Pioneers in Neuroendocrinology II

PERSPECTIVES IN NEUROENDOCRINE RESEARCH

Volume 1 • PIONEERS IN NEUROENDOCRINOLOGY
Edited by Joseph Meites, Bernard T. Donovan, and Samuel M. McCann • 1975

Volume 2 • PIONEERS IN NEUROENDOCRINOLOGY II
Edited by Joseph Meites, Bernard T. Donovan, and Samuel M. McCann • 1978

Pioneers in Neuroendocrinology II

Edited by

Joseph Meites

Department of Physiology
Michigan State University
East Lansing, Michigan

Bernard T. Donovan

Department of Physiology
Institute of Psychiatry
London, England

and

Samuel M. McCann

Department of Physiology
The University of Texas Health Science Center at Dallas
Dallas, Texas

PLENUM PRESS • NEW YORK AND LONDON

Library of Congress Cataloging in Publication Data

Main entry under title:

Pioneers in neuroendocrinology.

(Perspectives in neuroendocrine research; v. 1–)
Includes bibliographies and index.
1. Neuroendocrinology – History – Collected works. 2. Endocrinologists – Correspondence, reminiscences, etc. – Collected works. 3. Neurologists – Correspondence, reminiscences, etc. – Collected works. I. Meites, Joseph, 1913- II. Donovan, Bernard Thomas. III. McCann, Samuel McDonald, 1925- [DNLM: 1. Endocrine glands – Physiology – Personal narratives. 2. Neurophysiology – Personal narratives. W1 PE871H v. 1/WZ112 P662]
QP356.4.P46 599'.01'88 75-19075
ISBN-13:978-1-4613-4029-4 e-ISBN-13:978-1-4613-4027-0
DOI: 10.1007/978-1-4613-4027-0

Softcover reprint of the hardcover 1st edition 1978

A Division of Plenum Publishing Corporation
227 West 17th Street, New York, N.Y. 10011

Preface

This second volume of *Pioneers in Neuroendocrinology* differs from the first in several respects. First, with one exception, the present writers represent the second generation of pioneers, who began their work in the early or mid-1950s, at the onset of what became the most intensive and fruitful period of neuroendocrine research. Second, with few exceptions, the present authors are still active in work on many aspects of neuroendocrinology and show a keen appreciation of current as well as past problems in the field. Third, this volume covers the period when definite proof for the "chemotransmitter hypothesis" was first provided. This includes work by Saffran, Schally, Guillemin, McCann, Meites, Schreiber, and others demonstrating that "releasing factors" actually are present in the hypothalamus, early attempts to characterize them, development of assay methods, and studies on their physiology. All of this culminated in the dramatic announcements during the second half of 1969 of the isolation and synthesis of the first hypothalamic hypophysiotropic hormone, pyroglutamylhistidylprolinamide (TRH). Guillemin and Schally provide colorful accounts of their early work on this and other hypothalamic releasing factors. Both received Nobel prizes for their achievements.

This volume also deals with the early attempts to relate specific areas of the hypothalamus and of other brain regions to secretion of each of the anterior pituitary hormones; with the neuroendocrine control of reproductive functions during prepuberty, puberty, sexual maturity, and aging; with CNS bioelectrical correlates and endocrine functions; with effects of hormones on behavior; with the role of biogenic amines and other neurotransmitters in the CNS on hypothalamic and pituitary functions; with the comparative aspects of neuroendocrinology in nonmammalian species; with "short-loop feedback"; with negative and positive feedback of hormones on the CNS; with the "hypophysiotropic" region of the hypothalamus; with the relation of the CNS to development of mammary and pituitary tumors. Many other topics of interest in neuroendocrinology are also discussed.

The decision as to who should be invited to contribute to this volume was made by the three editors. Regrettably not all accepted. Lord Zuckerman had been invited to contribute to the first volume, but could not do so at the time because of illness, and hence his chapter appears here. He has provided a thoughtful and provocative analysis of his famous debate with Geoffrey Harris and explains why he still disagrees with the "portal vessel hypothesis." At least three of Harris's former students or postdoctoral fellows (Cross, Donovan, Reichlin) recount some revealing aspects of how their famous mentor operated, and it will interest the readers to see how many of the other writers of this volume were influenced by Harris.

Each of the chapters in this volume is to some extent autobiographical, and reveals much about each author's personality, ambitions, ideas, inspirations, disappointments, triumphs, sources of financial support, facilities available, research approaches, views of colleagues, and other interesting aspects of their careers. The reader will gain considerable insight—and, it is hoped, inspiration—in reading this book.

The Editors

Contents

1

Charles A. Barraclough

Charles A. Barraclough was born on July 13, 1926, in Vineland, New Jersey. He was educated in the public schools of Hammonton, New Jersey, and received a B.S. degree from St. Joseph's College, Philadelphia, Pennsylvania, in 1947. He received his M.S. degree in 1952 and his Ph.D. degree in 1953 from Rutgers University, New Brunswick, New Jersey. From 1953 to 1955 he was a postdoctoral fellow with Dr. Charles H. Sawyer at UCLA; in 1955 he was appointed instructor in anatomy at UCLA and in 1956 was appointed assistant professor of anatomy. In 1961–1962 he was awarded a special research fellowship and studied with Drs. Barry Cross and Marthe Vogt at Cambridge University, England. In 1962 he became an associate professor of physiology at the School of Medicine, University of Maryland, and was appointed full professor in 1965. In 1969–1970 he received a special research fellowship to spend a year at the Institute of Pharmacology, University of Milan, Italy, with Professor Luciano Martini.

Dr. Barraclough is a member of many societies including the Endocrine Society, the American Physiological Society, the American Association of Anatomists, the Society for Experimental Biology and Medicine, the American Association for the Advancement of Science, Sigma Xi, the Society for the Study of Reproduction, and the International Brain Research Organization. He has served on the editorial boards of *Endocrinology* from 1965 to 1972 and the *Proceedings of the Society for Experimental Biology and Medicine* from 1974 to the present.

He served as an original member of the Reproductive Biology Study Section, NIH, from 1967 to 1969. He was reappointed to this study section from 1970 to 1974 and became its chairman from 1973 to 1974. He also served on the board of directors of the Society for the Study of Reproduction from 1971 to 1973 and was Porter Lecturer of the APS at Howard University from 1967 to 1969.

Adventures with the Androgen-Sterilized Rat and Reproductive Neuroendocrinology

CHARLES A. BARRACLOUGH

THE FORMATIVE YEARS (1949–1961)

My route to a scientific career was rather serpentine—during my high school and undergraduate college years, my interests were directed more to music than to science. Like many, my parents had high hopes of my becoming a medical doctor, a career which would provide financial security in a respectable profession. I entered St. Joseph's College, a small Jesuit school in Philadelphia, in September 1944 as a premedical major, and I completed my B.S. degree in 1947. During these war years, St. Joseph's College had adopted an accelerated (3-year) curriculum to increase the number of students moving into the job market or into various professional schools. My one application for admission to Jefferson Medical College (Philadelphia) was rejected, to my parents' chagrin but to my delight, and for the next 2 years I worked as a professional musician in the Philadelphia area.

Then I entered the graduate program at Rutgers University, New Brunswick, New Jersey, to obtain a M.S. degree and to be sufficiently close to New York City to continue my musical profession. It was at Rutgers that I first encountered the human dynamo of the Zoology Department, Dr. James H. Leathem, and it was his contagious enthusiasm for endocrinology

CHARLES A. BARRACLOUGH • Department of Physiology, School of Medicine, University of Maryland, Baltimore, Maryland 21201

that subsequently led me into my present career. He was willing to accept me as one of his students (I was number four), and I discussed the possibility of doing a thesis with him. After rejecting several different research projects (all of which dealt with hormones and metabolism), I settled on a study of the effects of androgen administered as a single injection at select early postnatal ages on subsequent development of the reproductive systems of both male and female mice. The use of mice was predicated on the fact that all rats in the colony at the Bureau of Biological Research (Rutgers) were being used for dietary or biochemical studies, while the mouse colony was receiving little attention at this time. Also, it was unheard of during those years to buy animals; thus I learned to breed, sex, and care for my own mouse colony. This was particularly necessary since we were working with 5-, 10-, and 20-day-old animals. Also, during these years, federal research support was small and university support was almost nonexistent, so we designed our own cages, bread pans with wire tops and sawdust bedding collected from the floors of the local lumber companies. Fortunately, Dr. Leathem was able to obtain some financial support for his students, and the University waived tuition costs. I found I could survive on $90.10 a month and whatever erratic funds I could obtain by playing with local musical groups. As students, we purchased our own hypodermic syringes and needles from the local college pharmacy, and all my dissections were done with a pair of eyebrow tweezers and cuticle scissors.

In reviewing the literature in the late 1940s, it soon became apparent to me that androgenic hormones had been the subject of many investigations, and consequently numerous reports of their effects on the reproductive organs of a variety of species had been published. The results obtained were not always in agreement, because the manner in which the hormone was employed (multiple vs. single injections, variable dosages, routes of administration, etc.), the species studied, and the age of the animal all influenced the results. For such reasons, we designed our original studies to eliminate many of these variables, giving the hormones as a single injection in a standardized dose of 1.0 mg and varying only the age of the animal. Prepubertal mice of 5, 10, and 20 days of age were given testosterone propionate as a single injection and were autopsied at 10-day intervals thereafter to 60 days of age. The studies in male mice were not very rewarding, because early steroid exposure merely delayed the time of appearance of spermatozoa in the testes, and the growth rates of the seminal vesicles were somewhat sluggish between 20 and 60 days of life. Nevertheless, all animals were fertile when they reached adulthood (Barraclough and Leathem, 1959). Identical treatment of female mice, however, revealed some interesting information: the ovaries of animals which received testosterone at 5 days of life failed to form corpora lutea at any age studied, whereas the subsequent ovulation of animals injected at 20 days was not affected. Those mice treated at 10 days of age assumed an intermediate position, with some animals being sterile and others fertile (Barraclough and Leathem, 1954; Barraclough, 1955).

Some clues as to what effects androgen might have on the pituitary-ovarian or pituitary-testicular axes actually had been postulated as early as 1936 by Pfeiffer. He had observed that transplantation of testes into newborn female rats resulted in the ovaries of these animals, when adult, having only follicles, no corpora lutea, and persistent vaginal cornification. In contrast, if male rats were castrated at birth and an ovary was transplanted into the eye, follicular development and corpora lutea formation occurred when these animals reached adulthood. Pfeiffer proposed that the mechanism by which androgen (secreted by testes) produced the anovulatory persistent estrous condition was by "masculinization" of the adenohypophysis so that a permanent imbalance in gonadotropin secretion resulted. This investigator suggested that the pituitary gland of the newborn rat was undifferentiated. If it differentiated in the presence of androgen, only FSH was elaborated, whereas if it differentiated normally, both FSH and LH were secreted. This hypothesis was not unreasonable at the time it was published (1936), because little was known of CNS influences on adenohypophyseal function, and the negative feedback hypothesis of Moore and Price had only been published several years before. In our publication in 1955, I ventured a thought that perhaps androgen acted at levels above the pituitary gland, but without experimental evidence. Proof of this hypothesis remained dormant until 1961.

In 1952 I married Eleanor Kolakowski of South River, New Jersey. In June 1953 I received my Ph.D. degree, and in July 1953 my first daughter, Janet, was born. At this time I was offered the opportunity of studying with Dr. Charles H. Sawyer as a postdoctoral fellow at the University of California, Los Angeles (UCLA). I readily accepted this position since I was aware of the studies published by Dr. Sawyer in collaboration with Drs. Everett and Markee during his tenure at Duke University on the hypothalamic regulation of gonadotropic secretion in rats and rabbits. My training at Rutgers had been primarily in physiology and biochemistry but I knew little of the anatomy or physiology of the nervous system, nor was I particularly cognizant of the developing field of neuroendocrinology.

The primary research core of the Department of Anatomy, UCLA, was located in Long Beach, because construction of the medical school had only been begun by 1953. It was here that I was enveloped in an environment of CNS neurophysiologists, speaking glibly about such esoteric subjects as the ascending reticular activating system, evoked potentials, and electroencephalography. Only in Sawyer's laboratory did I hear the word "reproduction" mentioned. It was in Long Beach that my second daughter, Patricia, was born, and I shall never forget the wild ride to the hospital which preceded this event since Patti had decided to become a native Californian at 8 rather than 9 months of gestation. During this time I pursued some studies in androgen-sterilized female mice with the hope that the polyfollicular ovary might offer a target organ by which to measure changes in plasma LH/FSH without the need for using hypophysectomized animals. These studies were not very rewarding since ovarian responsiveness to either LH or FSH alone or in various combinations gave spurious results. Only recently have we learned that ovaries of androgen- or estrogen-sterilized animals are highly insensitive to LH. I also collaborated with Dr. Sawyer on the effects of morphine (Barraclough and Sawyer, 1955) and of reserpine and chlorpromazine (Barraclough and Sawyer, 1957) on blockade of ovulation in proestrous rats. It was our hope at that time to identify CNS systems which could be involved in the hypothalamic activation of pituitary LH/FSH release using a pharmacological approach. It soon became apparent that the multiple sites of drug action prevented us from reaching any clear understanding of the CNS systems involved in spontaneous gonadotropin release. We also observed that both chlorpromazine and reserpine, if administered at specific times during the estrus cycle, resulted in pseudopregnancy in the rat (Barraclough and Sawyer, 1959) and in galactorrhea in the rabbit. About this same time, Nikitovitch-Winer and Everett published their observations on the function of pituitary grafts autotransplanted to the kidney capsule. These glands, when removed from hypothalamic influences, secreted primarily prolactin. We reasoned that the long-acting tranquilizers such as reserpine had a similar action of inhibiting hypothalamic control of pituitary LH/FSH secretion and thus permitted prolactin to be released. Today we realize that reserpine is an effective depletor of hypothalamic catecholamines, including dopamine, and this transmitter might prove to be "prolactin-inhibiting factor."

During this interval, I was permitted to study both human gross anatomy and neuroanatomy-physiology with the medical students. At the end of my postdoctoral experience, I was appointed instructor and then assistant professor of anatomy at UCLA and fulfilled my teaching responsibilities by instructing in the gross anatomy course. When the new medical school building was completed, I moved my research laboratory to the UCLA campus, and it was here that the studies on the androgen-sterilized rat were begun (Barraclough, 1961). Vaughn Critchlow (1958), presently chairman, Department of Anatomy, University of Oregon, had begun his Ph.D. thesis studies with Sawyer at this time, investigating sites within the preopticohypothalamic system of Nembutal-blocked proestrous rats which, when stimulated, would induce the discharge of sufficient gonadotropin to cause ovulation. Using this type of preparation, I stimulated the same CNS regions in androgen-sterilized rats and their littermate controls (oil-injected only) and observed that only those areas in medial basal hypothalamus, arcuate-medium eminence region caused sufficient gonadotropin to be released to induce ovulation in sterile rats. Preoptic area stimulation, which readily induced ovulation in oil-injected controls, proved to be completely ineffective in this preparation. As a consequence of these studies, I proposed the theory of dual hypothalamic control of pituitary gonadotropin secretion (Barraclough and Gorski, 1961). A considerable amount of supportive evidence for this CNS regulatory system existed at this time, but it was muddled because the earlier studies of Hillarp and Greer in rats and Dey in guinea pigs had employed electrolytic lesions as a method of studying CNS regulation of pituitary function. I recall the frustration I encountered trying to correlate the exact neuroanatomical structures within the preopticohypothalamic system which were destroyed with the type of ovarian failure that developed with time after lesion production. Most lesions produced extensive damage to anterior, lateral, dorsal, and medial hypothalamic structures, and the lesion size and location varied considerably from animal to animal and from laboratory to laboratory. However, the data revealed that anterior lesions resulted in animals with polyfollicular ovaries and persistent vaginal cornification, whereas lesions in the median eminence region produced ovarian atrophy. The selective localization of the stimulation sites employed by Critchlow in normal rats and by me in cyclic and in androgen-sterilized rats permitted a more accurate delineation of the CNS regions involved in the ovulatory discharge of gonadotropin as compared to those areas necessary only for follicular development and steroid production. Consideration of the results of my studies together with those of the earlier workers led me to term the two respective regions "cyclic" (preoptic-suprachiasmatic area) and "tonic" (arcuate-median eminence region), and such terms seem to have become part of the accepted terminology in discussing CNS regulation of LH secretion. The logical extension of these observations was to separate the preoptic area from other hypothalamic structures to demonstrate the importance of this region in the cyclic release of gonadotropin. This was accomplished by the innovative studies of Halasz and Pupp (1965) some years later. In 1958 Roger Gorski arrived at UCLA from the University of Illinois, Urbana. His interests were in neuroendocrine control of pituitary function and he had obtained his M.S. degree with Dr. A. Nalbandov, studying some aspects of the hypothalamic regulation of pituitary function in sheep. Dr. Sawyer asked if I would be willing to accept Gorski as a graduate student; thus we became associated for several fruitful years, pursuing a variety of studies on the various types of abnormalities produced by early sex steroid exposure (Barraclough, 1967).

About this time, Dr. Albert Parlow had reported on the ovarian ascorbic acid depletion (OAAD) technique for measuring LH, and I readily implemented this methodology in my laboratory to examine changes in pituitary LH concentrations in normal, sterile, and progesterone-primed sterile rats. We noted that LH was reduced in the sterile animal but that progesterone pretreatment elevated the concentration to cyclic proestrus control levels. Further, it was this increased concentration which was released after preoptic stimulation. In rats in which low dosages of androgen (10 μg) had been used to produce anovulatory sterility, such progesterone treatment and preoptic stimulation resulted in ovulation. More recent studies by Barraclough and Turgeon (1975) and Chappel and Barraclough (1977), which employed radioimmunoassay procedures, revealed that preoptic stimulation induced significantly greater peak plasma LH concentration in progesterone-primed sterile rats than in untreated persistent-estrus rats. We now realize that these elevated plasma LH levels were required to induce ovulation because of the elevated ovarian ovulatory thresholds present in the sterile rat. Gorski and I also investigated the effects of prepubertal steroid exposure on hypothalamic regulation of pituitary FSH secretion employing the technique of ovarian compensatory hypertrophy as an index of the release of FSH. We concluded that at least one component of the FSH system in the sterile rat was not altered in this preparation. Recent studies by Chappel and Barraclough (1976*a*) reinforced this early observation. Consequently, as new techniques have been adopted and older observations have been reexamined, it becomes increasingly apparent that the biological end points we used for measurements of pituitary function were accurate and useful tools for their time in history.

I recall one particular series of studies that Roger Gorski and I initiated which, at the time, created a small incident in the laboratory. We decided to examine the mating behavior of rats sterilized with large (1.0 mg) or small (10 μg) dosages of androgen (Barraclough and Gorski, 1962). To this end, I had built eight boxes with glass fronts through which we could observe the behavior of the animals. They also served as the home cages for sexually active male rats. Our indications of mating behavior were not sophisticated; rather, we were satisfied only if the male mounted the female and achieved introitus. The experiment went well with cyclic female rats, but trouble began with the introduction of the high-dose sterile animals. The male approached the female, attempted to mount, and was rebuffed by a well-placed kick to the groin. Needless to say, the male animal lost interest in his mate. I stared at the male through the glass and the male rat reciprocated by sitting in front of the glass returning my stare. This went on for several hours. At this time, I had a laboratory technician named Terry Mangold, who had received her B.S. degree from a small Catholic girl's college in Los Angeles. In sheer frustration, I began to call on my males to try again (among other expressions). Needless to say, one should be aware of the personnel around him when doing mating behavior studies, for a very blushing and aggravated Terry departed the laboratory and was not seen again until these studies were concluded.

Meanwhile, the Brain Research Institute (BRI) was constructed and a small research laboratory was alloted to me which I thought would alleviate the severe space shortage in the medical school. Assistant professors were required to share both offices and laboratories, which led to some problems. However, by accepting BRI research space I also was required to accept into my laboratory visiting research scientists who wished to spend a year at the BRI. Thus the new space was rapidly occupied. During this time, I had the opportunity of meeting and collaborating with such interesting individuals as Dr. Juan Tramezzani from Argentina, Dr. Shaul Feldman from Israel, and Dr. Sergio Yrarrazaval from Chile. Dr. Tramezzani and I spent considerable time and effort attempting to localize EEG changes within the hypothalamus that occurred after vaginal cervix stimulation, which we hoped would shed some light on neural pathways involved in the release of prolactin. While we obtained highly suggestive data that activation of median forebrain bundle occurred after prolonged cervical stimulation, I was never sure that these results were specific to the system being studied and thus these data were never published.

During this time I also began studies which attempted to decipher the sites of action of progesterone in facilitating the ovulatory discharge of gonadotropin in normal rats. Some years earlier, Everett had reported that administration of progesterone to 5-day cyclic rats would advance ovulation by 24 hr. Further, in a strain of rats which became spontaneously anovulatory and exhibited persistent estrus when they reached adulthood, he observed that progesterone, when administered as sequential injections, would result in ovulation (Everett, 1940). I reasoned that perhaps the failure of the spontaneous gonadotropin surge to occur in androgen-sterilized rats was an absence of progesterone (due to absence of corpora lutea). However, after a considerable amount of effort it became evident that, regardless of the dosage or sequence of treatment, steroid-induced ovulation could not be induced in the sterile rat. The results of these studies created something of a dilemma for me since I had proposed that the region deleteriously affected by prepubertal androgen treatment was the preoptic area. Yet, in 1953, Greer reported that rats made persistently estrous by anterior hypothalamic lesions ovulated in response to progesterone. Why then, didn't androgen-sterilized rats respond the same way? One could rationalize that neonatal steroid exposure affected pituitary function by a direct action on the gland as proposed by Pfeiffer. Yet Harris and Jacobsohn (1952) (in rats) and Martinez and Bittner (1956) (in mice) had shown that male pituitaries (subjected to endogenous androgenization) revascularized when transplanted into proximity with the median eminence of hypophysectomized female rats or mice and ovarian cyclic behavior was restored. In rereading the Greer study, I noted that a certain percentage of his lesioned, persistent-estrus rats did not ovulate in response to progesterone therapy. Consequently, we decided to repeat the lesion experiments in normal cyclic rats but to be selective in the regions of the preoptic area to be destroyed. Lesions of the dorsal preoptic area failed to alter cyclicity, whereas selective destruction of the suprachiasmatic nuclei resulted in persistent estrus, and ovulation could be induced by progesterone treatment. Only when the suprachiasmatic nuclei and the periventricular portion of the medial preoptic area were destroyed was progesterone ineffective in inducing ovulation (Barraclough *et al.*, 1964). As far as I can determine, these were the first definitive studies which implicated the suprachiasmatic nuclei as having a role in the ovulatory discharge of gonadotropin. As well, they served to further define the important CNS regions involved in regulating this cyclic release of gonadotropin.

I should point out that during this entire historic interval I, like my colleagues, was continuously frustrated by the lack of methodology to accurately measure changes in plasma steroids or gonadotropins. Most conclusions were based on end points such as ovulation or changes in ovarian morphology or target organ weight.

In 1961, I requested permission from the UCLA administration to spend a sabbatical year at Cambridge, England. Within 3 months of my departure for England I also was offered an associate professorship within the Department of Physiology at the Medical School, University of Maryland. I decided to accept the position since it increased my salary substantially, tripled my research space, and decreased my teaching load by 90%. I was fortunate to be able to obtain my complete salary for the sabbatical leave year from the NIH and took a leave of absence from UCLA. In England, I spent my time collaborating with both Dr. Marthe Vogt at the ARC Institute of Animal Physiology, Babraham, and Dr. Barry Cross in the Subdepartment of Veterinary Anatomy at Cambridge. With Dr. Vogt we attempted to learn if the hypothalamic catecholamine content in the sterile rat differed from that in the normal animal. With Barry Cross we began what were perhaps the first studies on unit cell recording in the rat hypothalamus. Most of these latter studies examined changes in unit frequencies in the lateral hypothalamic area to a variety of stimuli in proestrus rats (Barraclough and Cross, 1963).

THE MIDDLE YEARS (1962–)

In September 1962 I arrived at the University of Maryland. At this time, the medical school was in the process of revitalization and the Department of Physiology, under the direction of Dr. William Blake, was being completely restaffed. Initially, we were jammed into old quarters with poor animal facilities and wholly inadequate space. Equipment was slow to arrive, technicians had to be trained, and consequently research progressed at a slow pace for the first 2 years. During this time, I reevaluated the direction in which I thought my research should go. While I continued some studies on the abnormal endocrinology of the sterile rat preparation, it became increasingly apparent to me that I could not fruitfully study reproductive abnormalities if I didn't understand the normal physiology of the system. As a consequence, I began to perform in-depth studies on various components of the preopticotuberal-pituitary-ovarian axis to learn how each component could regulate each other component, and our present research is following this direction. Periodically, we resume studies in sterile animals to see if our recently acquired knowledge can detect other abnormalities which heretofore were unknown because of a lack of basic understanding of normal processes.

In 1965 I was appointed full professor of physiology. Between 1962 and 1967 I worked mostly alone in a laboratory with one or two technicians. More and more frequently, when I am surrounded with an overwhelming morass of paperwork, I reminisce about the enjoyable aspects of this uncomplicated and rewarding type of life.

In 1967 Dr. Edwin Haller joined me and was the first of a series of postdoctoral fellows subsequently to pass through my laboratories. With him we reexamined the usefulness of unit cell recording techniques in deciphering the hypothalamic events which lead to the ovulatory discharge of gonadotropic hormones. I came to realize that this approach was not to my liking, and within 2 years I abandoned it in favor of more quantitative chemical methods. As a final study in our laboratory, which employed the OAAD method, Haller and I examined the changes in pituitary and plasma LH concentrations in normal and sterile rats to castration and estrogen replacement. From these studies, as well as with information provided by the literature, we diagrammed a scheme of the differential feedback actions of estrogen on pituitary LH synthesis, storage, and release (Barraclough and Haller, 1970).

In the fall of 1967 a graduate student from the biochemistry department arrived in my office and indicated her dissatisfaction with the graduate program she had selected for a career. Her husband, Dwight, a very bright medical student, recommended that she talk with me about transferring into our graduate program in physiology. Thus, in early 1968, Oneida Morningstar Cramer became my first graduate student at Maryland. Her work supplied us with basic information on changes in plasma LH following electrical stimulation of the preoptic area and, as well, a more concise picture of the neuroanatomical organization of the preopticotuberal system which was involved in LH release (Cramer and Barraclough, 1971, 1973).

In 1969 I took advantage of my sabbatical leave time to spend a year with Dr. Luciano Martini and his group at the Institute of Pharmacology, University of Milan, Italy. Here I pursued studies on the temporal changes in plasma progesterone in relation to the time of the LH surge. As well, among other things, I learned that you say *buon giorno* (good morning) until 4 p.m. More importantly, my years in England and Italy gave me a broader perspective and understanding of the differences and similarities, often great, between the problems and rewards encountered by scientists in these countries and scientists in the United States.

In 1969, during my sabbatical leave, Dr. Judith Turgeon arrived as a postdoctoral fellow, and on my return we began a highly rewarding collaborative series of studies on the role of the preoptic brain in inducing quantitative releases of pituitary LH depending on the amount of preoptic tissue stimulated (Turgeon and Barraclough, 1973). Judy remained in my laboratories as a colleague (assistant professor) until recently (1975) when she departed for a new position at the University of California, Davis. Prior to leaving, she completed a study on the possible short-loop feedback action of LH in regulating its own secretion (Turgeon and Barraclough, 1976). While this concept had been originally proposed by Kawakami and Sawyer (1959) and had received some attention thereafter, I was never truly convinced that such a system was operative, because data on castrated animals revealed a progressive elevation of LH to high plasma concentrations, which argued against this hypothesis. Perhaps one explanation for the failure of the LH feedback loop system to function in castrates is the absence of estrogen, which seems essential as a cofactor with progesterone and LH for activating hypothalamic peptidases which degrade LHRH.

In 1971 Dr. David Mann arrived as a postdoctoral fellow after having received his degree with Dr. James H. Leathem. It was at this time I realized that I had aged. I was also to learn that Leathem's students were still buying their syringes at the college pharmacy. With Dave we did an in-depth study of the role of progesterone in regulating the timing of the proestrous LH surge. From the data obtained, it seems that this steroid has an important function in setting of the time at which LH is released and perhaps synergizes with ovarian estrogen to alter thresholds of excitability within the hypothalamus (Mann and Barraclough, 1973). Apparently the adrenal glands are the source of this progesterone.

Meanwhile, two new graduate students had embarked on their Ph.D. programs with me in 1972–1973, Scott Chappel and George Barr. With Scott we pursued an extensive study of the hypothalamic regulation of FSH secretion. These studies revealed that stimulation of the dorsal anterior hypothalamic area in Nembutal-blocked proestrous rats resulted in the selective release of FSH but not of LH. In the same preparation, preoptic stimulation induced the pituitary release of both LH and FSH (Chappel and Barraclough, 1976*a*). These observations, coupled with other data obtained by Chappel (as yet unpublished), raise questions of whether LHRH is indeed also FSH-RH. With George Barr we have been pursuing studies on changes in medial basal hypothalamic LHRH in response to stimuli or other experimental maneuvers known to alter plasma LH concentrations.

In 1973 we had the honor of having Dr. Béla Flérko spend 6 months with us collaborating in some studies on sex steroid feedback on pulsatile LH release patterns in castrated rats. In 1974–1975 Wan-Pang Pi from Taiwan spent a year with my group for advanced training, and most recently (1975–1976) Dr. Lajos Tima from Hungary collaborated with me in further studies on the short-loop feedback action of LH.

This brings us to today. I suddenly find myself again surrounded by new faces of eager young investigators (Dr. Douglas Shander and Dr. Phyllis Wise) and students (Louis De Paolo, Naomi Rance, Patricia Camp, and Keith Lookingland) embarking on careers such as I did in 1949. I wonder what the future holds and what mysteries will be unraveled by these individuals. Clearly, all the work that I have accomplished to date would not have been possible without the collaborative efforts of those mentioned in this chapter. To them I express my deepest gratitude.

REFERENCES

Barraclough, C. A. (1955). Influence of age on the response of preweaning female mice to testosterone propionate. *Am. J. Anat.* **97**:493.

Barraclough, C. A. (1961). Production of anovulatory sterile rats by single injections of testosterone propionate. *Endocrinology* **68**:62.

Barraclough, C. A. (1967). Modifications in reproductive function after exposure to hormones during the prenatal and early postnatal period. In Martini, L., and Ganong, W. F. (eds.), *Neuroendocrinology*, Vol. 2, Academic Press, New York, pp. 61–69.

Barraclough, C. A., and Cross, B. A. (1963). Unit activity in the hypothalamus of the cyclic female rat: Effect of genital stimuli and progesterone. *J. Endocrinol.* **26**:339.

Barraclough, C. A., and Gorski, R. A. (1961). Evidence that the hypothalamus is responsible for androgen-induced sterility in the female rat. *Endocrinology* **68**:68.

Barraclough, C. A., and Gorski, R. A. (1962). Studies on mating behavior in the androgen-sterilized female rat in relation to the hypothalamic regulation of sexual behavior. *J. Endocrinol.* **25**:175.

Barraclough, C. A., and Haller, E. W. (1970). Positive and negative feedback effects of estrogen on pituitary LH synthesis and release in normal and androgen-sterilized female rats. *Endocrinology* **68**:542.

Barraclough, C. A., and Leathem, J. H. (1954). Infertility induced in mice by a single injection of testosterone propionate. *Proc. Soc. Exp. Biol. Med.* **85**:673.

Barraclough, C. A., and Leathem, J. H. (1959). Influence of age on the response of male mice to testosterone propionate. *Anat. Rec.* **134**:239.

Barraclough, C. A., and Sawyer, C. H. (1955). Inhibition of the release of pituitary ovulatory hormone in the rat by morphine. *Endocrinology* **57**:329.

Barraclough, C. A., and Sawyer, C. H. (1957). Blockade of the release of pituitary ovulating hormone in the rat by chlorpromazine and reserpine: possible mechanisms of action. *Endocrinology* **61**:341.

Barraclough, C. A., and Sawyer, C. H. (1959). Induction of pseudopregnancy in the rat by reserpine and chlorpromazine. *Endocrinology* **65**:563.

Barraclough, C. A., and Turgeon, J. L. (1975). Ontogeny of development of the hypothalamic regulation of gonadotropin secretion: Effects of perinatal sex steroid exposure. In Markert, C. L., and Papaconstantinou, J., (eds.), *The Developmental Biology of Reproduction*, Academic Press, New York, pp. 255–273.

Barraclough, C. A., Yrarrazaval, S., and Hatton, R. H. (1964). A possible hypothalamic site of action of progesterone in the facilitation of ovulation in the rat. *Endocrinology* **75**:838.

Chappel, S. C., and Barraclough, C. A. (1976*a*). Hypothalamic regulation of FSH secretion. *Endocrinology* **98**:927.

Chappel, S. C., and Barraclough, C. A. (1976*b*). Plasma concentration changes in LH and FSH following electrochemical stimulation of the medial preoptic area or dorsal anterior hypothalamic area of estrogen- or androgen-sterilized rats. *Biol. Reprod.* **15**:661.

Chappel, S. C., and Barraclough, C. A. (1977). Further studies on the regulation of FSH secretion. *Endocrinology* **101**:24.

Cramer, O. M., and Barraclough, C. A. (1971). Effect of electrical stimulation of the preoptic area on plasma LH concentrations in proestrous rats. *Endocrinology* **88**:1175.

Cramer, O. M., and Barraclough, C. A. (1973). Effects of preoptic electrical stimulation on pituitary LH release following interruption of components of the preoptico-tuberal pathway in rats. *Endocrinology* **93**:369.

Critchlow, B. V. (1958). Ovulation induced by hypothalamic stimulation of the anesthetized rat. *Am. J. Physiol.* **195**:171.

Everett, J. W. (1940). The restoration of ovulatory cycles and corpus luteum formation in persistent-estrous rats by progesterone. *Endocrinology* **27**:681.

Greer, M. A. (1953). The effect of progesterone on persistent vaginal estrus provoked by hypothalamic lesions in the rat. *Endocrinology* **53**:380.

Halász, B., and Pupp, L. (1965). Hormone secretion of the anterior pituitary gland after physical interruption of all nervous pathways to the hypophysiotropic area. *Endocrinology* **77**:553.

Harris, G. W., and Jacobsohn, D. (1952). Functional grafts of the anterior pituitary gland. *Proc. R. Soc. London Ser. B* **139**:263.

Kawakami, M., and Sawyer, C. H. (1959). Induction of behavioral and electroencephalographic changes in the rabbit by hormone administration or brain stimulation. *Endocrinology* **65**:631.

Mann, D. R., and Barraclough, C. A. (1973). Role of estrogen and progesterone in facilitating LH release in 4-day cyclic rats. *Endocrinology* **93**:694.

Martinez, C., and Bittner, J. J. (1956). A non-hypophyseal sex difference in estrous behavior of mice bearing pituitary grafts. *Proc. Soc. Exp. Biol. Med.* **91**:506.

Pfeiffer, C. A. (1936). Sexual differences of the hypophyses and their determination by the gonads. *Am. J. Anat.* **58**:195.

Turgeon, J., and Barraclough, C. A. (1973). Temporal patterns of LH release following graded preoptic electrochemical stimulation in proestrous rats. *Endocrinology* **92**:755.

Turgeon, J., and Barraclough, C. A. (1976). The existence of a possible short-loop negative feedback action of LH in proestrous rats. *Endocrinology* **98**:639.

2

Frank A. Beach

Frank A. Beach was born in April 1911 in Emporia, Kansas. After receiving a B.S. and M.S. in psychology from Kansas State Teachers College in Emporia, he attended the University of Chicago, where he received the Ph.D. in psychology in 1940. He had taught high school for 1 year after receiving his Master's degree. Even before receiving his Ph.D. he was assistant curator in the Department of Experimental Biology at the American Museum of Natural History in New York City. In 1942 he was made curator and chairman of the Department of Animal Behavior of the American Museum of Natural History and served also on the staff of the Department of Biology of New York University. In 1946 he moved to New Haven to become professor of psychology at Yale University and held an endowed Sterling Professorship there until 1958, at which time he moved to the University of California at Berkeley as professor of psychology. He has delivered many lectures at schools around the country and has received many honors. He received the Warren Medal for Excellence in Scientific Experimentation in 1951 from the Society of Experimental Psychologists, an Award in Psychiatry from the Association for Research in Nervous and Mental Disease in 1958, the Distinguished Scientific Contribution Award from the American Psychological Association in 1959, the annual award for the contemporary scientist who has made an outstanding contribution to the study of sex from the Society for Scientific Study of Sex in 1968, the Kenneth Craik Research Award from St. John's College at Cambridge in 1973, the Fifth Annual Carl G. Hartman Award from the Society for the Study of Reproduction in 1974, and many others. He is a member of numerous societies including the American Psychological Association, the American Philosophical Society, the American Academy of Arts and Sciences, and the National Academy of Sciences. He has served on many boards, panels, and major committees for the National Academy of Sciences, the National Science Foundation, and the National Institutes of Health. He founded the journal *Hormones and Behavior*, and his scientific contributions have spanned all areas of sex behavior and the effect thereon of hormones.

2

Confessions of an Imposter

FRANK A. BEACH

PROLOGUE

As I begin this autobiographical essay, I have just come from the laboratory, where, during the past few weeks, several graduate students and I have been making some interesting discoveries about the behavior of golden hamsters (*Mesocricetus auratus*). It is axiomatic that when females of this species are in physiological estrus they are sexually *receptive*, i.e., will readily mate with a male; but we have been testing the female's tendency to display a different manifestation of estrus, namely, to approach the male and "solicit" copulation. To differentiate such behavior from passive receptivity, I have recently proposed the term "sexual proceptivity" to designate those behaviors in which females of any species assume some degree of initiative in establishing and maintaining coital relations with the male (Beach, 1976).

Our experimental method is simple. The estrous female is placed in a rectangular arena (2 × 4 feet) which contains two small cages constructed of wire mesh. One cage confines a female hamster which is in diestrus, and the other confines a sexually potent male. For a period of 5 min, we record the amount of time the experimental female spends beside each cage and the kinds of behavior she shows.

We have found that when they are in estrus most females spend 5–10 times longer beside the male's cage than beside the female's. Furthermore, while she is standing near the cage containing the male the estrous female frequently assumes the position known as "lordosis," which involves flattening her back and raising her tail and head just as she does while the male is copulating with her. In more normal situations the female adopts this posture in response to the male's investigatory behavior as he noses her flanks

FRANK A. BEACH • Department of Psychology, University of California, 3210 Tolman Hall, Berkeley, California 94720

and licks her genitalia. The fact that she will seek out a caged male, remain in his vicinity, and occasionally exhibit lordosis without physical contact demonstrates that the estrous female is sexually attracted to males. She is *proceptive* as well as receptive.

When the same female is not in estrus, she shows no preference for a male over a member of her own sex and never displays lordosis, so we believe that proceptivity depends on estrogen and progesterone, the same hormones that are known to induce sexual receptivity. I believe that proceptivity is characteristic of female animals of many species when they are in heat, and that from a functional point of view this is as important as the fact that they are at the same time receptive to the male. The function of proceptive behavior is to arouse the male and elicit his sexual responses at a time when the female is susceptible to impregnation. Temporal contiguity of proceptivity, receptivity, and fertility is "orchestrated" by the ovarian hormones.

A different series of experiments in progress in my laboratory for more than 10 years has to do with the behavior of dogs. Working with a sequence of graduate student assistants and visiting postdoctoral fellows, I have been exploring some of the causes for sexual dimorphism in the behavior of male and female beagles. Thanks to the work of earlier investigators (Phoenix *et al.*, 1959; Harris and Levine, 1962), we already knew that female rats and guinea pigs which have been exposed to stimulation by testosterone before birth (guinea pigs) or within a few days after birth (rats) tend to show marked reduction or absence of female mating responses in adulthood. Furthermore, such females exhibit male mating responses if they are injected with androgenic hormone as adults.

Work carried on at our Field Station for Behavioral Research has revealed that in dogs, as in rodents, administration of testosterone to the mother during a critical stage of pregnancy modifies female fetuses so that when they are born their external genitalia resemble those of a male much more than those of a normal female. From the behaviorist's point of view, the most interesting question is whether exposure to androgen during prenatal development also alters functional characteristics of the developing brain so that the experimental females will exhibit behavioral as well as anatomical masculinization.

Studies completed so far show that two kinds of psychological or behavioral change have taken place (Beach, 1975). (1) Experimental females are *defeminized*, which is to say they fail to show certain patterns of behavior, e.g., sexual receptivity, characteristic of the normal female. (2) At the same time, they are psychologically *masculinized* in the sense that they exhibit typical male behavior under certain circumstances. For example, in adulthood, perinatally masculinized females are more likely than normal bitches to lift one hind leg while urinating. They tend to be socially dominant over normal females, although not over normal males. When confronted with a normal female in heat, the experimental females frequently mount her in male fashion and exhibit many elements in the normal male copulatory pattern.

A current investigation is aimed at determining whether female dogs exposed to androgen before birth will exhibit the normal masculine preference for female rather than male odors. I would also like to discover whether such females are deficient in their maternal responses, but that will involve an entirely new experiment and a source of foster puppies, because masculinized females probably are infertile, and even if they could conceive would have to be delivered of pups by caesarian section because they possess no vagina.

EARLY MOTIVATIONAL INFLUENCES

I have begun this essay with an unorthodox introduction because I am not a neuroendocrinologist, and my name will be familiar to very few readers of this volume. The truth is I cannot even qualify as either an endocrinologist or a neurologist; but the editors were kind enough to invite my contribution and I was immodest enough to believe that it might be a good thing if one autobiography in this series were contributed by a scientist whose primary interest lay in the study of behavior. Several contributors to Volume 1 conducted important experiments involving behavior as one variable, but I believe I am the only contributor thus far who has specialized in such research. A principal aim of this story, therefore, is to explain how a young, honest, clean-cut American psychologist could have strayed into a career in behavioral endocrinology, and to recount what happened when he did so.

The explanation involves a number of concatenated coincidences which began during the Great Depression of the 1930s. When I graduated from the State Teachers College of Emporia, Kansas, in 1929, I planned to earn my living by teaching English in high school, but jobs were not available for unmarried men without teaching experience, and so I stayed at the Teachers College and took a Master's degree in psychology, which had been my minor subject. The principal inducement consisted of a $600 fellowship, which was just about what I could have earned pumping gas in a filling station if I had been lucky enough to find such a job.

I regarded the extra year of college as a temporary diversion from my real goal of high school teaching, but at the end of the year there were still no openings in secondary schools, so I decided to continue in graduate study provided I could get some kind of assistantship and could get into a good department of anthropology, a subject in which I had taken one or two undergraduate classes. I had enjoyed one graduate year of psychology but did not think it exciting enough to pursue further. If I were going to spend time on a Ph.D., it might as well be in a subject that was really interesting!

On the farthest trip I had ever taken from Kansas, I went to the University of Chicago and asked the chairman of the Anthropology Department if there were any graduate assistantships available. The reply was negative, but before taking the train home I stopped by the Psychology Department to see if they had any openings for a potential graduate student willing to work for a Ph.D. in practically any discipline. There was one $400 scholarship which would just pay tuition for four quarters, and the chairman was willing to gamble on me. Since that was the best offer available, I accepted gratefully, thus greatly improving the future prospects of anthropology and unwittingly determining the course of my own life for the next 35 years.

My rather cavalier attitude toward choice of a career may seem strange to readers of a later generation. The fact is that most young college graduates of my age felt an almost compulsive need for "security," which meant a steady and dependable income. The economic crash and subsequent depression made a strong impression on all of us. Grown men with families had lost their jobs and were working at any menial job they could find. Some were selling apples on the street. A small number had even committed suicide. My father, a professor in the Teachers College, took a one-third cut in salary, and the family felt fortunate that even this reduced income was assured.

Added to these external events was the value system I inherited from my New England-raised parents. Financial success above and beyond necessity was not valued, but the ability to earn a living and support a family was an essential dimension of adult masculinity. Dependence on outside help was a sign of failure or weakness. My first goal was to earn a living so that I could support myself and eventually a family. The particular way I was going to achieve this objective was not irrelevant but was of secondary importance. My choice of teaching as a profession was influenced by the fact that as a teacher my father had been able to weather the depression, and also by his great enjoyment of and success at his job.

One other fact concerning personal motivation should be mentioned. As a youngster in school I did fairly well until the last 2 years of high school, at which time my interest in social pursuits (particularly those of a heterosexual nature) resulted in scholastic disaster. Despite stern parental warnings, I majored in fun and games during my freshman year in college, with the result that over half my grades were either D or F. This resulted in considerable embarrassment to my father and he bundled me off to Antioch College, where I was to "prove" myself. I proved that girls in Ohio were much like girls in Kansas.

My father's displeasure and my own sense of failure combined to convince me that a college education was not for me, and anyway I would be happier earning my own living starting immediately. Three summer months on a railroad construction gang modified that conclusion slightly. There simply had to be an easier way to support myself! That was the point at which the solution of high school teaching suggested itself and I returned to the Teachers College with new resolve.

This time it worked, and I sailed through the last 2 years with practically a straight A record. I think it was during those 2 years that I realized I could do something well; and I then became resolved to do it *very* well. A need to succeed became a need to excel, probably in some degree to prove something to my father, but mostly, I suspect, to emulate him, because he was a nationally recognized figure in his profession and undoubtedly my most important role model throughout childhood.

These personal introspections are relevant to the major purpose of this essay because for many years after leaving the university I retained the somewhat adolescent motivation to succeed in whatever I attempted and to do whatever I did better than anyone else. Of course, no one can achieve either of these unrealistic goals, but, fortunately, experience and maturity bring readjustments in one's measure of "success" and constraint in the choice of objectives.

DEVELOPMENT OF INTEREST IN RESEARCH

One year of graduate work at the Teachers College did not convert me to psychology as a profession but it did teach me that research could be fun. At least it was fun under the unique set of conditions in which I learned to conduct experiments without guidance, tuition, or supervision. This came about because I was allowed to do a thesis on vision in rats even though no faculty member had ever conducted any experiments with animals and there was no one to advise me as to how my subjects should be housed and fed, what apparatus should be used, what behavioral tests would be appropriate, etc., etc. In any graduate department of psychology these details would have been common knowledge to at least some of the faculty, but I had to learn what I could from journals and books, and then solve any remaining problems by trial-and-error improvisation.

The experience was exhilarating, and I even went so far as to teach myself how to remove parts of the rat's cerebral cortex and keep the patient alive for further behavioral tests. This was a gratifying achievement, because at the beginning I didn't know a trephine from a cautery and had never seen an animal anesthetized. The thesis was completed and approved, although it could never have been published in a reputable journal. In retrospect, I can see that this was my first step toward a career in research, even though I could not recognize it at the time. In some ways it was an unfortunate beginning because it led me to view all research as exploration of uncharted territory, without maps or expert guidance. At the same time, the aura of complete independence and necessity for initiative and imagination lent great appeal to the research enterprise.

During the next year at the University of Chicago, I took a seminar under Professor Karl S. Lashley, an outstanding physiological psychologist famous for research on the neural basis for learning. Problems of learning did not excite me, but study of brain function did, and Lashley became my model of a "real scientist." At the end of that graduate year, I had to leave school. Earning a living at various part-time jobs, going to classes, and conducting research were more than I could manage so I took a job for 1 year teaching English in the high school of a small Kansas town. With the money I was able to save, I returned to Chicago, but in the interim Lashley had been lured to Harvard by President Conant, who first got Lashley appointed as University Research Professor and then offered him the position.

No one on the faculty at Chicago was qualified or interested in sponsoring research on brain function, but despite this deficit I was allowed to conduct a Ph.D. thesis involving the neurophysiological approach. At that time, practically all neuropsychological research revolved about problems of learning and memory, but I chose to investigate the effects of neocortical lesions on unlearned or "instinctive" behavior as represented by the female rat's care of her young. Very few experiments had been done on maternal behavior and none had involved investigation of its neural mediation, so I had again maneuvered myself into a situation in which I had to devise my own behavioral tests, decide how the effects of brain injury might be measured, select types of lesions that should be inflicted, and determine the most meaningful ways of analyzing my results.

All of this was accomplished without faculty guidance, and many mistakes resulted, but the final product was a success because the thesis was approved and, even more important, it resulted in a job offer from Lashley. He proposed that I come to Harvard to work in his laboratory on any research I chose. The pay amounted to $75 a month, and I accepted without hesitation. If I could have gotten a teaching job I probably would have taken it, but in 1934–1935 new positions for unfledged Ph.D.'s without college teaching experience were few and far between and I counted myself lucky to find any kind of professional employment.

INCIDENTAL INTRODUCTION TO ENDOCRINOLOGICAL VARIABLES

One of the first experiments I began in the Harvard Biological Laboratories where Lashley had his quarters was a study of mating behavior in male rats before and after destruction of different parts of the cerebral cortex. The floor above ours was occupied by graduate and postgraduate students working with Professor Hisaw. In conversation with one of the endocrinologists I happened to mention that some of my rats ceased mating after destruction of parts of the cortex, and the question was raised as to whether brain injury might not indirectly produce functional castration. Knowing nothing at all about endocrinology, I could not follow the explanation, which involved possible disruption of gonadotropic stimulation, but I could understand the need for assessing potential deficits in testicular hormone. Accordingly, I obtained a supply of testosterone propionate and injected it into some of the brain-operated rats that had ceased mating.

The results were interesting in more ways than one. Some males showed renewal of normal mating behavior, whereas others were totally unresponsive to the exogenous androgen. When all animals came to autopsy, several additional facts were discovered. By examining operated males which had not received hormone treatment, I found that brain injury by itself had not produced regressive changes in histology of the testis or the accessory glands. This suggested that failure to mate after cortical destruction could not be referred to secondary withdrawal of testis hormone. Why, then, had some postoperative noncopulators resumed mating after a series of testosterone injections?

In all experiments involving neocortical lesions it is essential to measure precisely the amount and locus of tissue removed. When this was done, I found that the capacity of testosterone propionate to restore preoperative sexual performance was negatively related to lesion size. Individuals which stopped copulating after loss of relatively small amounts of cortex were likely to resume mating in response to androgen treatment, whereas postoperative noncopulators with extensive neopallial invasion could not be induced to mate despite administration of very large amounts of TP.

Original loss of mating responses was not due to indirect depression of testosterone production, nor did it appear to reflect sensory impairment or motor impediment. Accordingly, I could only speculate that removal of parts of the neocortex had reduced sexual excitability. If this were correct, then restoration of mating by TP treatment might have reflected a renewal of responsiveness to sexual stimuli. It is well known that male rats cease mating after castration not because they are physically incapable of copulation but because they are no longer aroused by the estrous female. Putting all of the evidence together, I made a guess that under normal conditions sexual arousal in male rats depends jointly on testis hormone and on some general excitatory function of the cerebral cortex.

After this initial experience in hormone–behavior research, I should have recognized the necessity of acquiring at least some background knowledge in basic endocrinology. This might have happened had I remained at Harvard and established closer relations with the Hisaw group, but instead I moved to another job in which I had no endocrinologically sophisticated neighbors.

HUNT-AND-PECK EXPERIMENTATION ON HORMONAL CONTROL OF BEHAVIOR

My next position was that of assistant curator in the Department of Experimental Biology at the American Museum of Natural History in New York City. I had a full-time research appointment with no responsibilities other than to conduct my own experiments and maintain viable working relations with my boss, a brilliant but emotionally labile biologist named Gladwyn Kingsley Noble. The research facilities were excellent and one very important "fringe benefit" of working in a large museum was the opportunity to learn from my peers in other departments something about various areas of specialization such as paleontology, ornithology, herpetology, mammalogy, ichthyology, arachnology, and the like.

While learning something about natural history and more about modern evolutionary theory during brown-bag lunches and coffee-break seminars, I was at the same time indulging in a veritable orgy of experiments dealing with the effects of gonadal hormones on mating responses in animals. Testosterone propionate, estradiol benzoate, and progesterone were to me no more than easily injectable "male" and "female" hormones conveniently packaged in glass vials. They were magical independent variables which I used to vary behavior without any knowledge of their chemistry, synthesis, or general physiological functions.

I did not concentrate exclusively on experiments with hormones, but over a period of 9 years I published 16 separate papers dealing with relations between hormones and behavior. Most of them were simple and straightforward studies of withdrawal and replacement, administration of ovarian hormones to males and testosterone to females, induction of precocious behavioral puberty, etc. I worked with a variety of species, including laboratory rats, cotton rats, rice rats, cats, pigeons, lizards, and one small alligator.

It may be retroactive rationalization, but perhaps for the first few years my endocrinological naivete had some beneficial effects. It protected me from the temptation to dabble with a wide variety of hormones, and it compelled me to concentrate on operational definition and quantification of behavioral or psychological variables. These were the points at which some earlier studies by endocrinologists had been methodologically and theoretically deficient. Although I was an endocrinological ignoramus, I did have training as a comparative psychologist plus practical experience in the design and interpretation of experiments with behavior as the dependent variable. Consequently, over a period of years I was able to work out several objective and reliable tests for specific behavioral effects of gonadal hormones, and some of these measures are in use today by psychologists, endocrinologists, and neuroendocrinologists in this country and abroad. Standardized and reliable behavior measures are of course just as important as standardized measures of endocrine factors.

FORMAL INSTRUCTION IN ENDOCRINOLOGY

In the graduate and early postgraduate years I had been so lucky in getting by as a self-taught experimentalist that I tended to discount the importance of basic training and of supervised research for the beginner. I made a few attempts to learn something about endocrinology by attending professional meetings, but was too lazy to read elementary textbooks, so the vocabulary and concepts were often incomprehensible and what I could understand was unexciting because it had no immediate relevance to behavior.

Nevertheless, I was publishing articles about hormonal function and eventually was forced to admit to myself that I simply had to acquire at least a rudimentary knowledge of endocrinology. My solution was to enroll as an auditor in an undergraduate course at New York University which was taught by Dr. Robert Gaunt. Attending lectures faithfully, I picked up some fundamentals but was quite disappointed by the almost complete lack of any references to behavior. When I complained about the deficiency, Bob Gaunt suggested that I prepare a lecture on the subject and deliver it to the class.

I did this and in addition wrote a lengthy term paper summarizing what I could find in the literature about behavioral effects of hormones. The term paper grew into a monograph which became my first book and was published in 1948 by Hoeber and Harper under the title *Hormones and Behavior*. It was the first comprehensive summary of the literature, and for more than 20 years stood as the only book in its field.

This is not to say that no one else was writing or conducting research on behavioral problems. An outstanding pioneer in the field was Dr. Curt Richter, a psychologist at Johns Hopkins University. Richter began publishing important papers on hormonal control of behavior in the 1920s and was the first to measure activity cycles in the female rat which correlated with the then recently described vaginal cycle (Richter, 1927). Two other psychologists, Dr. Calvin Stone at Stanford and Dr. Josephine Ball at Johns Hopkins, had reported several experiments dealing with endocrine influences on mating behavior in rats (Stone, 1923) and monkeys (Ball, 1936).

During the 1930s, one nonpsychologist was working consistently on hormone–behavior problems. William C. Young is the man I would nominate as the "father of behavioral endocrinology." Starting at Brown University in 1933, Young, with two of his students, Myers and Dempsey (Young *et al.*, 1933), published the first in a series of experiments which was to extend over a span of more than 30 years and expand into the most impressive single program ever conducted in this field.

Bill Young and many of his students, now eminent in their own right, contributed a very large proportion of what we know today about effects of hormones on behavior and the ways in which such effects are modulated by individual learning and experience. It was due to the editorial judgment of Young that six of the 24 chapters making up the third edition of *Sex and Internal Secretions* dealt exclusively with behavior. In earlier editions, one chapter on behavior had been deemed sufficient.

DOG DAYS AT YALE

After I had been at the American Museum of Natural History for several years, the chairman of my department died following a short illness and I was instructed to wind up current research so that the department could be discontinued. There was no thought that a 28-year-old assistant curator could take over the department after its founder had gone to his reward. I asked for a chance to keep the shop open, and with the moral and financial support of R. M. Yerkes, then chairman of the NRC Committee for Research on Problems of Sex, I won at least a temporary reprieve.

Things went well enough so that my appointment as chairman and curator was made permanent. For reasons of intrainstitutional politics I changed the name to the Department of Animal Behavior, and for a few years actually had great fun running the small department as well as continuing my own research. Nevertheless, a few years of administration were sufficient, and so, in 1948, 10 years after my arrival, I left the Museum to accept a professorship in the Psychology Department at Yale University. Upon my departure, Dr. Lester Aronson became chairman at the Museum, and in the ensuing 26 years the Department of Animal Behavior has been an important research and training center for behavioral endocrinology and animal behavior generally.

One of the first things I wanted to do at Yale was to start a program of research with dogs. At that time, it was generally agreed that some hormones affected behavior by altering functional activity of the CNS, but there was a dearth of direct evidence. It seemed to me the first task was to define the loci of hormonal action, and with this in view I planned an elaborate study of genital reflexes in male dogs. A sequence of steps was involved: (1) condition males to respond to masturbation with erection and ejaculation; (2) castrate the animals and await loss of the genital reflexes; (3) restore the reflexes by systemic administration of TP; (4) perform lumbar transection, which should spare the reflexes but permit survival if bladder function was maintained; (5) demonstrate that erection and ejaculation can be maintained in spinal males by systemic androgen treatment, which would indicate that neurons distal to the point of section were responding to the hormone; (6) finally, expose different segments of the cord, apply androgen topically, and identify the regions or circuits immediately sensitive to testosterone.

As a preliminary step, I examined the effects of castration on mating behavior, and the results were so surprising that the research plans were changed and no spinal operations were ever performed. It turned out that males continued to copulate with receptive females for as long as 5 years after castration without any replacement therapy. Even castration combined with adrenalectomy failed to eliminate mating in sexually experienced animals. It has since been shown that these results depended on unique aspects of the way in which my dogs were maintained and tested; under more normal circumstances, castration occasions fairly prompt loss of intromission and ejaculation, although it may never completely eliminate mounting behavior (Beach, 1970).

Nonetheless, the original findings were valid and important because they compelled me to recognize that dogs are not large rats which bark. They made me aware of the importance of species differences in dependence on gonadal hormones for sexual performance. This realization, combined with the interest in evolutionary changes stimulated by my years at the Museum, had a formative effect which eventually shaped my thinking into a definitely comparative mold. One outcome of this recognition of the obvious was a spate of quasi-theoretical papers based on what I now consider superficial comparisons among man, nonhuman primates, and "lower" mammals.

The evidence was scant and the reasoning was shallow if not spurious, but the papers attracted the attention of various psychiatrists and psychologists who knew little about endocrinology, and some endocrinologists lacking first-hand acquaintance with animal behavior. One result was that I began to receive invitations to lecture and to contribute to various national and international symposia. Psychologists regarded me as an endocrinologist, endocrinologists thought I was a psychologist, and specialists in each discipline forgave me for ignorance in their own area on the assumption that I must be an expert in the other.

DISCOVERIES I ALMOST MADE

Most of us who have been in the research game for a long time can look back and identify near misses that could have resulted in important contributions if we had been just a little bit smarter or if events beyond our control had taken a slightly different turn. I can think of one example of each. First, I will describe the "events-beyond-our-control" instance, and then the "if-I-had-only-been-smarter" example.

In 1951 I attended a Ciba Symposium on Steroid Hormones in London and talked about some experiments on mating behavior in animals. Geoffrey Harris was there, and although he did not deliver a formal paper he contributed to several discussions. In commenting on one of the papers, Harris mentioned his own unpublished research on effects of estrogen implanted into the hypothalamus of rabbits. He was interested primarily in hypophysial-gonadal relations but noted in passing that spayed females with cerebral implants of estrogen showed mating responses comparable to those of an animal in natural estrus.

I was currently studying sexual receptivity in female cats and sexual potency in males, and had worked out reliable, quantitative measures of both functions. I suggested to Harris that he come to New Haven so that we could combine our technical skills in studying the behavioral effects of steroids implanted in various brain regions. He was in favor of the idea, and when I returned to Yale I obtained Hugh Long's agreement to offer Harris a temporary appointment in the Physiology Department at the Medical School.

At the time of the London meetings, Harris was on the faculty at the University of Cambridge, but before our plans for collaborative research could be put into action he was offered a position as head of the laboratory at the Maudsley Hospital and so he never came to Yale. Years later, working at Maudsley with Richard Michael, Harris showed that spayed cats can be brought into behavioral estrus by stilbestrol implants in the hypothalamus (Harris and Michael, 1964).

I had no control over the turn of events in arrangements with Harris, but another near miss could have been prevented if I had been more imaginative. It occurred while I was still at the American Museum of Natural History. The experiment was planned to test the hypothesis that hormones secreted early in life might exert an organizing influence on the developing brain and thus influence the type of mating behavior shown in adulthood. Male rats were castrated on the day of birth and tested for sexual responses to receptive females in adulthood after receiving a series of testosterone injections.

Results showed that neonatal castrates cannot mate normally despite hormone replacement because the penis is too small to permit intromission, and ejaculatory responses are not evoked. Nevertheless, my day-1 castrates responded to TP in adulthood by displaying marked sexual excitement, vigorously mounting the female and showing all of the normal pattern of which they were physically capable. I concluded that brain organization was normal in the absence of postnatal testis secretion (Beach and Holz-Tucker, 1946).

What I did not know, and what was not discovered until 14 years later, was that neonatal castration has a "feminizing" effect on male rats which is reflected in their capacity to display female mating behavior if they are treated with estrogen and progesterone in adulthood. If I had injected my neonatally castrated males with ovarian hormones and tested them for female behavior, I could have anticipated by 14 years a discovery which, when it was finally made, introduced a whole new dimension into behavioral endocrinological research.

FROM THE IVY LEAGUE TO THE GOLDEN GATE

Ten years at Yale were stimulating and productive but after a decade in the same job I was ready for a change, so the offer of a professorship at the

University of California in Berkeley was welcome and I moved in 1958. One of the many attractions was the climate in the Bay Area which would allow me to maintain experimental animals in an outdoor facility and observe their behavior under seminatural conditions 12 months of the year. By 1961 we had our Field Station for Behavioral Research in the hills behind the University, and I was able to breed my own beagles, rear them in small groups, and test their sociosexual reactions in a large field or in small, fenced compounds.

One of the first projects begun after the dog colony was established was to replicate with a canine species those studies which had been done on rodents indicating that prenatal exposure to androgen can behaviorally masculinize and defeminize female animals. In the prologue to this account, I mentioned some of the results, and research along these lines still continues. Behavioral effects of early masculinization extend far beyond the realm of mating behavior. Treated females are socially dominant over normal females, although subordinate to normal males. They are attracted to female rather than to male odors. Their excretions are responded to by other dogs more like male than like female excretions.

The move to California had one unexpected benefit because at about that same time Bill Young moved to the Oregon Regional Primate Center near Portland, where he began a program of experiments on hormonal control of behavior in rhesus monkeys. Young and two associates, Charles Phoenix and Robert Goy, conducted an ambitious series of experiments to analyze the role of hormones in mediation of sociosexual behavior of these nonhuman primates. Their findings concerning effects of prenatal masculinization of females paralleled in many respects my own results with dogs, although there were, of course, important species differences.

The two laboratories kept in close touch and benefited mutually from frequent exchanges of ideas and data. Today, Charlie Phoenix is in charge of behavioral research at the Oregon Primate Center, and Bob Goy has become director of another primate facility at the University of Wisconsin. Behavioral endocrinology flourishes at both institutions.

FINANCIAL SUPPORT AND SOCIAL ACCEPTANCE OF RESEARCH

In the 1930s and well into the next decade, anyone who specialized in research on sexual behavior was apt to be regarded as somewhat odd. Most scientists are about as prudish as laymen. As late as 1947 an invited article which I prepared for *Physiological Reviews* was almost turned down because the editor judged certain passages "indelicate" and likely to embarrass the readership. Twenty years later, the editor of *Science* flatly rejected a paper (not mine) dealing with the effects of surgical denervation of the penis on mating in cats. His explanation was that many subscribers were chemists, physicists, and other kinds of specialists who would be offended by public discussion of genital anatomy and coitus. At a Laurentian Conference organized by Gregory Pincus, an eminent endocrinologist congratulated me after my talk on gonadal hormones and mating behavior and added, "But for the life of me I cannot understand how you worked up the courage to use all those sexual words in public."[1]

As the public's attitude toward open discussion of sexual matters became more tolerant, research into hormones and animal sexuality became more respectable. This change was undoubtedly related to another, a marked increase in funds available for the support of such research.

When I went to Harvard as a postgraduate worker in Lashley's laboratory, he paid my salary from a small grant provided by the NRC Committee for Research on Problems of Sex. From its formation in 1922 until the end of World War II, that small committee, funded by the Rockefeller Foundation, was practically the only source of support for research on sexual functions, especially on sexual behavior. Most of the grants were given for studies of reproductive physiology, and the majority of grantees were endocrinologists. In 1932 the "Sex Research Committee" sponsored publication of an impressive compendium entitled *Sex and Internal Secretions*, which was edited by Edgar Allen and included among its contributors such well-known scientists as E. A. Doisy, E. T. Engle, H. L. Fevold, W. U. Gardner, C. G. Hartman, F. C. Koch, C. R. Moore, W. O. Nelson, O. Riddle, A. E. Severinghaus, P. E. Smith, B. H. Willier, and E. Witschi. As mentioned earlier, there was one chapter on behavior and that was written by a psychologist, C. P. Stone.

From 1937 until about 1953, practically all of my own research depended on grants from the Sex Research Committee. Sometime in the late 1950s I became first a member and finally chairman of the committee. By that time, money for behavioral research was being supplied by the National Science Foundation and by the National Institute of Mental Health in amounts far beyond our committee's ability to compete. Our annual budget, including operating expenses, totaled $50,000. After a great deal of painful deliberation, we decided that the committee had fulfilled its function and should be discontinued because all the first-rate workers were receiving federal funds which were more abundant and provided supple-

[1] Many years later, after a lecture on dog behavior, a lady thanked me for my "very educational" talk. "You know," she gushed, "I just never thought of the term 'bitch' as applied to a dog!"

mentary salaries, publication costs, and more institutional overhead, which our committee could not pay for.

I shall always think of the Sex Research Committee and of the prolonged support given it by the Rockefeller Foundation as an outstanding example of the way in which a small group of individuals with a limited but uncommitted budget can, with foresight and courage, stimulate pioneering research in a difficult area, thus making possible eventual public support of a more massive attack on basically important scientific problems.

FUTURE DEVELOPMENTS

I had not meant to focus quite so heavily on sex-related research, but for many years this was the kind of behavior studied by the majority of behavior scientists with endocrinological interests. It was also the kind of research that eventually paid off so that more funds became available for studying the effects of hormones on other types of behavior. In the last 15 years there has occurred a tremendous increase in research dealing with the behavioral effects of hormones and the effects of behavior on endocrine function.

In 1946 it would have been difficult to list as many as ten Americans whose research dealt primarily with these topics. In 1976 I attended a regional meeting titled the East Coast Conference on Reproductive Behavior. Two-hundred people attended and at least 80% of the papers dealt in one way or another with hormones and their behavioral correlates. The average age of the speakers and listeners could not have exceeded 35 years. The majority of these young people were psychologists, but there were a number of zoologists and a few endocrinologists on the program and in the audience.

In 1970 two of my former students (Richard Whalen and Julian Davidson) and I founded a journal named *Hormones and Behavior*. Research in the area was rapidly increasing and results were being published in many different journals, each devoted to a different discipline and each read by a different group of specialists. It was our belief that much could be gained if reports of such research could be pulled together in one journal and made available to specialists in all areas. Fortunately, the Academic Press shared our views and agreed to publish the journal with no financial guarantees.

The first few years were meager ones but gradually the flow of good papers increased until at present we are fighting the inevitable battle-of-the-backlog and have just been granted a 100% increase in the number of pages for 1977.

All signs point toward continuing expansion of interest and research on behavioral problems. Tremendous strides have already been made toward better and better understanding of the role of hormones in cellular biology. As this and the preceding volume of the present series attest, we have learned a great deal about interdependence of the neural and endocrine systems in the past and stand certain to learn more in the very near future.

The line of advance is clear: from cells to systems to organisms and their behavior. Neuroendocrinology is a broad field encompassing many different levels of organization. At one extreme it deals with interactions between hormones and the neurons, and current research is yielding exciting results in this area. At the other end of the continuum stand those integrative activities of the brain which we know as thought, emotions, and behavior. Investigation of neurohormonal bases for these complex functions is imperative and, I am convinced, has been well begun.

REFERENCES

Ball, J. (1936). Sexual responsiveness in female monkeys after castration and subsequent estrogen administration. *Psychol. Bull.* **33**:811.

Beach, F. A. (1970). Coital behavior in dogs. VI. Long-term effects of castration on mating in the male. *J. Comp. Physiol. Psychol.* **70**:1.

Beach, F. A. (1975). Hormonal modification of sexually dimorphic behavior. *Psychoneuroendocrinology* **1**:3.

Beach, F. A. (1976). Sexual attractivity, proceptivity and receptivity in female mammals. *Horm. Behav.* **7**:105.

Beach, F. A., and Holz-Tucker, A. M. (1946). Mating behavior in male rats castrated at different ages and injected with androgen. *J. Exp. Zool.* **101**:91.

Harris, G. W., and Levine, S. (1962). Sexual differentiation of the brain and its experimental control. *J. Physiol.* **163**:42.

Harris, G. W., and Michael, R. P. (1964). The activation of sexual behavior by hypothalamic implants of estrogen. *J. Physiol.* **171**:275.

Phoenix, C. H., Goy, R. W., Gerall, A. A., and Young, W. C. (1959). Organizing action of prenatally administered testosterone propionate on the tissues mediating mating behavior in the female guinea pig. *Endocrinology* **65**:369.

Richter, C. P. (1927). Animal behavior and internal drives. *Q. Rev. Biol.* **2**:307.

Stone, C. P. (1923). Experimental studies of two important factors underlying masculine sexual behavior: The nervous system and the internal secretion of the testis. *J. Exp. Psychol.* **6**:84.

Young, W. C., Myers, H. I., and Dempsey, E. W. (1933). Some data from a correlated anatomical, physiological and behavioristic study of the reproductive cycle of the female guinea pig. *Am. J. Physiol.* **105**:393.

3

Howard A. Bern

Howard A. Bern was born on January 30, 1920, in Montreal, Canada. He attended the University of California, Los Angeles, receiving his B.A. and M.A. in zoology in 1941 and 1942, respectively. During World War II, he saw active service with the US Army Medical Department partly in the Pacific theater and Captain Bern returned to UCLA in 1946 and received his Ph.D. with Professor Boris Krichesky in zoology (endocrinology) in 1948, at which time he joined the staff of the Zoology Department at Berkeley, attaining the full professorship in 1960. In the early 1950s he helped Professor K. B. DeOme establish the Cancer Research (Genetics) Laboratory, and continues to serve as research endocrinologist in the Laboratory. In the late 1960s he assumed the chairpersonship of the Group in Endocrinology.

Professor Bern has been a visiting professor at the universities of Bristol, Kerala, Tokyo, and Puerto Rico. His honors include a Guggenheim Fellowship in 1951, National Science Foundation Senior Postdoctoral Fellowships in 1958 and 1965, fellow at the Center for Advanced Study in Behavioral Science at Stanford in 1960, Nieuwland Lecturer at the University of Notre Dame in 1972, Eli Lilly Lecturer of the Endocrine Society (U.S.) in 1975, and Walker-Ames Professor at the University of Washington in 1977. He received a Distinguished Teaching Award from the University of California in 1972 and is a member of the National Academy of Sciences.

Professor Bern has served on the editorial boards of the *International Review of Cytology* and a variety of journals, including *Endocrinology*, *General and Comparative Endocrinology*, *Neuroendocrinology*, and *Cancer Research*. He is the author or coauthor of more than 250 papers in his several areas of research interest, coauthor of a *Textbook of Comparative Endocrinology* with Aubrey Gorbman in 1962, and coeditor of *Progress in Comparative Endocrinology* with W. S. Hoar in 1972. Almost three dozen students have received their Ph.D.'s with him. He has served as president of the American Society of Zoologists (1967) and is currently on the Council of the International Society of Neuroendocrinology.

3

A Second-Generation Neurosecretionist Looks at His Field

HOWARD A. BERN

This highly personal essay will attempt to outline the development of what I can only call a "second-generation" neurosecretionist. My entrée into the field of neurosecretion began at a time when the fundamental concepts had been soundly established, not only by the recurring discovery of stainable secretory-appearing neurons in representatives of one animal group after another but also by the delineation of the neurosecretory pathway—the obligatory connection of these stainable neurons with a tract that led to a neurohemal organ wherefrom the products of these cells could be released.

I had been intrigued by the phenomenon of neurosecretion since early years of teaching my course in comparative endocrinology at Berkeley. In 1953, Robert B. Clark, now professor of zoology at the University of Newcastle-upon-Tyne, arrived as a visiting assistant professor in the Berkeley Zoology Department and was soon immersed in his study of polychaete worm brains and their varying structure in a variety of species (especially nephthyids). He became entranced by glandular areas in these brains (*cf.* Clark, 1966), and I had the pleasure of raising the specter of neurosecretion in regard to his findings. Polychaete annelids, indeed all annelids, do have neurosecretory cells in their brains, but—as a cautionary lesson to us—much of what Clark was studying really concerned the presence of exocrine glands in the brain: integumentary structures that had become incorporated into the neural tissue during brain development. Clark went on to open up physiological experimentation on polychaete neuroendocrinology at Bristol, a field later pursued avidly by his student David

HOWARD A. BERN • Department of Zoology and Cancer Research Laboratory, University of California, Berkeley, California 94720.

W. Golding (*cf.* 1972), who was a postdoctoral with me at Berkeley in the mid-1960s, and by Denis G. Baskin (*cf.* 1976), then a student of R. I. Smith's at Berkeley. I remain indebted to Bob Clark for stimulating me to think about neurosecretion as a truly ubiquitous phenomenon in the animal kingdom.

Our own studies of neurosecretion began simultaneously at both ends of the vertebrate organism. My undergraduate honors students of the time, Lowell D. Wilson and J. Arthur Weinberg, lent their talents to defining the hypothalamic neurosecretory system in a local treefrog (Wilson *et al.*, 1957). Captivated by the ability of paraldehyde-fuchsin to pick out special neurons from an otherwise largely incomprehensible mass of nervous tissue, we sent our first findings to Ernst Scharrer for his expert confirmation. Scharrer encouraged us both gently and genteelly to proceed, scratching his head, I suspect, at our naive excitement at seeing in our own sections, finally, what he had been looking at for 30 years or more in a variety of vertebrates. Nevertheless, our study was a thorough one and provided a "good learning experience." If the published paper was ever cited to any extent by anyone other than my student Carolyn G. Smoller, who later used it as the light-microscope basis for her detailed ultrastructural analysis of the treefrog hypothalamic neurosecretory system (Smoller, 1966), I have yet to learn of it.

The other end of the nervous system in fishes is equipped also with secretory-appearing cells, albeit not staining with "standard" neurosecretion stains. C. G. Speidel in 1919 had opened the field of neurosecretion by clearly characterizing these caudal neurosecretory cells as glandular neurons and by contradicting U. Dahlgren's dogmatic earlier contention that these neurons were involved in the innervation of electric organs in certain skates. The reward for Speidel's virtue was the immortalization (temporarily) of Dahlgren by naming the cells after him, and the passing of Speidel's fine study (it is instructive to reread this paper as an example of careful analytical observation) into virtual oblivion (also temporarily). It was Ernst Scharrer's revival of the Speidel paper that led Irvine Hagadorn and me to start our own examination of the "Dahlgren" cells in an elasmobranch fish (Bern and Hagadorn, 1958). We chose the shovelnose guitarfish because (1) it was big and (2) it did not bite. Both of us spent hour after hour of a summer month at the Scripps Institution of Oceanography at La Jolla painstakingly removing muscle and opening neural arches to expose the caudal spinal cord, to learn weeks later that two cuts through the vertebral column would allow removal in a few minutes' time of the intact spinal cord in the region in which we were interested. We spent more time swimming after this devastating discovery.

These early intrusions into the field of neurosecretion were characterized, as was most of the work at that time, by a preoccupation with morphology. The esthetics of neurosecretory neurons are not to be underestimated, and, along with many others, we indulged ourselves fully in our appreciation of the brilliant staining reactions of these extraordinary neurons (the ordinary neurons—called *neurones banales* by our French and Belgian colleagues—had little to offer by comparison). This domination by morphology admittedly extended too long; it enjoyed a second ascendancy with the expansion of ultrastructural observations. Again, the micromorphology of neurosecretory neurons rivaled in esthetic value the images of light microscopy, despite the retrogressive replacement of color with black-and-white. Some of the electron micrographs of neurosecretory neurons are among the classical contributions of the then-emerging field of electron microscopy. My association with Richard S. Nishioka, which began in 1959, was most fortunate in regard not only to ultrastructural aspects of our research program but also to the later emergence of a substantial functional orientation. His continuing collaboration has been indispensable to whatever contributions our laboratory has been able to make to the field of neurosecretion.

Preoccupation with stains and electron-dense granules undoubtedly delayed the onset of physiologically oriented studies, if our own laboratory was any reflection of the field as a whole. Nevertheless, the morphological orientation invited the investigation of a whole series of putative neurosecretory and neuroglandular organs. It became evident with electron microscopy that staining images could be deceptive, that the "specific" stains were specific neither for neurosecretory phenomena nor for secretory products of any kind (*cf.* Bern, 1966). Staining could be due to accumulations of pigments or to a variety of cell organelles—mitochondria, lysosomes, and rhabdomes included—or even to viral particles. Among the possible neurosecretory organs that evanesced during electron microscope examination were rudimentary insect ocelli, octopus epistellar bodies and squid parolfactory vesicles, and gastropod follicle glands. The octopus and squid organs, long suspected of neuroglandular activity, proved to be non-visual photoreceptors, a serendipitous discovery raising major questions about photoregulation in these animals (Bern, 1967*a*). It was also recognized that the presence of granules and vesicles with diameters in the 100–300 nm size range did not *ipso facto* define neurosecretion. Such granules, especially in gastropod molluscs, could be associated with neurotransmitters (peptidergic?).

It was also true that staining might be due to granules/vesicles not in the usual elementary granule size range. Concern with this problem led to comparative studies of the two parts of the neurohypophysis in birds: the median eminence and the neural lobe. Beginning with Hideshi Kobayashi's visit to our laboratory in 1960 and our analysis of the parakeet neurohypophysis, and continuing in collaboration with Donald S. Farner and L. Richard Mewaldt on the whitecrowned sparrow, Nishioka and I raised questions about the cytological basis for the various paraldelyde-fuchsin-staining regions of the neurohypophysis (*cf.* Bern *et al.*, 1966). It seemed wise to conclude that this nonspecific "specific stain" can color a variety of membrane-limited entities. For some years, my laboratory was distracted by granule and vesicle measurements, as we developed an intimate understanding of the bird median eminence. This dedication led to Professor Farner's categorization of me as a biologist whose picture of the whitecrowned sparrow was that of a hypothalamus with feathers.

Particular attention at this time (around 1960) was focused on the mode of elaboration of neurosecretory granules. At about the same time, Ernst Scharrer (Scharrer and Brown, 1961) and Nishioka, Hagadorn, and I (Bern *et al.*, 1961) began to recognize the strict association of the elementary neurosecretory granules with the Golgi apparatus. Despite some fits and starts, the neurosecretory neuron emerged as a classical example of a protein-exporting cell, and autoradiographic and biochemical studies of the rat supraoptic neurons have confirmed this status (*cf.* Nishioka *et al.*, 1970).

R. B. Clark joined H. Heller in organizing the Third International Symposium on Neurosecretion in Bristol in 1961 (Heller and Clark, 1962), and Hagadorn and I found ourselves on a podium with the masters of the field. E. deRobertis was there, heretically lumping neurotransmission with neurosecretion, while most of the rest of us, led by the Scharrers, W. Bargmann, and H. Heller, defended the special status of neurosecretion. This defense continues to the present; despite the similarities between neurosecretory and ordinary neurons, the hormone-secreting neuron remains a special contribution of the nervous system to classical endocrine function. The argument will also obviously continue; W. F. Ganong (1975) has again recently questioned the special status accorded the neurosecretory neuron. At Bristol, when H. M. Gerschenfeld stoutly presented some studies on *Aplysia* neurons, Ernst Scharrer more than implied that studies of the electrophysiology of such "ordinary" neurons detracted from the special business at hand (there was then no electrophysiology of neurosecretory neurons). In retrospect, in view of the burgeoning interest in gastropod neurosecretory neurons (Strumwasser, Gainer, Davis, Coggeshall, Kandel, Kupfermann, Arch, *et al.*), it would be interesting to know whether Gerschenfeld was not indeed studying neurosecretory neurons even then!

The state of appreciation of the art and science of neurosecretion by biologists in the early 1960s was not overwhelmingly positive. I was plainly shocked when I learned that in their colossal project, *Structure and Function in the Nervous Systems of Invertebrates*, T. H. Bullock and G. A. Horridge (1965) had no intention of including any discussion of neurosecretion. For neurobiologists of the time, this topic was simply not germane. I remember mounting a campaign to recruit Ted Bullock to the faith: Hagadorn and I put up a roomful of demonstration microscopes equipped with explanatory cards, covering the animal kingdom as best we could. Bullock was on sabbatical leave in Berkeley and in a mood not to be bothered by anyone. Nevertheless, we convinced him to attend the special viewing, and the need for a chapter on neurosecretion became apparent. *Venit, videt, vincitur*. It also became suddenly apparent that if there was going to be such a chapter Hagadorn and I were going to have to write it (better candidates had shown a remarkable lack of enthusiasm). It was written (Bern and Hagadorn, 1965).

In the late 1950s, while Hagadorn and I were looking for and at caudal neurons, M. Enami in Japan had made a substantial contribution by establishing the existence of a caudal neurosecretory system in teleost fishes analogous to the cranial hypothalamoneurohypophyseal system. The caudal system is also composed of aggregates of neuron somata leading by a definable tract to a variably developed neurohemal organ, the urophysis. The "neurosekretorische Bahn" of Bargmann was being traveled again in the tail end of fishes. Professor Enami was invited to present his important studies at the Second International Symposium on Comparative Endocrinology at Cold Spring Harbor in 1958. The death of this brilliant scientist, still a young man, en route to the Symposium was a sad event. In my own case, it led to the reintensification of my interest in the caudal system and to the decision not to let the essential contribution of Enami remain unappreciated.

To this day, I remain an unashamed advocate of the field of "urophysiology." In the early 1960s Noboru Takasugi added a new pretty color to the battery of stains available for studying neurosecretory systems—acid violet (Takasugi and Bern, 1962). A stain was certainly needed for the caudal system since it was not reactive to the usual "Gomori" methods then available. Katsutoshi Imai, Enami's last student, joined us in some physiological studies, and Gunnar Fridberg (who was looking at elasmobranch cells in the snows of Stockholm during the same period that Hagadorn and I were on the sands of southern California) added much good functional morphology to our efforts, including the still-significant finding that the caudal system, if completely removed from the fish, may regenerate from remaining caudal ependymal elements (Fridberg *et al.*, 1966). The special regenerative properties of neurosecretory neurons invite further study; regrettably, they are still receiving too little.

The caudal system provided us with an opportunity to contribute electrophysiological information to our knowledge of neurosecretory neurons. Kinji Yagi arrived as a postdoctoral to penetrate neurosecretory neurons of this system (and also of the leech brain) and to record their responses to functional manipulations (Yagi and Bern, 1965). He was joined later by Shizuko Iwasaki. Earlier, H. Morita, T. Ishibashi, and M. V. L. Bennett had provided us with the first definitive recordings from any neurosecretory neurons, and in all cases these workers chose the caudal neurons. Shortly after, E. Kandel gave us information on goldfish hypothalamic neurons. It began to emerge that there was a distinctive electrophysiological feature of neurosecretory neurons in both vertebrates and invertebrates: the presence of a long-duration action potential (too often confused with more rapid conduction, which does not occur). Although information from mammals is not unequivocal, the long-duration action potential is a feature of the neurosecretory cells of at least nonhomeothermic animals (*cf.* Bern and Yagi, 1965), and may be related to the special problems of release of large granules at terminals by exocytic processes.

The caudal neurosecretory system of fishes provides a model for the study of a strictly neural endocrine system uncomplicated, as is the neurohypophysis, with any association with an epithelial organ. It is present in almost all groups of fishes and is best developed in the teleosts—the higher bony fishes (*cf.* Fridberg and Bern, 1968; Bern, 1969). Research on this system has again been dominated by morphological studies, but a few hardy "urophysiologists" (and "urobiochemists") persevere. Regrettably, so far it is not physiology but pharmacology that has largely been investigated; succinctly, we really do not know what this system and its hormones, the urotensins, do—in a functional sense—in the fishes that possess them.

There are few areas of modern physiological-biochemical study where the number of workers is so few that a single investigator can feel himself as occupying an "umbilical" position in the research that is going on. Immodestly, this is the role that seems to be mine to play at the present. My association with Daniel K. O. Chan began when I left an unwanted gift of urophysial material collected from mullets in Naples to be tested on cardiovascular functions in eels in I. Chester Jones's laboratory at Sheffield, where Chan was still a graduate student (in 1966). Karl Lederis, then at Bristol with Hans Heller, entered urophysial research during a year in my laboratory at Berkeley when he discovered the general smooth-muscle-contracting action of urotensin II (Lederis, 1969). Lederis had been warned by Professor Heller not to waste his time in the seductive but unrewarding banlieu of "urophysiology." At the same time, Flor Lacanilao, then my graduate student from Manila, discovered the arginine vasotocin-like properties of urotensin IV (Lacanilao, 1969). Shortly thereafter, my postdoctoral, Allan Berlind, found evidence for carrier proteins ("urophysins") possibly analogous to the neurophysins of the hypothalamic neurosecretory system (Berlind *et al.*, 1972), and further support for urophysins is coming both from Lederis's group and from Robert Gunther in my laboratory.

Professor Kobayashi and his student Tomoyuki Ichikawa have pressed forward with their original description (along with Yasumasu and Ootani) of a factor depressor and antidiuretic in rats (presumably urotensin I), and also with the proof of the existence of large quantities of acetylcholine in the urophysis (substantially more than in any other tissue yet investigated) (Ichikawa and Kobayashi, 1978). Kobayashi and I have recently solidified our collaborative efforts on the urophysis with the help of a United States/Japan Cooperative Science Grant. Meanwhile, the studies of Lederis and Chan and their colleagues on urotensins I and II have done much to characterize these two putative neurohormones (*cf.* Lederis *et al.*, 1974; Chan and Bern, 1976; Bern and Lederis, 1978).

The laboratories of the investigators mentioned above represent the active centers of functional studies on urophysial principles. Chan at Hong Kong, Kobayashi at Misaki, Lederis at Calgary, and members of my own laboratory at Berkeley in continuing association with Geschwind at Davis and Pearson (Geschwind's former postdoctoral) at Duarte not only correspond regularly but also meet to talk in person. The international aspect of the efforts is clearly evident, and the friendly and open exchange among the several of us is generally a happy one indeed. I emphasize this highly positive situation because, again immodestly, I think it is exemplary of how scientific collaborations can work. The branches branch again and intertwine. Enthusiasm and a wish to advance a small and largely unrecognized field provide the necessary motivation. And, whereas the ubiquitous caudal neurosecretory system must be of some fundamental importance to fishes as well as to fish biologists, there are other rewards. Urotensin I, which is hypotensive in mammals (but not in fishes), is a peptide of potential major pharmaceutical importance, as Lederis correctly continues to emphasize. The little urophysis in the tail of fishes may yet prove to be a storehouse for a biological agent long sought by clinicians for counteracting hypertension and the visceral vasoconstriction characterizing the phenomenon of shock. Comparative neuroendocrinology has its inherent justification as a field of investigation; the additional positive benefits are as unexpected specifically as they are predictable generally (*cf.* Bern, 1977).

As the 1960s proceeded, I found myself giving general presentations on the properties of neurosecretory neurons to a variety of biological audiences: to the comparative endocrinologists at Oiso, Japan, in 1961 (Bern, 1962); to cell biologists in Eugene, Oregon, in 1962 (Bern, 1963); to mammalian endocrinologists in London in 1964 (Bern and Yagi, 1965); and to neurobiologists at St. Andrews, Scotland, in 1965 (Bern, 1966). This "exposure" led to a certain pontifical status, which was brought into challenging perspective by productive discussions with Francis G. W. Knowles in 1965. It became evident that there could develop a literature of positions and counterpositions by Knowles and me, in which we would clutter the field without clarifying it. We decided with enthusiasm to become nonpolemicists. The years 1965 and 1966 found us jointly proposing an expanded definition of neurosecretion in *Nature* (Knowles and Bern, 1966), writing an expansion of our views as a chapter for Martini and Ganong's *Neuroendocrinology* (Bern and Knowles, 1966) and surveying the field before the world's neurosecretionists at Strasbourg by judicious division of the subject matter (Knowles, 1967; Bern, 1967*b*). I remain indebted to my deceased friend Francis for his forthright conversion of both of us to a policy of arguing out our differences in conversations and letters rather than in the scientific literature.

One area of initial conflict between Knowles and me concerned the problem of neurosecretory neurons projecting directly to other organs over which they exerted control, rather than secreting neurohormones into vascular pathways as typical endocrine structures and exerting control from a distance (the difference between anatomical addressing and chemical addressing of the specific messages). Terms such as "neurosecretory innervation" and "neurosecretomotor control" had been used. Knowles and I compromised our views by admitting to the province of neurosecretion such innervation only when it extended to other endocrine tissues, underlining the central endocrine function of neurosecretion. Innervation of the teleost pituitary would be a classical example of such neurosecretory activity (*cf.* Zambrano *et al.*, 1972). In retrospect, I feel this was an error. Just as one can speak of local hormonal phenomena (histamine mediated, for example, or as occurs in embryonic inductions), one should be able to speak of local neurohormonal phenomena, whether the responding tissue be an endocrine effector or a nonendocrine effector: exocrine gland, muscle, sensory organ, even other central neurons. The basic issue is still one of the nonsynaptic relationships—of chemical as opposed to anatomical addressing. Berta Scharrer (*cf.* 1972) has written extensively and incisively in the area of conventional and nonconventional junctions, and I need not discuss the issues further here. Suffice it to say that there are many ways in which neurons communicate with each other and with other cells. Knowles and I had oversimplified the problem, I feel, but it was an heuristic solution, as are scientific doctrines generally. The "heure" lasts longer in some cases than in others, but change is certain.

It has been my fortune as a member of the second generation to know well and to profit from contacts with many of the real pioneers and the international parentage of the field of neurosecretion. Berta and Ernst Scharrer, Wolfgang Bargmann, Manfred Gabe, Lucie Arvy, K. Karunan Nayar, Francis Knowles, Hans Heller, Henriette Herlant-Meewis, John Sloper, Lewis Kleinholz, Fred Stutinsky, Baldassare de Lerma, John Welsh, and John Green I number among my friends and my teachers. As should be true in any friendly teacher-student relationship, it was occasionally possible for the student to instruct the teacher. I remember the strength of argument needed to convince Manfred Gabe to include electron micrographs in his unequaled encyclopedic monograph, *Neurosecretion* (1966). Gabe relied only on his own illustrations whenever possible. I am not sure that he considered the electron microscope to be a frivolous toy, but he never used it himself. Our compromise was an appendix to his volume, with electron micrographs provided as best I could—and at the last moment—as a result of Richard Nishioka's expertise in my own laboratory. The volume suffered from the uniqueness of source, but it would have been more unique to have included no ultrastructure!

A most impressive aspect of the field of neurosecretion which the third generation now dominates is the growth in the quantity and quality of the literature. Modern students of the field have added their sophisticated techniques at all levels from the histological (wholemount staining, as Chandran Unnithan did in my laboratory, and neuron injection and filling, as Carol Mason did in her studies with Hugh Rowell and me) through the ultrastructural (enzymological) and electrophysiological to the biochemical (radioisotopic) and molecular (granule isolation, "dissection," and characterization). Organ culture of intact neurosecretory systems is a reality, and analytical studies of control at both systemic and intracellular levels are now possible. At the same time, physiological elucidation of systems barely thought to exist 15 years ago has been dramatic. There are positively flourishing fields of polychaete worm neuroendocrinology, gastropod mollusc neuroendocrinology, and echinoderm neuroendocrinology (*cf.* Clark and Olive, 1973; Golding, 1974). Unlike the more optimistic Durchon (1967), I would have visualized the expansions of these fields that occurred as possible but hardly likely. The Berkeley contributions of Irvine Hagadorn, David Golding, Paul Schroeder, and Ralph Smith's student Denis Baskin to annelid endocrinology and of Leonard Simpson, Richard Nishioka, and Walter Miller to mollusc endocrinology were real and pioneering, but only suggest what has developed in several laboratories around the world. Haruo Kanatani also did some of his echinoderm work at our Bodega Marine Laboratory.

In the early 1960s, Hagadorn and I could summarize invertebrate neuroendocrinology in part of a chapter; a few years later, Gabe did this in encyclopedic fashion in a major part of his book. One can predict that no one will attempt any encyclopedic compilation again. As for advances at the cellular and subcellular level, the few pages devoted to this material in our chapter for Bullock and Horridge can be compared with the depth and breadth of information which Carol A. Mason and I had to summarize 10 years later for the *Handbook of Physiology* (Mason and Bern, 1977). Once the property of a few comparative endocrinologists, the biology of neurosecretion has come of age: not only does it provide the central substrate for the whole field of neuroendocrinology, but also it has become essentially inseparable from the realm of modern neurobiology.

ACKNOWLEDGMENT

It seems only proper even in a nonresearch paper to acknowledge the continued support by the National Science Foundation for our program of study on neurosecretion. Beginning in 1959, we have received considerable encouragement in both a financial and a general professional sense, and this has allowed us not only to pursue our descriptive and experimental investigations but also to contribute to the development of the changing concept of neurosecretion.

REFERENCES

Baskin, D. G. (1976). Neurosecretion and the endocrinology of nereid polychaetes. *Am. Zool.* **16**:107.

Berlind, A., Lacanilao, F., and Bern, H. A. (1972). Teleost caudal neurosecretory system: Effects of osmotic stress on urophysial proteins and active factors. *Comp. Biochem. Physiol.* **42A**:345.

Bern, H. A. (1962). The properties of neurosecretory cells. *Gen. Comp. Endocrinol. Suppl.* **1**:117.

Bern, H. A. (1963). The secretory neuron as a doubly specialized cell. In Mazia, D., and Tyler, A. (eds.), *The General Physiology of Cell Specialization*, McGraw-Hill, New York, pp. 349–366.

Bern, H. A. (1966). On the production of hormones by neurones and the role of neurosecretion in neuroendocrine mechanisms. *Soc. Exp. Biol. Symp.* **20**:325.

Bern, H. A. (1967*a*). On eyes that may not see and glands that may not secrete. Presidential Address to the American Society of Zoologists. *Am. Zool.* **7**:815.

Bern, H. A. (1967*b*). The hormonogenic properties of neurosecretory cells. In Stutinsky, F. (ed.), *Neurosecretion* (Fourth International Symposium on Neurosecretion), Springer-Verlag, Berlin, pp. 5–7.

Bern, H. A. (1969). Urophysis and caudal neurosecretory system. In Hoar, W. S., and Randall, D. J. (eds.), *Fish Physiology*, Vol. II, Academic Press, New York, pp. 399–418.

Bern, H. A. (1977). Some possible contributions of comparative endocrinology to mammalian and human endocrinology. *Dobutsugaku Zasshi* (*Zool. Mag.*) **86**:1.

Bern, H. A., and Hagadorn, I. R. (1958). A comment on the elasmobranch caudal neurosecretory system. In Gorbman, A. (ed.), *Comparative Endocrinology* (Proceedings of the Columbia University Symposium on Comparative Endocrinology), Wiley, New York, pp. 725–727.

Bern, H. A., and Hagadorn, I. R. (1965). Neurosecretion. In Bullock, T. H., and Horridge, G. A., *Structure and Function in the Nervous Systems of Invertebrates*, Vol. 1, Freeman, San Francisco, pp. 353–429.

Bern, H. A., and Knowles, F. G. W. (1966). Neurosecretion. In Martini, L., and Ganong, W. F. (eds.), *Neuroendocrinology*, Vol. 1, Academic Press, New York, pp. 139–186.

Bern, H. A., and Lederis, K. (1978). The caudal neurosecretory system of fishes in 1976. *Seventh International Symposium on Neurosecretion*, Leningrad (in press).

Bern, H. A., and Yagi, K. (1965). Electrophysiology of neurosecretory systems. *Proc. II Intl. Congr. Endocrinol.* **83(I)**:577.

Bern, H. A., Nishioka, R. S., and Hagadorn, I. R. (1961). Association of elementary neurosecretory granules with the Golgi complex. *J. Ultrastruct. Res.* **5**:311.

Bern, H. A., Nishioka, R. S., Mewaldt, L. R., and Farner, D. S. (1966). Photoperiodic and osmotic influences on the ultrastructure of the hypothalamic neurosecretory system of the white-crowned sparrow, *Zonotrichia leucophrys gambelii. Z. Zellforsch. Mikrosk. Anat.* **69**:198.

Bullock, T. H., and Horridge, G. A. (1965). *Structure and Function in the Nervous Systems of Invertebrates*, Vol. 1 and 2, Freeman, San Francisco.

Chan, D. K. O., and Bern, H. A. (1976). The caudal neurosecretory system—a critical evaluation of the two-hormone hypothesis. *Cell Tiss. Res.* **174**:339.

Clark, R. B. (1966). Secretory activity in the nervous system. In Barnes, H. (ed.), *Some Contemporary Studies in Marine Science*, Allen and Unwin, London, pp. 129–154.

Clark, R. B., and Olive, P. J. W. (1973). Recent advances in polychaete endocrinology and reproductive biology. *Oceanogr. Mar. Biol. Annu. Rev.* **11**:175.

Durchon, M. (1967). *L'Endocrinologie des Vers et des Mollusques*, Masson, Paris.

Enami, M. (1959). The morphology and functional significance of the caudal neurosecretory system of fishes. In Gorbman, A. (ed.), *Comparative Endocrinology*, Wiley, New York, pp. 697–724.

Fridberg, G., and Bern, H. A. (1968). The urophysis and the caudal neurosecretory system of fishes. *Biol. Rev.* **43**:175.

Fridberg, G., Nishioka, R. S., Bern, H. A., and Fleming, W. R. (1966). Regeneration of the caudal neurosecretory system in the cichlid teleost *Tilapia mossambica. J. Exp. Zool.* **162**:311.

Gabe, M. (1966). *Neurosecretion*, Pergamon, New York.

Ganong, W. F. (1975). Book review: *Neurosecretion: The Final Pathway* (eds. F. Knowles and L. Vollrath). *Science* **190**:44.

Golding, D. W. (1972). Studies in the comparative neuroendocrinology of polychaete reproduction. *Gen. Comp. Endocrinol. Suppl.* **3**:580.

Golding, D. W. (1974). A survey of neuroendocrine phenomena in non-arthropod invertebrates. *Biol. Rev.* **49**:161.

Heller, H., and Clark, R. B. (eds.) (1962). *Neurosecretion. Mem. Soc. Endocrinol.*, Vol. 12.

Ichikawa, T., and Kobayashi, H. (1978). Acetylcholine in the urophysis and the release of urophysial hormones by neurotransmitters *in vitro. Seventh International Symposium on Neurosecretion*, Leningrad (in press).

Knowles, F. (1967). Neuronal properties of neurosecretory cells. In Stutinsky, F. (ed.), *Neurosecretion*, Springer-Verlag, Berlin, pp. 8–19.

Knowles, F., and Bern, H. A. (1966). The function of neurosecretion in neuroendocrine regulation. *Nature (London)* **210**:271.

Lacanilao, F. (1969). Teleostean urophysis: Stimulation of water movement across the bladder of the toad *Bufo marinus*. *Science* **163**:1326.

Lederis, K. (1969). Teleostean urophysis: Stimulation of contractions of bladder of the trout *Salmo gairdnerii*. *Science* **163**:1327.

Lederis, K., Bern, H. A., Medakovic, M., Chan, D. K. O., Nishioka, R. S., Letter, A., Swanson, D., Gunther, R., Tesanovic, M., and Horne, B. (1974). Recent functional studies on the caudal neurosecretory system of teleost fishes. In Knowles, F., and Vollrath, L. (eds.), *Neurosecretion: The Final Neuroendocrine Pathway*, Springer-Verlag, Berlin, pp. 94–103.

Mason, C. A., and Bern, H. A. (1977). Cellular biology of the neurosecretory neuron. In *Handbook of Physiology* Sec. 1: *The Nervous System*, Vol. I, *Cellular Biology of Neurons*, Part I, American Physiological Society, Bethesda, pp. 651–690.

Nishioka, R. S., Zambrano, D., and Bern, H. A. (1970). Electron microscope radioautography of amino acid incorporation by supraoptic neurons of the rat. *Gen. Comp. Endocrinol.* **15**:477.

Scharrer, B. (1972). Neuroendocrine communication (neurohormonal, neurohumoral and intermediate). *Progr. Brain Res.* **38**:7.

Scharrer, E., and Brown, S. (1961). Neurosecretion. XII. The formation of neurosecretory granules in the earthworm *Lumbricus terrestris*. *Z. Zellforsch. Mikrosk. Anat.* **54**:530.

Smoller, C. G. (1966). Ultrastructural studies of the developing neurohypophysis of the Pacific treefrog, *Hyla regilla*. *Gen. Comp. Endocrinol.* **7**:44.

Speidel, C. G. (1919). Gland-cells of internal secretion in the spinal cord of the skates. *Carnegie Inst. Washington Publ.* **13**:1.

Takasugi, N., and Bern, H. A. (1962). Experimental studies on the caudal neurosecretory system of *Tilapia mossambica*. *Comp. Biochem. Physiol.* **6**:289.

Wilson, L. D., Weinberg, J. A., and Bern, H. A. (1957). The hypothalamic neurosecretory system of the treefrog *Hyla regilla*. *J. Comp. Neurol.* **107**:253.

Yagi, K., and Bern, H. A. (1965). Electrophysiologic analysis of the response of the caudal neurosecretory system of *Tilapia mossambica* to osmotic manipulations. *Gen. Comp. Endocrinol.* **5**:509.

Zambrano, D., Nishioka, R. S., and Bern, H. A. (1972). The innervation of the pituitary gland of teleost fishes. In Knigge, K. M., Scott, D. E., and Weindl, A. (eds.), *Brain–Endocrine Interaction. Median Eminence: Structure and Function*, Karger, Basel, pp. 50–66.

4

Emanuel M. Bogdanove

Emanuel M. Bogdanove (known to many as "Manny") was born February 20, 1925, in New York City. He attended Hunter College Model and Townsend Harris High Schools, and then City College, CCNY (1940–1943 and 1945–1946), where he majored in biology and chemistry, with time out to serve in the Army of the United States. His formal education continued at Wayne University (1946–1948) and the State University of Iowa (1948–1953), where he received the M.S. degree in physiology (1950) and the Ph.D. degree in anatomy (1953). In 1953 he joined the Department of Anatomy at Albany Medical College, where he remained until 1961, leaving to serve as associate professor and then professor of anatomy and physiology at Indiana University in Bloomington (1961–1971). He has served on the editorial boards of *Endocrinology* (1961–1973) and *Endocrine Research Communications* (1973 to the present), and on the Endocrinology Study Section, NIH (1969–1973). Since 1971 he has been professor of physiology and head of endocrine physiology at the Medical College of Virginia.

4

"*Gullible's Travails*": *or How I Eventually Discovered the "Implantation Paradox"*

EMANUEL M. BOGDANOVE

The editors asked me, in writing on my role in the history of neuroendocrinology, to focus mainly on the "implantation paradox." In that frame of reference, I must discuss three periods of my scientific life. These might be described as "preconceptive" (in several senses), "conceptive," and "gestational" phases in the development of a view of brain-pituitary-gonadal interplay (Fig. 1) which I guess I "pioneered." Although I shall emphasize the conceptive phase, the flavor of that adventure can be savored best in the light of what has so far followed, and insight into how I managed to become a "pioneer" requires some understanding of my preconceptive period.

PRECONCEPTION (*1950–1958*)

I had not been a notably clever boy and, as a young man of 25, I still lacked confidence in my own intelligence. I relied heavily on the seemingly awesome intellectual equipment of my teachers, particularly the three who were perceptive (or kind) enough to help me over the hazards of education to the Ph.D.: N. S. Halmi, W. R. Ingram, and W. O. Nelson. I did have confidence in my brawn—I never doubted that, for me, endurance was

EMANUEL M. BOGDANOVE • Department of Physiology, Medical College of Virginia, Virginia Commonwealth University, Richmond, Virginia 23298.

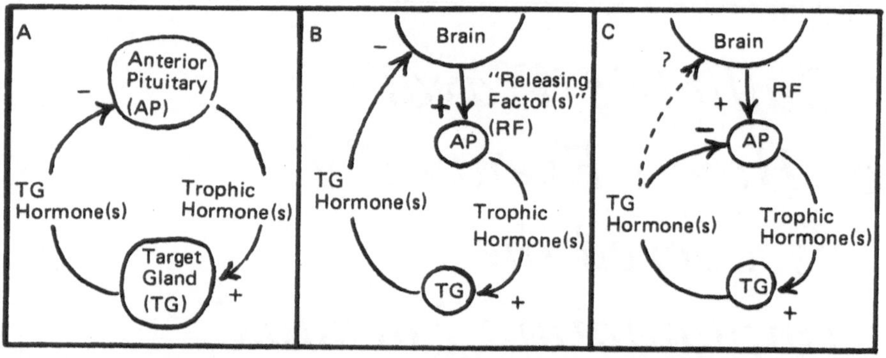

FIGURE 1. Some historic concepts of long-loop feedback. A: Moore-Price hypothesis (see *Pioneers in Neuroendocrinology*, Vol. 1, pp. 229–235). Note that only a "direct" feedback is shown. This idea was proposed 5 years before the first suggestion (Friedgood, 1970) of neurohumoral involvement in anterior pituitary regulation. B: Flerkó-Szentágothai hypothesis (as stated in Flerkó and Szentágothai, 1957) which invoked only "indirect" feedback. This idea was based on a belief that direct feedback could not occur. C: Bogdanove hypothesis (Bogdanove, 1962, 1964) which attempted to integrate known neural and feedback influences to the extent possible at the time, and was based on a belief that the evidence for "indirect" feedback was equivocal (see Fig. 3).

merely persistence. It would never have occurred to me (in those days) that I could not always run or swim farther, or lift more, or collect more data, simply by *continuing*.

The preceding quarter century of my life certainly was not without some relevance, but I have dated this period from the time I first became aware that the hypothalamus might be involved in the control of the anterior pituitary gland. The idea that the hypothalamic-pituitary portal venous system might be a functional link between the two great regulatory systems of the body, the nervous system and the endocrine system, appealed to me. For one thing, it smacked of revolution, because I had been taught that these two great controllers held independent hegemony over different sets of physiological adjustments. (Anyone who did not also thrill to the idea of revolution at the age of 25 should not read this chapter.) There was another, peculiarly local, reason.

In 1950 I was a moderately advanced graduate student in anatomy at the University of Iowa. I had done stints elsewhere (in pharmacology at Wayne University and in physiology at Iowa), but had been drawn to this department by the twin brilliances of the two full professors, Ingram and Nelson. However, mindful perhaps of Daedalus and Icarus, I did not dare to work directly under either. I greatly admired being "at the top," but did not aspire to it.

Ingram, although he had had his start in endocrinology (*Pioneers in*

Neuroendocrinology, Vol. 1), had subsequently advanced far in the neural sciences and now gave little thought to what he called the "infernal" secretions. Nelson, one of the early giants in the study of sex endocrinology, viewed the nervous system as something enormously complex, and quite beyond either his own purview, or that of anyone who hoped to get somewhere with a research problem within a single lifetime. A graduate student who chose to work under either man effectively cut himself (or herself) off from the other. I wanted to learn what I could from both. Investigating this newly discovered brain-pituitary linkage made that possible.

During my "prepreceptor" phase of graduate education in this dichotomous department, I had somehow become closely allied with a young instructor, "Nick" Halmi. Halmi had arrived in the department a few months after I had, following an 11-month stint in Chicago. The University of Chicago had helped him to immigrate after he and his father had left their native Hungary (appropriately, by stealing[1] an airplane). At Chicago, he had enormously enjoyed his contacts with a fellow Hungarian, George Gomori, but little else. He brought with him to Iowa City a fantastic knowledge of the literature, a marvelous eye for histological detail, and an armamentarium of Gomori's most useful stains. With these, he set out to do for the pituitary of the rat what Romeis (1940) had done for the human hypophysis.

I soon was earning $60 a month (and a place, the lab, in which to eat and sleep, albeit illegally), and keeping too busy to get into *too* much mischief, by assisting with Halmi's efforts to identify the pituitary cells of origin of the trophic hormones. The method was to trick the rat pituitary into selective hyperplasia or depletion of its various (then Greek-lettered) distinguishable cell types and then to attempt to relate these cytological change(s) to inferred changes in secretion of the two kinds of glycoprotein hormones (TSH and the gonadotropins) or to bioassayable changes in pituitary ACTH content. Later, I also helped Halmi with his early studies of iodide trapping by the rat thyroid. These technologies naturally came into play when I "discovered" the hypothalamus.

One day in 1950, the departmental seminar series was graced by a barnstorming Geoffrey Harris. It was probably more the way Harris presented his findings than the evidence itself which influenced me, but I left his talk boiling with enthusiasm for the idea that the two great regulatory systems of the body might not be discrete after all, but linked by some sort of neurohumoral mechanism. I got out Harris's review (1948) and fine-combed the evidence he cited. It was clear that there was some sort of

[1] I refer here to the Hungarian gypsy recipe for chicken soup, which begins "First, you steal a chicken. . . ."

hypothalamic control over the gonadotropins, but the available evidence that secretion of TSH, or any of the other anterior lobe hormones, depended on the postulated neurohumoral link was not yet convincing. I talked this over with Harry Lipner, who was then doing his own doctoral dissertation, and ultimately approached Halmi with the idea of possibly investigating, for my Ph.D. thesis, whether TSH secretion was partly, wholly, or not at all under neural control. I thought this could be done by testing, in stalk-sectioned rats, the extent to which the pituitary "β" cells would undergo transformation in response to thyroidectomy. Halmi suggested that I try hypothalamic lesions instead of stalk section, but I (mindful of Nelson's views) felt that I didn't want to hunt all over the central nervous system for a "TSH center" before I knew for sure that such a center existed. Since stalk section would cut the "final common pathway," even negative findings would allow me to make a statement of value (and get a degree), whereas negative findings with any number of lesions might only mean that I hadn't made the right lesions. Halmi, grudgingly conceding that this was logical, pointed out that he could not be my official advisor on the project since he was still only an instructor. According to the rules at that time (which had merit), someone of professorial rank would have to serve as my mentor. This proved to be no problem; both Nelson and Ingram agreed to provide the "clout" if Halmi would accept the day-to-day responsibility for supervising my activities.

Lipner remained interested in the project for a while, and he and I spent many a long night[2] nibbling away rat crania, elevating temporal lobes, and trying unsuccessfully to sever the pituitary stalk completely without simultaneously cutting the carotid arteries or injuring the hypothalamus. The few rats that survived our butchery commonly displayed signs of hypothalamic trauma, but we never did succeed in cutting the stalk! No matter how carefully we restudied Jacobsohn's account of this operation, neither of us had her hands. Halmi was too canny to try to show us how to do it, and Nelson avoided the problem by being almost always out of town, so at last I timidly approached Ingram for help. He shrugged, and blinked a bit, then said "Sure!" The next morning (the daylight arrangement being solely in honor of "the Chief"), he walked up to the rat we had prepared and, within a few seconds, simply lifted the brain and cut the stalk. The rat recovered quietly and autopsy revealed that the transection had been total and clean. It was so demoralizing that Lipner quit!

[2] Days were spent, for the most part, teaching, taking courses, doing Halmi's chemistry, or (for one brief but intense period) playing chess (which Emil Steinberger called "chest"). Nelson halted the chess by presenting me with a Christmas gift of a small chess set designed for traveling. The message was clear: if I continued seducing my colleagues into what had virtually become the chief departmental activity, I could expect to travel!

I plugged along alone (still with no success) for quite a while, until Monte Greer (who had taken a chance where I had feared to tread) reported that anterior hypothalamic *lesions* could markedly, if not totally, impair pituitary TSH "reserve" (Greer, 1952).

We judged from Greer's abstract that his index of TSH reserve had been exclusively two extrapituitary signs (goiter development and thyroid iodide "trapping") of the TSH hypersecretion he had produced by treating the rats with propylthiouracil (PTU), one of the then-new goitrogens. Consultation with my triumvirate of mentors led to agreement that (even though I had been "scooped") extension of Greer's work, using impairment of thyroidectomy cell development as a parameter of lesion effectiveness, would be an acceptable thesis because (if it worked as expected) it would reinforce Halmi's investigations on the cell of origin of TSH.

Nelson, it turned out, had a marvelous relic—a Horsley-Clarke stereotaxic apparatus for rats. (Ingram worked exclusively with cats.) It weighed about 20 lbs, and I had to hang it by wires from the overhead surgical lamp to keep the rat from becoming a crêpe suzette, but I was able to use it (and Ingram's d.c. lesion maker; see *Pioneers in Neuroendocrinology*, Vol. 1) to burn holes in the hypothalamus. After that *I*, and I alone, injected the rats with PTU! (The reason for this was simple. Neither Halmi, nor anyone else in his right mind, would have opened those rat cages. The lesions had a number of effects, including the production of obesity, testicular atrophy, transient hyperthermia, and, most importantly, a well-coordinated hypothalamic *rage* which was neither transient nor "sham." Only an idiot who desperately wanted to get a degree would have undertaken to inject those rats![3] Years later, I learned to put PTU in the diet.)

The work went slowly. Ingram patiently provided moral support and, at several points in the project, even money. One of these points was on the occasion of a trip to St. Louis, during which Halmi and I wanted to meet Greer. After hunting for Greer at the meetings for 2 days and finally deciding that he had not come, we gave up and proceeded to Halmi's other desired rendezvous, with Edward Dempsey, who was then editor-in-chief of *Endocrinology*. In the midst of this visit a round-faced young man in a bow tie came in unannounced, to be warmly greeted by Dempsey as "Monte!" The mountain had come to Mohammed.

Greer was delighted to hear of our work, which so far confirmed his. His arrival shifted the conversation (in which I had been a very silent

[3] This is not strictly true. The department maintained for many years some cats that Wheatley (1944) had rendered savage with similar lesions. Ingram, to see if the emotional imbalance had yet worn off, would regularly attempt to pet these cats. He did this late at night, so only we "night hawks" ever got to see the bloody consequence.

partner) away from the staining characteristics of β and Δ cells and onto the role of the hypothalamus in the control of the anterior lobe (a subject on which I could comment). At that stage of what would later become my "career," I was greatly encouraged by the evident interest with which Dr. Dempsey listened to my views. This helped me a lot, and I am still very grateful that Ingram (who supplied the cash for the trip) had thought it a good idea for me to go along.

The other major money outlay provided by Ingram was for the Krieg-Johnson stereotaxic apparatus. The Horsley-Clarke machine was terribly awkward to use, and, when Halmi saw an ad for the Krieg machine, I got so excited I jumped into my 1929 Model A and raced to Berwyn, Illinois, to get one. It was about 8:00 p.m. when I clanked up to the address listed for the Johnson Apparatus Company. It didn't look much like a factory. It looked more or less like any other house on Maple Avenue. However, it was the right place, and even though it was way past business hours Roy and Ernie Johnson invited me in. Yes, they (Roy) were making the new machines (in the basement) and, yes, they (Ernie) were selling them (all over the world). However, they were not a stock item. Each was made to order. Roy couldn't afford to build any on speculation, since he was only charging what seemed to him a fair price of about $200. (After Roy died and Ernie sold the business, the price jumped immediately to around $700.) If I would go home and have an institutional purchase order sent, they said, I could have a machine *in about 6 months*!

I think my disappointment, and the fact that I had forgotten to eat during my 8-hr drive to Chicago, must have been evident. Ernie and Roy started feeding me home-baked Swedish cake and coffee in large amounts. Between bites and gulps, I told them all about Harris, and Greer, and the hot project I needed to complete so I could graduate and get a job before the Model A fell apart. (I probably spilled out a lot more about my personal cosmos, too.) Roy was thoughtful during most of this, while Ernie, being the socially adept member of the team, drew me out as she kept bringing more cake and coffee to the table. Suddenly Roy spoke up, "Those other guys can wait!" By dawn's early light I was home, dismounting from an overheated Model A with the Krieg-Johnson machine in hand (and the entire factory delivery of a dozen or more round-the-world orders set back by a month or so)!

Roy Johnson not only had let me take the machine with me, he also had trusted me to pay for it. A few days later, payment was authorized by Ingram (whom I had not thought to consult before I left and who certainly was not obligated in the slightest to undertake this payment). My disorderly approach distinctly bothered Ingram, and it was very fortunate for me that he finally decided to bail me out, since $200 was about twice what I had (with much pain) assembled to buy the Model A.

With the new machine, the lesion work went about three times as fast. Soon Halmi and I had a big chart and, as the findings accumulated, we began to distinguish not only "goiter block" but also several other "hypothalamic" syndromes.

The thesis work was done, as far as Halmi was concerned. The lesions which had prevented goiter formation had, as predicted, also prevented thyroidectomy cell development. I presented a paper at the 1952 Anatomy meetings in Providence, to which several of us (including Jack Davies) drove in Spirtos's brother's car. The report included Halmi's discovery that certain lesions seemed to have removed an inhibitory neural influence on the pars intermedia. I was terribly nervous, but I got through the (well-rehearsed) talk without misadventure. However, when I was asked a question I had not anticipated, namely, how much my rats weighed, the veneer of *savoir-faire* dissolved. "Oh," I replied, "between 250 and 450 milligrams."

Among the laughers was Willy Etkin, my old comparative zoology teacher from CCNY, who had been so delighted to read of findings that related to his early observations pointing to an "MSH-inhibitory factor" that he had traveled all the way to Providence to hear me talk. This was quite a heady compliment!

The work was *not* done as far as Ingram and Nelson were concerned. Ingram wanted me to analyze the lesions as to size and location. Nelson was particularly interested in knowing which lesions caused testicular atrophy. I was given the distinct impression that I would do well to furnish the neuroanatomical part of the study before presuming to present my thesis. Consequently, I spent many weeks sectioning and staining brains, in and around my other duties as a teaching assistant, embalming assistant, and Sunday morning animal caretaker.

Christmas 1952 came, and with it a midsemester break of 3 weeks or so. Both Ingram and Nelson left to seek some brief escape from the Iowa winter, but the sorcerer's apprentice stayed on! In my loneliness, I suddenly realized that I could do the microscopic work fully as well in New York as in the lab, and get to see my family besides. I packed two heavy suitcases with slides, a microscope, and a sheaf of the mimeographed charts of the rat diencephalon which I had prepared for coding the lesions for analysis. I arranged with someone (Roger Boynton, I think) to feed the animals for the next two Sundays and escaped the Iowa winter, too!

I enjoyed the vacation, which included a vist to Hamilton's department at what was then the Long Island College of Medicine, where I met Jack

Gross and Don Ford, who both seemed excited about my findings. I also charted every single lesion, and gorged on ethnic foods (which in those days were not available in Iowa City), and ingested unhealthful amounts of sodium bicarbonate as a consequence. When I got back, it was quite warm (and I am not referring to the Iowa weather)! Ingram would only mutter over his shoulder something about irresponsibility and Nelson wouldn't even look at me at all. I may never know whether they really were more anxious for the results than I had realized or simply outraged at my being AWOL. I rather think the fact that I did come back with the results well analyzed (as well as with the borrowed microscope) was a major determinant of my fate at that point. Both Ingram and Nelson eventually forgave me, but it took time.

I was (and still am) rather impressed with my lesion analysis, which clearly correlated destruction of the arcuate nucleus of the hypothalamus with testicular atrophy. I felt (even if I was too cautious to say so) that I had found the "sex center." McCann (1953) came to about the same conclusion at about the same time. (In the intellectual climate of that period, pursuit of "centers" in the hypothalamus was a normal goal, even though it was modeled on the way people were then organizing such things as civil defense, which is not necessarily the way God made the central nervous system.)

Over the next several years, having gotten a degree, a job at Albany Medical College, a new car (all in 1953), a wife (1955), and a daughter (1957), I continued trying to map the "endocrine-influencing centers in the hypothalamus." I even got a grant for that purpose (1954). (At this writing, I still have the same grant, with the same title, but I do not know how long I will retain it in the intellectual climate of *this* period.)

During the rest of my "preconceptive" phase, most of my independent efforts were unimaginative. I wasn't discovering, I was cataloguing. The most advanced idea I had along these lines was that it might be clever to do the converse of the lesion experiments and stimulate the hypothalamus to secrete the individual pituitary hormones selectively. In retrospect, I was quite a pedestrian. I don't really like to admit this, but facts are facts. Two events, perhaps more, brought about a change. (There must have been a change, or I wouldn't have been asked to contribute to this book.)

For several years, I had done basically the same experiment repeatedly, sometimes in long-distance collaboration with Nick Halmi and Bill Spirtos. In the prototype, I had castrated rats with and without arcuate nucleus lesions. When the lesions prevented pituitary gonadotropin content from increasing, I concluded that the lesions, since they also caused testicular atrophy, had interfered with both the synthesis and the release of gonadotropin(s). I was trying very hard to use a similar approach for ACTH and, to that end, had started some extensive collaboration with Murray Saffran and Guy Rochefort in Montreal. (During my several visits there—in those days the only safe way to convey frozen rat pituitaries through Canadian customs was to drive them there personally—I also met Andy Schally, who asked my advice about accepting a job offer from Guillemin in Texas. I forget whether I gave him any advice, but do recall that he took the job.) These ACTH studies *never* panned out. I later got into similar collaboration with Parlow (this time *re* lesions that seemed to increase gonadotropin release), who had just developed a specific bioassay for LH. I had lots of energy, but little sophistication.

Perhaps the first insight which led me to change the direction of my research was a chance meeting with S. A. ("Sam") D'Angelo, a colorful man of enormous energy and rare wisdom. It was at the 1956 Anatomy meetings in Milwaukee (where I also visited the Miller Brewing Company with Roy and Ernie Johnson, and showed them that I had not yet lost my capacity to take advantage of a free lunch—this time a liquid one). D'Angelo had a bioassay going for TSH, and, perceiving a chance to do my favorite experiment again, this time with TSH, I tried to sell him on collaboration. Wouldn't he like to find out if lesions which prevent goiter formation would also prevent the associated rise in pituitary TSH stores? D'Angelo looked at me in genuine dismay and then kindly explained that, in the rat, TSH stores do not rise—but *fall*—during the thyroidectomy reaction. [He went on to point out that this was not the case in the guinea pig, a fact which led me to become a "comparative neuroendocrinologist" since, with D'Angelo's help, I was able to do my then-favorite experiment in that species (Bogdanove and D'Angelo, 1959).] My embarrassment led me to realize, for the first time, the very important fact that the direction of a change in pituitary hormone stores has no necessary fixed relationship to the directions of any concomitant changes in the rates of synthesis and release of that hormone (interested readers may refer to Bogdanove *et al.*, 1975, p. 570).

The second event was that a biochemist friend of mine, Rudy Anker, brought me a paper he had run across (Flerkó, 1957*a*) which he thought I might find of interest. His guess was correct. I became very interested (after he had helped me to translate it). I became so interested that I even managed to learn to read the German myself, with a dictionary,[4] and I had

[4] I cannot speak for all Ph.D. language exams in this country but I had really stretched a point on mine by translating from Spalteholz's *Handbuch des menschlichen Körpers*. It was almost all pictures. The meager text was a cinch since the nouns were mostly Latin cognates of the English nouns and the verbs and adjectives were obvious from the pictures.

no trouble with the French (Flerkó, 1957*b*) or English (Flerkó and Szentágothai, 1957) papers, which I immediately obtained. This was in the spring of 1958.

CONCEPTION (1958–1963)

My readers may have deduced by now that I was (and am) a relatively slow thinker. I am poor at social repartee and generally can muster only one or two points, which may not be the key points, during a verbal scientific exchange. (However, I should add that I have generally continued to think when such an exchange is over and thus have often ended up with a written response which did not miss these key points. Consequently, although I can cite many examples of my own fallibility, which is of perfectly normal dimensions, they tend to be *unpublished* examples.)

The relevance of this perhaps unseemly admission is that I initially perceived no flaw in Flerkó's elegant experiments. At first blush, I believed in all of them, particularly the one I could read without a dictionary, which Szentágothai still seems to think was "conclusive" (see *Pioneers in Neuroendocrinology*, Vol. 1, p. 303). It took an experiment I did for a bad reason to lead me to begin to suspect that there might be flaws.

Since I did not initially question Flerkó's hypothesis (that negative feedback by the ovaries is exerted via the hypothalamus), I merely noted it and continued my ongoing frenetic activities, in which Ed Crabill had joined me. The principal energy drain was the ill-fated ACTH problem. We also began a collaboration with Parlow on LH, and I was busily involved in preparation for an exciting year abroad with my family (to which a son was about to be added). I had been able to secure a USPHS traineeship to finance a year's stay in Harris's lab, where I had hoped to learn why he and Woods had been able to use electrical stimulation of the hypothalamus to release TSH in the rabbit, whereas Greer and his colleagues, in studies in over 500 rats, had come up with nothing but a "drinking center." (I did learn why, but it isn't relevant here.) We were scheduled to leave for England at the end of June 1958.

At first, the most pressing of these projects was ACTH assays, using the adrenal ascorbic acid depletion method. (This quite outraged my boss, Jack Wolfe, who felt that I should be using "*anatomical*" methods, but he managed to tolerate this impropriety, *inter alia*.) The study was due to be finished by late June, after which Ed was to keep the lab going for a year, extending the work in guinea pigs (Bogdanove and D'Angelo, 1959) and pursuing the collaboration with Parlow, while I took off for my leave of absence abroad.

Ed and I were pushing very hard to make the June deadline when a medical student showed up wanting to do a summer research project. Since he was to receive money from the dean for this, and since the department supplied what rat care there was in those days, I decided to accommodate him, even though at that point I had only enough money left in my annual budget to finish the ACTH work. However, Crabill had a small grant and was able to buy a shipment of 60 rats.

The project I proposed for the student was simple enough, and not really conceptually original at all. It was merely a translation of a study by von Euler and Holmgren, who had looked at inhibition of thyroid hormone release after microinjection of thyroxine into the rabbit pituitary or hypothalamus and observed that the intrapituitary injection site was the effective one. This allowed them to conclude that thyroxine feedback was *direct*. I merely wanted the student to do, with thyroid, what Flerkó had done with ovary; transplant fragments of it to the brain or the pituitary. Having done that, all he had to do was thyroidectomize the rats, kill them 3 weeks later, make slides, and see whether intrapituitary or intrahypothalamic thyroid grafts, or both, interfered with the thyroidectomy reaction. The experiment seemed worth doing because, although I tended to believe von Euler and Holmgren's conclusion, I had some reservations due to the possibility of leakage of thyroxine from the microinjection site. The graft approach seemed safe from this criticism.

I am not often uncharitable any more, but the student proved to be completely irresponsible. First of all, he took a second job, but assured me that he would be able to handle that at night, so I went ahead and ordered the rats (using Ed's money). Then, when it came time for him to start operating, he said he couldn't do it yet because the other job required him to have 2 weeks of day training in addition to his night training. We couldn't defer this part of the study because I already had my steamship tickets, so Ed and I agreed to do the operations at night, for 2 weeks, after which the boy would complete the experiment. We continued our ACTH work in the daytime and then came back to the lab every night after supper, to alternate between washing glassware for the next day's ACTH run and operating on the rats. I figured we could stand it for 2 weeks.

I was more than right. We stood it for nearly 3 months! The student's sense of responsibility to the project (if he ever had one) diminished progressively and he ended up being paid (fortunately, by the dean) solely to *feed and water* the operated rats. After about 2 more weeks, when I discovered that he hadn't even done that right, his association with the lab terminated abruptly! (I'm afraid I wasn't very nice.)

This meant that, if the project was to continue, Crabill and I would have to do it all. To make matters worse, the rats got sick and it became clear that many of them would not survive to autopsy. I had no grant money left to spend for replacements. (Neither did Ed, since Wolfe had discovered that he had "misappropriated" about $100 of his hair follicle research money, and wouldn't let me buy another shipment of rats for our projects. Ed and I didn't exactly agree with Wolfe's views on morality, but he was the boss.) It was a situation in which one could easily have found a reason to quit.

We did not quit. Mr. Holtzman sent me a free order of rats to replace the sick ones. I negotiated with Harris, my landlord, and the Holland-America line to defer my trip for 3 weeks, and Ed and I continued our swing-shift existence. Again the rats developed "sniffles" and again (with Mr. Holtzman's help) I extended the time. Eventually, with the second and third waves running (and still doing the ACTH assays in the daytime), it came time to section the first group of brains and pituitaries. I had only a half-time technician, Mrs. Esther ("Suki") Senning, so Ed and I did most of the work. Finally, dog-tired, we were able to get to the staining. We had gotten some radioiodine uptake data on graft viability and I was able to select, on this basis and gross inspection of graft placement, which animals to look at first. The three of us worked late that night and, by about 11:00 p.m., with Suki staining and Ed coverslipping, I was able to play the job of inspector as the slides came off the assembly line.

The first few slides told the story! Not only had a feedback action of the thyroid been demonstrated, but also the demonstration that it was *direct* was unmarred by the question of possible leakage from the implant site. Each engrafted pituitary gland was its own control for that consideration (Fig. 2). (Had I expected Wolfe to be awake, I might have phoned to tell him that "anatomical" methods had certain merits, after all!) Suki, Ed, and I celebrated the fact that *something* had come of our summer's arduous efforts (by this time the ACTH project had at last been discarded) with a toast, consisting of orange soda and 95% ethanol, in a ratio of 2:1 (if I recall correctly, and I do). We then finished the night's quota of staining and, after Suki and Ed had gone home, I packed up the newly stained slides and the microscope and also went home.

In retrospect, this study (which Ed Crabill and I would probably not even have initiated on our own, and in which we had persevered mainly because my sense of respect for research had been royally outraged by the cavalier attitude of the student) was the foundation of the "implantation paradox." The key observation I made that night was that the *spread* of hormone from an implant (manifested only by histological distribution of hormonal effects) was obviously and markedly affected by what appeared to be a very strong posterolateral current of blood traversing the anterior lobe (Fig. 2). However, the relevance of this simple fact to Flerkó's negative

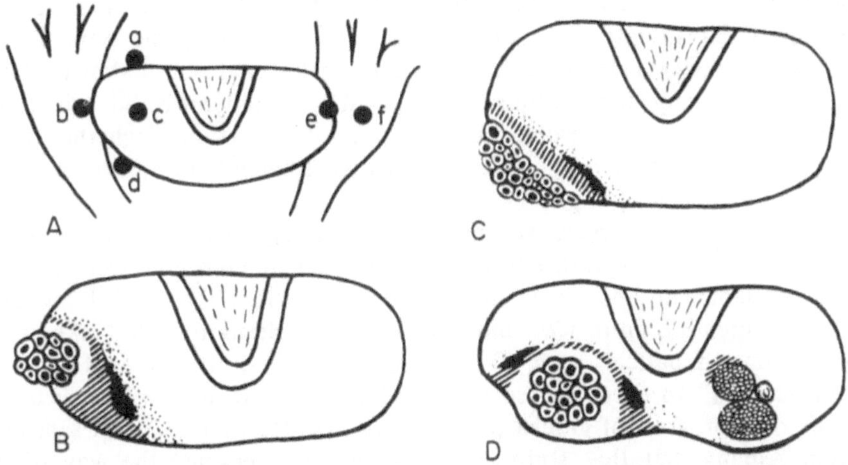

FIGURE 2. A: Scheme of rat pituitary in horizontal section. Various locations in which grafts were found are indicated by black dots labeled a–f. In B, C, and D, thyroid grafts are shown at left. In D, an ovarian graft (with two corpora lutea and a small follicle) is shown on the right. The ovarian graft did not appreciably modify the thyroidectomy reaction in adjacent pituitary tissue. Concentric zones (I–IV) of decreasingly severe inhibition of the thyroidectomy reaction were seen around the thyroid grafts. In zone I (clear area adjacent to graft) the histological signs of thyroid hormone (TH) effect were maximal. in zone IV (stippled) they were minimal. Note that no TH effect was evident in regions remote from the grafts (clear areas) so that the thyroidectomy reaction was ***not inhibited throughout most of the pituitary.*** Note also that the distribution pattern of TH effect was not spherical, but distorted (as discussed in text). Reproduced from Bogdanove and Crabill (1961) with permission.

findings with ovarian grafts in the anterior lobe (Flerkó and Szentágothai, 1957) did not occur to me until much later.

I have never fully analyzed my exaltation on that hot summer night in Albany, but I suspect the major esthetic appeal of the findings shown in Fig. 2 lies in the low complexity: decisiveness ratio of the evidence. A single slide (Fig. 2D) provided an unequivocal answer. Also, I had recently been having some frustrating discussions with a colleague in the physiology department who used "blackbox" philosophy (?) to justify his ignorance of anatomical structure. This night's work would put an end to such inanities, I thought! (I was wrong. Similar inanities persist today, in less-than-glorious isolation from the facts.)

In any event, I had to share the excitement and, shortly after 2:00 a.m. I entered the bedroom where Janet and Jonny (then about 2 months old) were sleepily engaged in an intercorporeal transfer of milk. I woke the former up as follows: "Madam, I have here a handy-Andy, Jim-dandy little optical device which is going to revolutionize the practice of neuroendocrinology! For only one thin dime, the tenth part of a dollar, etc. etc." I went on for a few minutes, paraphrasing (for the Ortholux) the pitch of the country huckster peddling a "new" potato peeler, and then I showed Janet the slides. Jonny slept through it all, and eventually, we did too. (Not, however, until we had resolved a recent domestic question of whether it had been worth while to twice defer our trip to England. For once, we both agreed I had been right.)

Over the next weeks, we finished the histological work and I took the cream of the slides with me to England. There I found myself working with Bernard Donovan and David El Kabir on a technique for which I saw only limited potential, an *in vitro* bioassay for TSH. Bernard soon ducked it by going on sabbatical with Dora Jacobsohn, but David and I were "encouraged" to carry on, so we did. I managed to use the assay to get some data on a set of guinea pig pituitary extracts which Crabill sent over from Albany, but that study went (for various reasons) the way of the ACTH study, and has no bearing on the "paradox." I sometimes also got to sneak a peek at Harris's electrical stimulation work (but was not invited to join in these experiments), and for a while Bernard and I played about with efforts to implant estrogen in beeswax into the rat pituitary. (That game ended when the stereotaxic gadget broke.) We also did some other rather unfocused studies on the side (Harris didn't mind, as long as the TSH work kept going at a reasonable pace) to confirm our belief (Donovan and van der Werff ten Bosch, 1959; Bogdanove and Schoen, 1959) that estrogen could not inhibit gonadotropin except via neuronal "receptors" in the preoptic-anterior hypothalamic region. These too came to naught, for various reasons (not least of which may have been that we were trying to *prove* a hypothesis, rather than *test* it).

What I *did* accomplish, mainly late at night so I wouldn't be in the way, was to teach myself photomicrography in Harris's darkroom, where I prepared the figures for the paper with Ed Crabill (Bogdanove and Crabill, 1961). In the process, I learned (see the mosaics in Bogdanove and Crabill, 1961) that, although wrinkled microscopic sections don't bother the eye much (since the viewer reflexly readjusts the fine focus), they play havoc with a photographic plate! (The other thing Janet and I learned something about was ourselves. We arrived in England as naive provincials from Albany, New York. I don't think we were "Ugly Americans," be we may have seemed to be. When we left, we had become, if not citizens of the world, at least Americans who understood some of their own barriers to empathy and communication with people from other cultures.)

Albany (1953–1958) had been a good, small medical school, but not a scientific Mecca. In contrast, Harris's lab was a touchdown point for almost everyone in the new field of brain-pituitary interplay. Visitors were frequent, and those who did not come were gossiped about with those who did. I loved it. Toward spring (1959), Harris suggested that if I planned to make the expected "tour of the Continent" I should do so *before* summer, since many of the civilized continentals tended to shut down their labs later on. (This was pre-air conditioning, but I think the custom persists.) I accepted the advice, but requested (and received) permission to set up a small experiment which would run itself in my absence. Having done that, off we went (armed with the Crabill slides, RAC maps of the continental jungle, two duffel bags of disposable diapers, 10 gallons of English drinking water, and a yellow plastic baby bathtub and other survival gear—most of which was loaded on, or dangling from, two roof racks on our tiny A35 van). Left at the Maudsley were a few hooded rats which had been castrated and given intrapituitary and intrahypothalamic implants of their own testis or ovary.

It is of crucial relevance to this story that we spent our first evening on the Continent in Leiden (actually Oegstgeest), where "Koos" van der Werff ten Bosch was our host and the other guest was none other than Belá Flerkó. After I had diplomatically drunk some Dutch tap water (and even taken a sorely needed shower in it) and discovered the fallacy of the English myth that "the water isn't safe on the Continent," and after the commotion of getting Bethie and Jonny to sleep had died down, Koos and Belá and I got down to business. They were both as delighted as I with the thyroid implant study (which Harris had not been) and as approving as I of the analogous study I had just set up with testicular and ovarian grafts. We all expected that, *in contrast to the thyroid grafts*, intrapituitary grafts of gonadal tissue would not inhibit castration cell development! I was still under the spell of Flerkó's brilliant hypothesis (Fig. 1B) and, although I had formulated my own (analogous to Fig. 1C) for brain-pituitary-thyroid interplay, I believed his for brain-pituitary-gonad interplay.

The rest of our month abroad was devoted mainly to nursing the wee A35 over 5000 miles of continental roads (including the St. Gotthard pass), swimming in both the Mediterranean and the Baltic, and keeping pace with the peak of the strawberry season as it progressed from *Sud de France* to *la belle Belgique*. Almost everywhere we touched down, the Crabill slides were my entrée into a delightful conference: Munich (Romeis), Milan (Martini and his group), Paris (Benoit), Brussels (Desclin), Lund (Dora Jacobsohn and Bernard Donovan), and Stockholm (Curt von Euler and Granit). Toward the end of the month, we visited Koos again. Flerkó had returned to Pécs but I asked Koos to assure him, in his letters, that I would let him know the outcome of the experiment as soon as possible.

As soon as I got back to the lab, I took a survey. Most of the rats had died and been thrown out, with no postmortem investigation. Harris had forgotten that I had an experiment in progress, and when the technician who (occasionally) fed and watered the rats had asked what to do with their carcasses he had been told to do anything he wished. This upset me some, but not enough to reproach "the boss," which would have been pointless anyway, since he had far more important concerns than my trivial experiment (the existence of which he evidently forgot again, once I had left his lab). Besides, some rats had somehow survived and since this was really to be only a confirmatory ("me too") type of report, I figured that one or two good rats were really all I needed.

I killed the surviving rats, inspected the grafts and resorption scars, fixed and embedded the appropriate chunks of tissue, and got to work with microtome and staining pots. In due course, I found three healthy-looking grafts of ovary in the anterior lobe (no testis implants had survived). The situation differed from that in the Crabill study only in that my first view of the result came in the early afternoon, rather than toward midnight (as detailed above). Otherwise, the result was parallel: castration cells were not seen in the regions close to the implanted ovarian tissue!

This, of course, was *not* what I had expected!

My "me too" experiment had suddenly raised a serious challenge to the most important new concept in the field since Friedgood (1970) and Harris (1948) had postulated neurohumoral control of the anterior lobe. There were questions, of course. These did not rest on the paucity of findings in only three rats[5] but on the fact that I did not know whether the inhibitory effect of the grafts was due to estrogen or to some other factor. The experiment was therefore inconclusive as far as Flerkó's indirect estrogen feedback hypothesis was concerned. Still, it was highly suggestive and I got quite excited. I immediately ran over to get Harris and Billy Nikitovich-Winer (who had recently joined Harris's group) and they became the second and third people, respectively, to become aware of this evidence that ovarian-pituitary feedback could be direct.

Janet was the fourth and, some days later, Flerkó (to whom I wrote at once) became the fifth. His hypothesis, I wrote, might require some modification since the *pattern* of effect (which I sketched for him, much as I drew it on the left in Fig. 3) clearly showed the influence of blood flow I have already commented on above (in connection with the Crabill study). This meant that much of an engrafted pituitary *could escape* the direct effect, allowing sufficient release of gonadotropin so that the deficit would *not be seen at the periphery* but would be apparent only if one looked within the gland itself, by using histology. From this, it followed (I wrote) that his (Flerkó's) brilliant study with Szentágothai (1957) was *not* conclusive after

[5] Halmi once said, with regard to the use and abuse of statistics, "If a single chicken lives with his head cut off, it *proves* that a chicken *can* live with his head cut off."

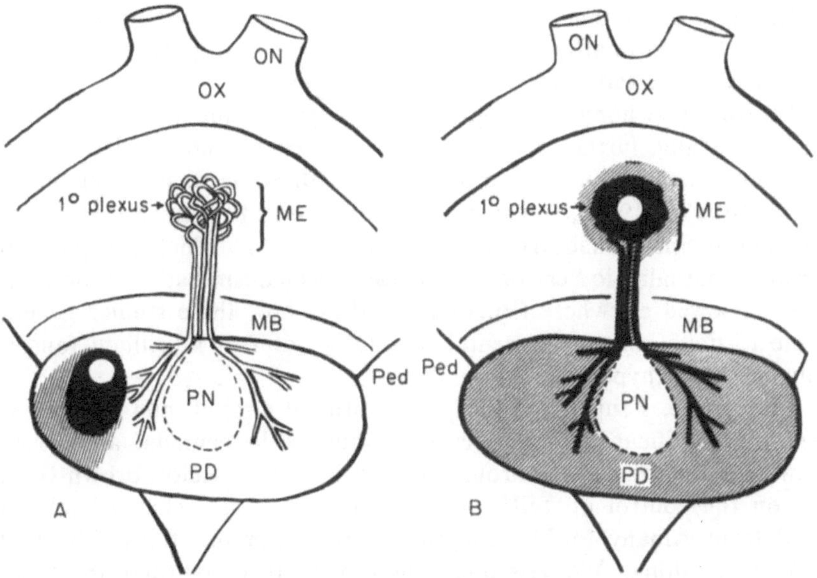

FIGURE 3. Diagrams of rat hypothalamus, hypophysis, and pituitary portal vessels. ON, OX, optic nerve and chiasm; ME, median eminence; MB, mammillary body; PD, pars distalis; Ped, cerebral peduncle; PN, pars nervosa. Intrapituitary (A) and intrahypothalamic (B) hormone depot sites are shown in white. Postulated regions of high and low concentrations of hormone from the depots are shown by heavy and light shading, respectively. The *implantation paradox*. A: An intrapituitary source cannot distribute hormone to the entire gland. Trophic hormone released from unaffected regions may be adequate to mask the effect of inhibition of a localized region on peripheral indices of trophic hormone release (such as uterine weight or serum hormone assays). B: An intrahypothalamic source can supply hormone to the median eminence, whence it may be distributed (via hypophysial portal vessels) to the *entire* anterior lobe. Since the resultant decrease in trophic hormone release could be due either to an *indirect* action of the hormone at the median eminence or to a *direct* action at the pituitary, the effects of intrahypothalamic hormone depots are not interpretable. The distribution pattern shown in A is the same as those shown in Fig. 2. From Bogdanove (1963*a*) with permission.

all! (Researchers of these histories may note, in connection with p. 303 of Vol. 1, that this was in June 1959).

Flerkó replied promptly, raising at once an argument which has not even yet been completely disposed of.[6] He asserted that even though estrogen might have a direct effect, which he and Szentágothai could surely have missed, the direct effect might occur only with supraphysiological concentrations of hormone. In other words, *indirect* feedback (which I had not detected) might be *more sensitive* than *direct* feedback, and thus be the mechanism which operates under physiological conditions. I replied that I

[6] Maybe it has (Bogdanove *et al.*, 1977; Nansel and Trent, 1977).

was happy to allow this possibility, and that I would investigate further. I was not yet ready to attack Flerkó's hypothesis, even if part of the evidence on which it rested had now been rendered equivocal.

We returned home in August 1959, to plunge into another round of cadaver teaching, further collaboration with Parlow, and a series of studies of the other arm of Flerkó's argument for indirect feedback—the assorted findings (Flerkó, 1957*a,b*) which he had interpreted to mean that estrogen could not inhibit gonadotropin secretion in rats in which the postulated anterior hypothalamic "estrogen receptor" mechanism had been destroyed. I have reviewed elsewhere (Bogdanove, 1964) how these studies generally led me to doubt the experimental foundation of Belá's brilliant (and certainly heuristic!) hypothesis.

Meanwhile, I submitted the paper with Ed Crabill, and debated with Greer the significance of his recent studies with Yamada, and wrote a review (Bogdanove, 1962) around this debate and Keith Brown-Grant's ideas on the control of TSH.[7] During all this, I also changed jobs. We moved from Albany to Bloomington shortly after our third child (Julie, 1960) came along. We remained there for 10 years (and three more children). A number of adjustments were involved.

Albany had been a relatively uncomplicated medical school and my job there had been merely to teach gross anatomy. If I cared to, I was free to do some research also. I had had no graduate students, but I had had a lot of fun with several inquisitive medical students who, like Herman Schoen, had transiently participated in what was going on in the lab. I had never had reason to view either research or research training as part of my job, only as opportunities for adventure my job made possible.

At Indiana I discovered that research *was* considered to be my job, or at least the main part of it. I still took medical student teaching seriously, and worked conscientiously to do a good job of it, but I'm afraid I was too ambitious in this respect. In retrospect, I was swimming vainly against a national tide of reducing medical *education* to medical *training*. (This ultimately influenced graduate education too—*viz*. "Ph.D. *training* programs.")

I had a fair amount of trouble with the administrators who did not share my aberrant philosophy. However, either Christian charity or my growing prominence in neuroendocrine research allowed me to be given tenure anyway (after which I promptly grew a beard!) and eventually I even became a full professor. Although my teaching responsibilities dwindled and I was kept off of committees, I was left alone to do my work. I was even

[7] In this review (1962) I proposed a model of thyroxine-TRH interaction at the pituitary which Magoun flattered me by including in his book *The Waking Brain*. I was amused just recently to see this concept referred to as the "classical" model, but attributed to someone who published the same idea in 1966. *Caveat lector*!

helped, in a major way, when the University, in 1968, actually funded and built a rat lab to my specifications. On balance, Indiana University was good to me, whether or not my bosses there ever understood what I was trying to do in the lab. (One or two of them did, I think.)

Among the early studies at Indiana, Kirby Tarry (one of my new herd of graduate students) and I made and implanted micropellets of estradiol and cholesterol. These showed what the ovarian grafts had shown 2 years earlier and allowed me finally to write the papers in which the "implantation paradox" (Fig. 3) was presented (Bogdanove, 1963*a*) and discussed (Bogdanove, 1964). This required some courage, since many important people had by now become enthused (as I had been earlier) with Flerkó's notion of obligatory indirect feedback of sex steroids. Some of them heartily resented my persistent pronouncements that *all* of the experimental evidence for this idea (including theirs) was uninterpretable!

I had not started to speak out until my own experiments had convinced me (Bogdanove, 1964) that there were possible alternative explanations for *every one* of Flerkó's findings, including the apparent inability of estrogen (or testosterone) to inhibit gonadotropin secretion in rats in which the anterior hypothalamus had been destroyed. Flerkó himself, visiting my new lab in 1961, verified the lesion *loci* for the paper (Bogdanove, 1963*b*) which showed that anterior hypothalamic lesions did not prevent estrogen from inhibiting castration cell development. The last of these experiments, using pregnancy urine (Bogdanove, 1964, p. 228), exploited the imminent arrival of Sam, our fourth child (1962).

In the process of speaking out, I became an "important person" too, but I think I did so rather reluctantly. I remember several instances in which I offered to refrain from public criticism if the person in question would only include the alternative possibilities I had raised (Figs. 1C and 3) in his own discussions. On one occasion, I sat on a bench in Atlantic City with an eminent protagonist of Flerkó's idea until 1:00 a.m., trying to eliminate any necessity to criticize his colleague's presentation the next day. However, my efforts at diplomacy failed, and I was forced to repeat my arguments during discussion of the paper. (This was viewed as unfriendly, which it was not.) This sort of thing repeated a number of times. I would send someone my edited reply to an unsound comment he had made at a meeting and offer to drop publication of my reply if he would only, having seen the logic of my rebuttal, amend his own original comment. Incredibly, these offers were never once accepted! I had to either ignore the truth or undertake to spend much of my time replaying, *ad nauseam*, the solitary role of devil's advocate in discussions about the site of long-loop negative feedback. I think it is the fact that I chose the latter course which led the editors of these volumes to consider me a "pioneer" (Fig. 4).

Gradually, others began to recognize the possibility of direct gonad-

FIGURE 4. The author as a "pioneer."

pituitary feedback. Ramirez (Palka *et al.*, 1966) was one of the first. Hilliard *et al.* (1966) followed. By 1971 nearly everyone could agree (Schally *et al.*, 1972; Naftolin, 1973) that direct feedback actions of sex steroids were probably the major determinant of how the pituitary responds to the neurohumoral influence(s) which impinge on it. Nearly everyone but me, that is (Bogdanove, 1972*a,b*).

GESTATION (1963-   )

This is not the place to delve into detailed analysis of the possible roles played by direct and/or indirect long-loop feedback in the regulation of gonadotropin and thyrotropin secretion. I did that elsewhere long ago (Bogdanove, 1962, 1964). This chapter deals (I hope, for younger generations of investigators) with how I happened to encounter certain facts, and through these facts a few key insights. I was asked to focus here on the "implantation paradox," and I have. However, I must add that becoming an "authority" on feedback did not lead me to cast in concrete my views about the marvelously intricate system of which the "paradox" is a manifestation. Much of what I have learned since I first glimpsed some of the intricacies of this system on that hot summer night in 1958 has recently been reviewed elsewhere (Bogdanove *et al.*, 1975). Since then, we have perceived new complexities (Bogdanove *et al.*, 1977; Nansel and Trent, 1977). The rest of the story has yet to be lived.

I have tried to convey some of the excitement, some of the disappointment, some of the joy, and some of the social difficulties inherent in a life in which research has remained the central focus. I have also touched on factors involved in my choice of this kind of life. However, motivations are elusive. I may never have had a choice.

Dedications

To Janet, who helped with the major work reviewed here. To Nick Halmi, who taught me, among other things, to be my own man. (*Alterius non sit qui suis esse potest.*) To A. J. Goldforb, who taught me that the admission of ignorance is the first step toward wisdom, and finally, to Adam (1964) and Eve (1968), who joined the family too late to get cited in this history.

Acknowledgment

Successive phases of the research reviewed here were made possible by support through three grants from the National Institutes of Health, U.S.P.H.S. (B-708, NB-03371, and HD-06600).

REFERENCES

Bogdanove, E. M. (1962). Regulation of TSH secretion. *Fed. Proc.* **21**:623.

Bogdanove, E. M. (1963*a*). Direct gonad–pituitary feedback: An analysis of effects of intracranial estrogenic depots on gonadotrophin secretion. *Endocrinology* **73**:696.

Bogdanove, E. M. (1963*b*). Failure of anterior hypothalamic lesions to prevent either pituitary reaction to castration or the inhibition of such reactions by estrogen treatment. *Endocrinology* **72**:638.

Bogdanove, E. M. (1964). The role of the brain in the regulation of pituitary gonadotrophin secretion. *Vit. Horm.* **22**:205.

Bogdanove, E. M. (1972*a*). Hypothalamic-hypophyseal interrelationships: Basic aspects. In Balin, H. and Glasser, S. (eds.). *Reproductive Biology*, Excerpta Medica, Amsterdam, p. 5.

Bogdanove, E. M. (1972*b*). Current knowledge of gonadotrophin releasing factor(s). *Med. Coll. Va. Q.* **8**:5.

Bogdanove, E. M., and Crabill, E. V. (1961). Thyroid–pituitary feedback: Direct or indirect? A comparison of the effects of intrahypothalamic and intrapituitary thyroid autotransplants on pituitary thyroidectomy reactions in the rat. *Endocrinology* **69**:581.

Bogdanove, E. M., and D'Angelo, S. A. (1959). The effects of hypothalamic lesions on goitrogenesis and pituitary TSH secretion in the propylthiouracil-treated guinea pig. *Endocrinology* **64**:53.

Bogdanove, E. M., and Schoen, H. C. (1959). Precocious sexual development in female rats with hypothalamic lesions. *Proc. Soc. Exp. Biol. Med.* **100**:664.

Bogdanove, E. M., Nolin, J. M., and Campbell, G. T. (1975). Qualitative and quantitative gonad–pituitary feedback. *Recent Progr. Horm. Res.* **31**:567.

Bogdanove, E. M., Nansel, D. D., Aiyer, M. S., and Trent, D. (1977). Evidence for physiological androgen-LH feedback being both direct and indirect. *Proc. 59th Ann. Mtg. Endocrine Soc.*

Donovan, B. T., and van der Werff ten Bosch, J. J. (1959). The hypothalamus and sexual maturation in the rat. *J. Physiol.* (*London*) **147**:78.

Flerkó, B. (1957*a*). Einfluss experimenteller Hypothalamusläsion auf die durch Follikelhormon indirekt hervorgerufene Hemmung der Luteinisation. *Endokrinologie* **34**:202.

Flerkó, B. (1957*b*). Le rôle des structures hypothalamiques dans l'action inhibitrice de la folliculine sur la sécrétion de l'hormone folliculo-stimulante. *Arch. Anat. Microsc. Morphol. Exp.* **46**:159.

Flerkó, B., and Szentágothai, J. (1957). Oestrogen sensitive and nervous structures in the hypothalamus. *Acta Endocrinol.* (*Kbh.*) **26**:121.

Friedgood, H. B. (1970). The nervous control of the anterior hypophysis. Harvard University Tercentenary Celebration, September 15, 1936. *J. Reprod. Fert. Suppl.* **10**:3.

Greer, M. A. (1952). The role of the hypothalamus in the control of thyroid function. *J. Clin. Endocrinol. Metab.* **12**:1259.

Harris, G. W. (1948). Neural control of the pituitary gland. *Physiol. Rev.* **28**:139.

Hilliard, J., Croxatto, H. B., Hayward, J. N., and Sawyer, C. H. (1966). Norethindrone blockade of LH release to intrapituitary infusion of hypothalamic extract. *Endocrinology* **79**:411.

McCann, S. M. (1953). Effect of hypothalamic lesions on the adrenal cortical response to stress in the rat. *Am. J. Physiol.* **175**:13.

Naftolin, F. (1973). Remarks by the Chairman of the Clinical Session on Gonadotropin Releasing Hormones. In Gual, C., and Rosemberg, E. (eds.), *Hypothalamic Hypophysiotropic Hormones*, Serono Research Foundation, Excerpta Medica, Int. Congr. Ser. No. 203, Amsterdam, pp. 424–425.

Nansel, D. D., and Trent, D. (1977). Frequency modulation of GnRH can modify direct feedback effectiveness of androgen. *Proc. 59th Ann. Mtg. Endocrine Soc.*

Palka, Y. S., Ramirez, V. D., and Sawyer, C. H. (1966). Distribution and biological effects of tritiated estradiol implanted in the hypothalamo-hypophysial region of female rats. *Endocrinology* **78**:487.

Romeis, B. (1940). Die Hypophyse. In von Mollendorf (ed.), *Handbuch der mikroskopischen Anatomie des Menschen*, Vol. 6, bd. 3, Springer, Berlin.

Schally, A. V., Kastin, A. J., and Arimura, A. (1972). FSH-releasing hormone and LH-releasing hormone. *Vit. Horm.* **30**:84.

Wheatley, M. D. (1944). Hypothalamus and affective behavior in cats; study of effects of experimental lesions with anatomic correlations. *Arch. Neurol. Psychiat.* (*Chicago*) **52**:296.

5

Alvin Brodish

Alvin Brodish was born June 11, 1925, in Brooklyn, New York. He received his undergraduate degree from Drake University in 1947 and his Master's degree in physiology at the University of Iowa in 1950. He then entered Yale University and received his Ph.D. in physiology in 1955. After serving as a postdoctoral fellow, he was appointed an instructor in physiology at Yale in 1957, assistant professor in 1959, and associate professor in 1965. In 1967 he moved to the University of Cincinnati where he was appointed associate professor of physiology. In 1968 he was appointed professor and director of the medical school physiology course at Cincinnati. In 1975 he became chairman of the department of Physiology and Pharmacology at the Bowman Gray School of Medicine in Winston-Salem, North Carolina.

Dr. Brodish is a member of the American Physiological Society, the Endocrine Society, the International Society of Neuroendocrinology, the International Brain Research Organization, and the American Association for the Advancement of Science, among others. He served on the editorial boards of *Neuroendocrinology* (1965–1977), *American Journal of Physiology* (1970–1977), and *Journal of Applied Physiology* (1970–1977). He has served as an advisor and research grant reviewer for NIH and NSF, and was on the Veterans Administration Research and Education Committee from 1970 to 1975. He has refereed manuscripts for *Endocrinology* and *Science*, and has been a consultant for several NIH site committees.

His honors include the American Physiological Society Porter Fellowship award and a National Science Foundation Postdoctoral Fellowship award while at Yale, a U.S. Public Health Service Research Career Development Award, and travel awards by the Endocrine Society, the National Academy of Sciences, and the Federation of American Societies for Experimental Biology. His major research interests include regulation of anterior pituitary secretions, particularly the hypothalamic-pituitary-adrenal axis. He has a special interest in the education of medical students.

Tracking the Elusive CRF

ALVIN BRODISH

From earliest recollections, I was fascinated by studies of the nervous system, mind, emotions, and behavior. I intended to become a neurophysiologist and devote my life to unraveling the secrets of the nervous system. As is so often the case, a series of unexpected circumstances detoured me from my original goals and seduced me into neuroendocrinology, a discipline that did not even exist at the time I was involved in career direction decisions.

After being liberated from the mass education system of New York City by enrolling in a small Midwestern university, Drake University, in Des Moines, Iowa, I was so pleased with the student-teacher relationships that I decided to continue my education in Iowa. It might be difficult for some to understand how a product from a large Eastern city could be captured by a rural Midwest cornfield atmosphere; suffice it to say that the change was a welcome and refreshing one for at least that period of my life.

Following the undergraduate degree from Drake University, I went on for graduate work to the State University of Iowa in Iowa City. Although I chose to work in the nervous system, it was at the State University of Iowa that I was first introduced to endocrinology, and became acquainted with several individuals who were later to become well-known leaders in endocrinology.

Sam Barker was a young faculty member in the Department of Physiology at Iowa and among his graduate students at that time were Emanuel Bogdanove and Harry Lipner. Sam Barker was a meticulous investigator who demanded of his students exacting experimental design and technique. How clearly I remember the late evenings put in by Barker's "poor" graduate students. In retrospect, I would conclude that some of Barker's "unreasonable" demands probably played an important role in developing the critical analytical minds that we see in Bogdanove and Lipner today.

ALVIN BRODISH • Department of Physiology and Pharmacology, Bowman Gray School of Medicine of Wake Forest University, Winston-Salem, North Carolina 27103.

After 2 years in Iowa City, I decided to seek another institution where I could receive in-depth training in the central nervous system. This led me to Yale University, where John Fulton had established an international reputation for research and training in neurophysiology. I arrived at Yale in 1950 and was to remain there for the next 16 years.

From my graduate student days at Yale in the early 1950s, some vivid impressions come to mind that I am certain influenced many attitudes and characteristics that I bear today. John Fulton was a scholar with a warm, outgoing personality. The Department of Physiology in particular, and Yale in general, was steeped in British tradition. Accents acquired by some of the faculty were cherished and nurtured; mannerisms were imitated and the entire atmosphere was contagious.

John Fulton was strongly influenced by Sir Charles Sherrington, and the Department of Physiology at Yale continued much of the work on the integrative action of the nervous system that had originated in Sherrington's laboratory. In fact, the department at Yale was not called a Department of Physiology, but was referred to as the Laboratory of Physiology, reflecting Fulton's attitude that the entire department was indeed a laboratory for research and teaching scholars.

In the early 1950s the Department of Physiology in particular and Yale in general brought together a group of individuals that today would read like a *Who's Who* in neuroscience. Among the faculty members in physiology were John Brobeck, Paul MacLean, Robert Livingston, Karl Pribram, Jose Delgado, Harold Lamport, and Donald Barron. Paul Munson was in pharmacology, Frank Beach and Neal Miller were in psychology, and Harold Burr, Sanford Palay, and a young newcomer, Jerome Sutin, were in anatomy. With an array of such talent around me, it is no wonder that a strong appreciation for the nervous system just had to rub off.

Shortly after C. N. H. Long, a well-known endocrinologist, became chairman of Physiology, I had to make a decision concerning my graduate research program and thesis advisor. I would say that C. N. H. Long, more than any other individual, was instrumental in influencing my choice of a career in endocrinology. C. N. H. Long was a most interesting person—distinguished and aristocratic in appearance and manner so that one might hesitate to seek his counsel; when he spoke in public the audience listened attentively to his generally concise and perceptive analysis expressed in a slight British accent. Long's hair was prematurely white, contrasting with dark eyebrows and a rather long, thin face with a slightly deviated pointed nose. He conversed easily and was a restless talker, leaning back in his chair or pacing about the room, stopping periodically to assume a characteristic pose—bent at the waist, one hand on hip, the other busy with pencil or chalk.

C. N. H. Long and I talked at length concerning my interest in neurophysiology and possible research programs in the department. After considerable discussion, it was decided that we would try to develop a program of research that would blend my interest in the nervous system and Long's interest in ACTH secretion. At the time, Long considered the autonomic nervous system (particularly the sympathetic nervous system) as one of the important regulations of ACTH secretion. Catecholamines and stress was Long's concept, and the lines were being drawn among the "believers and doubters." In my discussions with Long, it was finally decided that I would concentrate my studies on the reflex control of ACTH secretion. I suppose that this decision made me a neuroendocrinologist in current terminology, but that particular designation was not in vogue at the time, and I became, instead, an endocrinologist, and had to throw myself into the endocrine literature.

After searching the literature on ACTH, I realized, to my chagrin, that ACTH in blood could not easily be measured in 1952, and therefore studies on the dynamic changes in ACTH secretion could not readily be carried out. My dilemma was apparent—how could I investigate reflex control of ACTH secretion when ACTH in blood could not be reliably quantitated?

Since a direct method was not available for determination of ACTH, previous investigators had to resort to indirect indices such as adrenal ascorbic acid depletion. However, the depletion of ascorbic acid in the adrenal of intact rats could only indicate that ACTH secretion had occurred; it could not be used to infer the pattern of ACTH secretion—i.e., magnitude or duration of ACTH release. Hypophysectomized animals, however, could, under certain conditions, be used to quantitate ACTH. One method of blood ACTH assay, employed by Sayers and his colleagues, was to inject a quantity of blood (usually limited to 4 ml) directly into hypophysectomized rats at intervals following an experimental stimulus, and to determine its effect on the adrenal ascorbic acid. This method was limited by the small amounts of ACTH normally present in the blood and the small volume of blood that could be infused as a single injection. In fact, the direct blood injection could detect ACTH levels of adrenalectomized rats, but the level of sensitivity was insufficient to detect ACTH in the blood of intact rats. Other investigators pooled large quantities of blood from groups of rats subjected to the same procedure, then extracted and concentrated the ACTH prior to injection into the hypophysectomized donor for assay. However, loss or inactivation or even fragmentation into active units of hormone activity was possible as a consequence of exposure to chemical procedures, and investigators employing different extraction procedures reported results that were not in agreement.

Before I could investigate the regulation of ACTH secretion, I had to develop a method for assay of ACTH in blood that would avoid the problems of extraction, yet provide for the assay of large volumes of blood, because ACTH concentration of intact animals was so minute. A cross-circulation technique was developed in the rat whereby the blood level of ACTH in its natural state in the circulation could be measured. This method of cross-circulation in the rat permitted accurate control of blood volume transfers between two animals, and the apparatus was designed so that the blood volumes of two rats could be kept in balance. Using this method, volumes as large as 100 ml were exchanged without detriment to the partners (Brodish and Long, 1956*a*), and large volumes of blood were quantitatively assayed for ACTH activity (Brodish and Long, 1956*b*).

The cross-circulation apparatus was a "home-made" setup that would look primitive compared to the shiny apparatus we see in contemporary laboratories. Graduate students in the early 1950s were expected to fabricate their own specialized instrumentation, because this was the period prior to the "golden era of NIH support" when machine shops and electronics facilities either were not present or were certainly unavailable to graduate students. In retrospect, I am convinced that I am less fearful of troubleshooting present-day equipment because I had to learn to use the drill press, band saw, lathe, etc., in order to fabricate necessary equipment. What is even more important, I am able to modify, improvise, and design laboratory items that cannot be purchased.

Graduate student days at Yale in New Haven were hectic but exciting. Long hours were spent in cramped quarters, not only overcoming the experimental hurdles of the unknown but also persisting despite the local environmental frustrations. In those days, only the animal quarters were air-conditioned and temperature-regulated, whereas the investigator laboratories varied with weather, and experiments progressed according to the stamina of the researcher. Nevertheless, these were exciting days of youthful energy and inquiry. Interesting experiences come to mind. Because I had to use hypophysectomized rats for ACTH bioassay, these animals could be employed only for approximately 3 days after hypophysectomy in order for the adrenal ascorbic acid measurements to be reliable. Therefore, I arranged for rats to be hypophysectomized at the Charles River Breeding Laboratories in Boston on Mondays and shipped via railroad from Boston to New Haven, where I would meet the rats at the railroad station and bring them to the animal quarters Monday night so that I could experiment with them Tuesday through Thursday of each week. My vigil consisted of a Monday night stroll from the medical school to the New Haven railroad station, where I would meet my passengers and transport them via taxi to the animal quarters at the medical school. In those days, few graduate students owned automobiles, and the comments of the taxi drivers concerning their unexpected cargo was always amusing. They were curious to know how the rats would be used; they wanted assurances that none would escape from the cartons; but most frequently this event was a first experience that they couldn't wait to tell friends and co-workers.

Interactions with C. N. H. Long as my graduate student advisor were exceedingly stimulating and rewarding. Long was extremely imaginative and aggressive in the kinds of experiments that he proposed. As a relatively conservative novice in endocrinology, under Long's influence I became more assertive and somewhat more unorthodox in the experiments that I contemplated. He helped me develop a self-confidence that allowed me to express myself more creatively and boldly. Long's enthusiasm for our work was evident; he would encourage me to move in a variety of directions, but he was relatively impatient and at times could hardly wait to assess our findings. Being somewhat systematic and thorough, I frequently resisted new ventures until I had had an opportunity to evaluate the results of our past experiments. I learned early in my career to satisfy myself that experiments could be replicated in my own laboratory before any reports of experimental findings were transmitted externally. This has been characteristic of much of my work—assurance that the results could be obtained more than once and that the experimental groups were reasonably adequate. Shortcuts were never consciously employed and conclusions were based on several lines of evidence wherever possible.

Development of the cross-circulation procedure for the assay of ACTH in blood made it possible to study the dynamics of ACTH secretion in response to stress. The magnitude of the change as well as the duration of the change in ACTH secretion could now be determined by uniting the stressed animal with a hypophysectomized assay partner at various times after the application of the stress. Variations in the secretion of ACTH by the intact rat were reflected by proportional changes in the adrenal ascorbic acid of the hypophysectomized partner. The results of these studies demonstrated a transient high rate of ACTH secretion after application of stress, followed by declining levels of ACTH 4–6 hr later. The decline in ACTH secretion was observed even after bilateral adrenalectomy stress, suggesting that elevated adrenocortical hormone levels were not solely responsible for the declining ACTH release. Additional stress could release ACTH during the period of reduced ACTH secretion, implying that the capacity for release of ACTH was intact but that reduced sensitivity was present. Control ACTH levels were reestablished 12–24 hr after the stress and, in the case of adrenalectomy, hypersecretion of ACTH was demonstrated 2 weeks after removal of both adrenals. These studies suggested to us a dual phase of ACTH secretion in response to stress—an initial phase of rapid but short-duration ACTH hypersecretion, probably under neural control, followed by a second phase which developed more slowly and possibly involved humoral feedback regulation (Brodish and Long, 1956*c*).

The studies on blood levels of ACTH estimated by means of a cross-circulation technique were a major part of my dissertation for the Ph.D. degree. The ability to determine the dynamic changes in ACTH levels in the blood of intact rats at different times after stress or experimental manipulation was a major advance at the time and received favorable recognition from the external scientific community.

In 1955, I was prepared to defend my dissertation prior to being awarded the Ph.D. degree from Yale. As is usually the case, a Ph.D. defense is a traumatic hurdle for graduate students—the culmination of years of effort now subjected to judgment and final decision. I well remember that awesome event, not only because so much hinged on its outcome but also because I learned unexpectedly that a famous endocrinologist would be in attendance. Bernardo Houssay, a Nobel prize winner, happened to be visiting with C. N. H. Long at Yale, and was invited to sit in on the dissertation defense, and to ask questions if he wished. My first meeting with Professor Houssay, under somewhat stressful conditions, proved to be an experience that I will always remember. He was warm and reassuring, and commented that I had performed better than he had on his examinations many years before. He commented favorably on my work and wished me well in my career. That was one of the special days in my life—not many graduate students can boast of the attendance of a renowned Nobel laureate at their dissertation defense.

The concept of neurohumoral regulation of anterior pituitary secretions was widely accepted, although the evidence for this idea at the time was largely circumstantial. Hypothalamic-hypophyseal portal blood was inaccessible; therefore, direct evidence for secretion of releasing factors from the hypothalamus to regulate pituitary hormone release was not available. During the course of our experiments on ACTH in blood, we inadvertently obtained evidence for corticotropin-releasing factor (CRF) in peripheral blood. In the cross-circulation technique for ACTH assay, rapid changes in ACTH could not be detected because the cross-circulation itself required anywhere from 10 to 30 min for completion. In order to assess the rapid changes in ACTH secretion, we used a modification of a procedure employed by Sayers (1957) and his colleagues in which blood was withdrawn from an intact rat and then infused into a hypophysectomized recipient animal. When the two methods of ACTH assay (cross-circulation vs. infusion-withdrawal) were compared, it was found that the cross-circulated animals consistently exhibited higher levels of ACTH in blood than animals assayed by the infusion-withdrawal technique but otherwise subjected to the same experimental procedure. In the cross-circulation technique an exchange of blood occurred between the normal rat and the hypophysectomized rat during the assay for blood ACTH, whereas in the infusion-withdrawal method the normal rat was not exposed to the blood of the hypophysectomized rat. It seemed entirely possible that the elevated blood ACTH levels following cross-circulation might be due to circulatory transfer of CRF from the hypophysectomized animal to the normal animal. Indeed, an ACTH-releasing substance (CRF) was demonstrated in the peripheral circulation of rats and a technique was devised for its detection under various experimental conditions. Previous investigators (Porter and Jones, 1956; Porter and Rumsfeld, 1956, 1959; Schapiro *et al.*, 1956) had reported the presence of ACTH-releasing substances in hypophysial portal blood or in blood draining the brain, but they did not show conclusively that the substance(s) originated in the hypothalamus. Others, particularly Hume and Wittenstein (1950), suggested the possibility that a hypothalamic humoral agent could enter the systemic circulation and be carried to the anterior pituitary to cause ACTH secretion, but they did not establish the neural origin (i.e., hypothalamus) of this substance. In our studies, the hypothalamic origin of the CRF in the peripheral blood of hypophysectomized rats was clearly demonstrated, because it disappeared following placement of discrete lesions in the hypothalamus of these animals. Our findings raised serious questions concerning the attempts to "isolate" the pituitary gland from the central nervous system by transplantation to distant sites. Some of the controversial reports could possibly be explained by the transfer of hypothalamic releasing hormones via the systemic circulation to influence a pituitary transplant at a remote site (Brodish and Long, 1962).

Our demonstration of CRF in the systemic blood of hypophysectomized animals was the first definitive evidence of hypothalamic releasing factor activity in the peripheral circulation. Since that time, other investigators have reported the existence of several other hypothalamic releasing factors in the peripheral circulation of hypophysectomized rats—i.e., luteinizing hormone releasing factor (LRF) (Nallar and McCann, 1965), follicle-stimulating hormone releasing factor (FRF) (Saito *et al.*, 1967; Negro-Vilar *et al.*, 1968), and growth hormone releasing factor (GRF) (Falconi, 1966; Müller *et al.*, 1967). The relative inaccessibility of the hypothalamic-hypophysial portal circulation (except for the brilliant skill of John Porter) made it difficult to study hypothalamic CRF secretion. Changes in CRF content of the hypothalamus (commonly employed) were more difficult to interpret since they were reflections of both synthesis and release. We therefore reasoned that studies of CRF in peripheral blood might provide insight into the dynamic responses of hypothalamic neurons as regulatory influences on ACTH secretion.

Before comprehensive studies of CRF release could be carried out, a suitable test system had to be devised for CRF assay. *In vitro* pituitary incubation assays for CRF were being criticized incessantly for lack of specificity; therefore, we were interested in developing an *in vivo* preparation for CRF assay, particularly one that could accommodate relatively large volumes of blood. Although a most suitable test animal for the study of hypothalamic releasing factors would be an animal with an appropriate hypothalamic lesion which presumably destroyed the nuclei concerned with the synthesis and release of specific releasing factors, neither the extent nor the precise location of these essential hypothalamic structures had been defined, at least for CRF. For ACTH secretion, there was no agreement as to the location of hypothalamic structures essential for ACTH release. Certain investigators reported that lesions in the posterior hypothalamus effectively blocked a stress response, whereas equally prominent investigators reported that lesions placed in the anterior hypothalamus produced the same result. In order to reconcile these controversial reports and to provide a reliable test animal for CRF assay, we attempted to define precisely the region of the hypothalamus that controls ACTH release from the anterior pituitary gland. To determine the extent of a hypothalamic area that regulates ACTH release, the hypothalamus of the rat was arbitrarily divided into four zones extending from the optic chiasm to the mammillary bodies, and the effects on ACTH release of bilateral lesions of the individual zones, as well as on several combinations thereof, were determined. In addition to varying the location of these lesions, the degree of hypothalamic damage also was varied so that the effects of small and large lesions in the same area could be compared. By using a standard animal population and by standardizing the stereotaxic placements in the same laboratory, it was possible to control lesion size and location, and therefore to make reliable comparisons. Furthermore, the nature of the stress imposed on these animals was essentially the same, so that the stress-induced test for ACTH release was also standard.

Certain investigators (McCann, 1953; Bouman *et al.*, 1957) reported that lesions in the anterior or posterior region of the hypothalamus did not impair a pituitary-adrenal response to unilateral adrenalectomy stress, but when at least 80% of the median eminence was destroyed the response to stress was blocked and no adrenal ascorbic acid depletion occurred. In our studies, by using ether stress as the stimulating agent and plasma corticosterone levels as evidence of pituitary-adrenal activation, we were able to grade the response to a given stress. Therefore, the effectiveness of a particular lesion was evaluated in terms of the percentage of normal response which remained rather than whether or not a response to stress was possible. Consequently, as a result of being able to grade the response, it was shown that a small lesion, placed in any one of the four zones of the hypothalamus, significantly impaired the stress-induced response. These findings implied that the entire region of the ventral hypothalamus, extending from the optic chiasm to the mammillary bodies, was involved in ACTH release, even though the median eminence-tuberal region seemed particularly important. Graded hypothalamic destruction resulted in graded pituitary ACTH response. A small lesion permitted an attenuated corticosterone release in response to stress, whereas a larger lesion resulted in greater attentuation of response. A lesion encompassing the entire ventral hypothalamus produced the greatest deficit in terms of ACTH release, resulting in corticosterone levels that remained essentially at basal control values.

The results of these investigations established that a small, discretely localized hypothalamic nucleus did not exist for the control of ACTH release. To the contrary, there appears to be a diffuse area at the base of the hypothalamus, extending from the optic chiasm to the mammillary bodies, that influences the secretion of ACTH in response to a variety of stimuli (Brodish, 1963). This study unequivocally resolved the problem of localization of a precise hypothalamic region for control of ACTH secretion and was instrumental in dispelling the concept of compartmentalization of hypothalamic control of anterior pituitary activity. There seem to be overlapping areas in the hypothalamus for the regulation of anterior pituitary secretions. These conclusions have since been confirmed by others employing stimulation techniques (D'Angelo *et al.*, 1964, for TSH; Redgate *et al.*, 1973, for ACTH).

Further studies on lesioned rats led us to an unexpected but extremely significant observation. Most of the previous studies by other investigators on lesioned or chemically blocked animals consisted of examinations of stress-induced responses within a period of 1 hr. An experimental design confined to observations within the first hour after application of stress could only detect failure of rapid responses in effectively lesioned animals, and obviously gave no opportunity to detect delayed hormone secretion. In our laboratory, we carried out studies that were directed toward evaluating the impairment in ACTH secretion that followed placement of hypothalamic lesions in rats. The effects of small and large lesions in various regions of the hypothalamus were investigated with respect to the rapidity of stress-induced ACTH release. One of the consequences of a hypothalamic lesion was to delay, rather than to prevent entirely, the ACTH secretory response to stress. The duration of the delay was proportional to lesion size; small lesions delayed for 1 hr whereas larger lesions delayed for 2 hr the increased secretion of ACTH following an appropriate stimulus. These studies showed that "effective" hypothalamic lesions prevented the rapid stress-induced release of ACTH, but they did not prevent a substantial delayed response in which plasma corticosterone concentrations reached levels comparable to those observed in control rats exposed to the same stress (Brodish, 1964*a*).

The delayed response of lesioned animals was unexpected and did not fit the dogma of the time, which assumed that "effective" lesions prevented increased secretion of ACTH in response to stress. Our studies contradicted the assumption that animals bearing hypothalamic lesions could not further increase ACTH secretion. Although lesioned rats did not release ACTH soon after stress, under certain conditions they did show delayed hypersecretion of ACTH and elevated plasma corticosterone concentrations comparable to those of intact animals. The delayed hypersecretion of ACTH and corticosterone was, however, not sustained and returned to initial levels 24 hr later (Brodish, 1969). Thus a delayed transient elevation of ACTH was possible despite considerable hypothalamic destruction. It was not until several years later that an explanation was forthcoming to explain these findings.

The experiments that led me into neuroendocrinology were carried out at Yale, first as a graduate student, then as a National Science Foundation Postdoctoral Fellow, and finally as a faculty member in the Department of Physiology at Yale. The years at Yale were exciting, not only from the standpoint of the rapid pace of neuroendocrine developments, but also because of the outstanding individuals who came through the department as brief visitors or as research fellows for varying periods of time. Arimura came to work with C. N. H. Long and we subsequently became good friends as well as scientific colleagues. Arimura was one of the first to bring the transauricular method of hypophysectomy (originally developed by Tanaka, 1955) to the United States, and I had the good fortune to learn the technique from him. Another research fellow from Japan, Hiroshige, also came to Yale during my residence there. Again, warm friendships were established that exist to this day. Both Arimura and Hiroshige excelled in surgical skills and patience that allowed them to develop preparations that few others could master.

Roger Guillemin visited Yale around 1960, when he was developing the methodology and validation of plasma corticosterone determinations. Guillemin's visit helped us set up the corticosterone procedure in our laboratory and allowed us to "lay to rest" the older adrenal ascorbic acid methodology. More importantly, however, was the warm friendship that developed from that time, and the high regard that I have for him as a scientist, scholar, and human being.

Geoffrey Harris visited Yale during the time I was becoming more involved in neuroendocrine pursuits. He was an exciting, dynamic speaker who obviously was completely committed to the concept of hypothalamic involvement via the portal vasculature. At that time, Harris was having a "running battle" with Zuckerman and his stalk-sectioned monkeys; Harris attributed Zuckerman's findings to vascular regeneration of the stalk-sectioned portal system. Harris had not completely convinced C. N. H. Long that the hypothalamus was the sole regulator of ACTH secretion; Harris was persuasive, but Long remained skeptical.

Another visitor to Yale during that era was Hans Selye. He was invited to address the medical school body on his concept of the general adaptation syndrome. Selye was an extremely effective speaker who enraptured his audience and left them spellbound.

In 1960, while attending the first International Congress of Endocrinology in Copenhagen, I was fortunate to have an opportunity to visit endocrine laboratories in the Soviet Union, particularly in Leningrad and Moscow. During a visit to one of the laboratories in Moscow, I was greeted enthusiastically by a Russian professor and with great excitement I was ushered into a laboratory to view an experiment in progress. To my pleasant surprise, I found them using a cross-circulation setup patterned after the procedure I had developed for rats. Through the assistance of interpreters, I learned that the system was built according to the descriptions in my publications but, instead of polyethylene tubing for cannulation of femoral blood vessels, they had to use glass capillary tubing because plastic was not yet available in Russia. Somehow it made me feel particularly good to know that I had truly made an international contribution.

Periodically, it is healthy to assess our characteristics as scientists and our goals for achievement. One of my basic characteristics is an unwillingness to blindly accept current dogma simply because it is fashionable or because it is expounded by prestigious individuals. Throughout my research efforts, my laboratory has frequently challenged the current dogma and has on a number of occasions been instrumental in revising certain concepts of endocrine control. Appropriate methodology has been developed to improve existing methods so that valid results could be obtained. For example, by developing a cross-circulation technique, quantitative measurements of changing ACTH levels in individual rats became possible without employing artifact-producing extraction procedures. Demonstration of a diffuse ventral hypothalamic system for ACTH release dispelled the concept of discrete compartmentalization that was in vogue at the time and resolved much of the controversy regarding the effectiveness of lesions placed in different parts of the hypothalamus. Application of the concept of diffuseness permitted the development of a reproducible and effective assay for CRF in lesioned rats. The studies that demonstrated a delayed secretion of ACTH in lesioned rats raised questions concerning the universality of hypothalamic control of ACTH secretion. Questions concerning the "absoluteness" of blockade had to be reconciled, and therefore directed attention to other pathways for pituitary ACTH release.

At this time I relocated to Cincinnati, where investigations continued on the lesioned rat and CRF assay. In the course of experiments in which lesioned rats were used as recipients for CRF bioassay, it became apparent that extracts of either hypothalamic or nonhypothalamic tissues could, under certain conditions, induce ACTH release, whereas ether stress alone was without effect. The mechanism of this "breakthrough" was not known; nevertheless, the effects of nonhypothalamic tissue extracts were eliminated by avoiding "prior sensitization" of the recipient while retaining the response to hypothalamic material. Conditions were therefore developed for the reliable use of lesioned rats for the assay of CRF in tissue extracts (Witorsch and Brodish, 1972*a*).

When conditions were established for the reliable use of lesioned rats to assay CRF, it was shown that hypothalamic extracts evoked rapid release of ACTH in rats bearing hypothalamic lesions. ACTH release was obtained from extracts of nonhypothalamic tissue only after lesioned recipients had been stressed ("sensitized") prior to the injection of the extract. Neither the previous stress itself nor injection of nonhypothalamic extract, when applied separately, was effective in releasing ACTH. Hypothalamic extract, on the other hand, evoked the release of ACTH in lesioned rats even in the absence of prior stress. Because the recipient animals had extensive hypothalamic lesions that consistently blocked stress-induced ACTH release and because the effect of nonhypothalamic tissue extract depended on prior manipulation of the recipient, experiments were designed to determine whether hypothalamic or extrahypothalamic pathways were involved. We therefore expanded the hypothalamic lesions to include virtually all of the hypothalamus, but the response persisted, although it was somewhat attenuated. Furthermore, removal of all brain tissue anterior to the mesencephalon (pituitary isolation) did not eliminate an evoked increase in plasma corticosterone. The observation of rapid, stress-induced increases in plasma corticosterone in rats bearing ventral hypothalamic lesions, expanded hypothalamic lesions, and pituitary islands provided evidence for the existence of extrahypothalamic pathways for acute ACTH release in rats (Witorsch and Brodish, 1972*b*).

Although it is well accepted that CRF from the hypothalamus regulates ACTH release, Egdahl in 1960 jolted the neuroendocrine community by demonstrating pituitary-adrenal function in dogs after removal of the entire nervous system rostral to the hindbrain. Egdahl (1960, 1961, 1962) observed elevated levels of adrenal corticosteroids in dogs with "pituitary islands" and also showed that these animals responded to additional stresses such as inferior vena cava constriction or hemorrhage. Explanations of Egdahl's findings ranged from "pituitary leakage" to removal of an "inhibitory input" from the central nervous system. Our findings of delayed responses in lesioned animals suggested the possibility that an extra-hypothalamic humoral CRF might be involved.

At the present time, we do not have purified CRF and, in fact, we do not know what CRF is because little information on its chemical nature or structure is available. We have therefore taken the position that a reasonable approach at this time is to attempt to describe "what CRF does" by characterizing the time course of the response of lesioned rats to substances that contain CRF. By characterizing the elicited response, we might be in a better position to determine whether one or more physiological CRFs exist.

CRF in the blood of hypophysectomized rats showed a time course of response (i.e., corticosterone elevation) similar to that observed after administration of hypothalamic extract (ME-CRF). Therefore, it was reasonable to conclude that the CRF in the blood of hypophysectomized rats represented CRF released from the median eminence of the hypothalamus. On the other hand, when lesioned animals were subjected to laparotomy stress and hypophysectomy, and 5 hr later had blood withdrawn for CRF assay, it was found that the delayed response (i.e., elevated corticosterone levels) was due to a potent CRF that produced an action that was distinctive compared to the effects normally observed with median eminence CRF (ME-CRF). On the basis of its extreme potency and unique time course of action, we proposed the term "tissue-CRF" to distinguish it from ME-CRF of hypothalamic origin (Brodish, 1973; Lymangrover and Brodish, 1973*a*,*b*).

Tissue-CRF has been distinguished from CRF of median eminence origin on the basis of its physical-chemical properties, its extreme potency, its prolonged action on the pituitary-adrenal system, and its existence even after the entire hypothalamus has been removed. Recent reports from our laboratory clearly demonstrate a humoral factor in the peripheral blood of stressed rats which can activate the pituitary-adrenal axis, resulting in a massive and prolonged secretion of corticosterone. Because the substance appeared in animals with extensive ventral hypothalamic lesions that were also hypophysectomized 5 hr earlier, it seems unlikely that this potent pituitary-adrenal activation was evoked by ME-CRF.

Nonhypothalamic CRF (tissue-CRF) seems to stimulate the pituitary-adrenal system in a sufficiently unique manner to suggest that it is not identical to hypothalamic ME-CRF. The term "tissue-CRF" characterizes the nonhypothalamic humoral substance that can directly stimulate the secretion of ACTH. There are a number of reasons why such terminology is suggested. This substance is probably not of hypothalamic origin; therefore, it is not unreasonable to assume that it is released from a more peripheral site in response to stress. Our studies, as well as those of other investigators, suggest that the types of stress that presumably activate an extrahypothalamic mechanism are extensive surgery, caval constriction, laparotomy, prolonged cold exposure, and high altitude; all seem to produce, in common, tissue anoxia and/or tissue destruction. We have postulated that intense and prolonged stress, which results in tissue damage, may evoke the release of tissue-CRF to supplement the rapid ME-CRF mechanism and thereby produce a prolonged output of adrenal cortical steroids in time of need.

What is the possible physiological role of tissue-CRF? Transient secretion of ACTH, presumably by hypothalamic activation, may be an appropriately rapid response to acute stress. The subsequent feedback suppression of the hypothalamic-pituitary system by the secreted hormones may be a means of preventing overstimulation and excessive secretion of the system.

In cases of severe trauma associated with extensive tissue damage, hypothalamic-pituitary suppression by the secreted hormones may be premature and inappropriate. Therefore, a mechanism may exist whereby the affected tissues themselves can sustain adrenocortical activation by releasing tissue-CRF as a signal to the pituitary for continued need. Tissue-CRF release would represent a valuable mechanism for damaged tissues to sustain the signals for the hormones or metabolites that are needed. When the tissue damage has been repaired or stabilized, then tissue-CRF release would cease.

One can further speculate that the mechanism whereby corticosterone suppresses tissue-CRF secretion is similar to the membrane stabilization theory proposed by Weissman and Thomas (1964). If tissue-CRF is a common factor which is found in all tissues and is released as a result of tissue destruction, then the suppression could be mediated through corticosterone acting at the cell membrane of lysosomes to stabilize the cellular lysosomes and prevent the release of enzymes that could disrupt the cell and release tissue-CRF. Further clarification must await purification of tissue-CRF and studies of its secretory process.

The significance of this new finding is that it suggests that tissue-CRF is under physiological regulation and thus may play a role in an animal's ability to adapt to its environment. What previously appeared to be a possible artifact of the lesioned animal preparation now appears to be related to a possible fundamental mechanism for the control of ACTH secretion.

Normally, a transient response to stress is observed which, if sufficient, may be terminated by neural adaptive mechanisms that prevent overstimulation and oversecretion of the pituitary-adrenal system. During continued stress of relatively high intensity, the needs of the organism may not be met by the hypothalamic (ME-CRF) mechanism, and another system (tissue-CRF) may be brought into play to sustain pituitary-adrenal secretions. Obviously, these studies require confirmation because of their significance in our comprehension of pituitary-adrenal regulation. The conclusion is inescapable that CRFs are produced at sites other than the central nervous system (Lymangrover and Brodish, 1973/4). These studies were first reported at an international symposium on Brain-Pituitary-Adrenal Interrelationships held in Cincinnati in 1972. E. S. Redgate and I organized this symposium, which included an international array of active investigators in pituitary-adrenal regulation.

In 1975 I became chairman of Physiology and Pharmacology at Bowman Gray School of Medicine of Wake Forest University in Winston-Salem, North Carolina. Our research efforts continue on understanding tissue-CRF and the overall regulation of the pituitary-adrenal system. Our laboratory has been known to involve itself in difficult procedures and in experiments that do not always have "guaranteed payoffs." It is somewhat unfortunate that the demands for publication in our current climate direct many into research programs assured of success in terms of publishable data, instead of wrestles with conceptual aspects.

My philosophy and attitude has always been to investigate and to pursue in sufficient detail so that a complete understanding would be forthcoming. I have attempted to approach my studies with enthusiasm, organization, and innovation. I try not to take untested shortcuts or to make frequent assumptions; the experiments have to be systematic, complete, and reproducible.

The laboratory is a place of hard work but the work should provide an inner satisfaction for the investigator—an investigator who does not enjoy his work does not belong in the laboratory. To my knowledge, no one who has limited his activities to a 9 to 5 schedule has ever become known or respected as a scholarly investigator. A good investigator finds it difficult to stay away from the laboratory. The most important piece of equipment in any laboratory is the scientist himself.

REFERENCES

Bouman, P. R., Gaarenstroom, J. H., Smelik, P. G., and de Wied, D. (1957). Hypothalamic lesions and ACTH secretion in rats. *Acta Physiol. Pharmacol. Neerl.* **6**:368.

Brodish, A. (1963). Diffuse hypothalamic system for the regulation of ACTH secretion. *Endocrinology* **73**:727.

Brodish, A. (1964*a*). A delayed secretion of ACTH in rats with hypothalamic lesions. *Endocrinology* **74**:28.

Brodish, A. (1964*b*). Role of the hypothalamus in the regulation of ACTH release. In Bajusz and Jasmin (eds.), *Major Problems in Neuroendocrinology*, Karger, Basel, p. 177.

Brodish, A. (1969). Effect of hypothalamic lesions on the time-course of corticosterone secretion. *Neuroendocrinology* **5**:33.

Brodish, A. (1973). Hypothalamic and extrahypothalamic corticotropin-releasing factors in peripheral blood. In Brodish, A., and Redgate, E. S. (eds.), *Brain-Pituitary-Adrenal Interrelationships*, Karger, Basel, p. 177.

Brodish, A., and Long, C. N. H. (1956*a*). A technique of cross-circulation in the rat which permits accurate control of blood volume transfers. *Yale J. Biol. Med.* **28**:644.

Brodish, A., and Long, C. N. H. (1956*b*). Estimation of blood ACTH by means of a cross-circulation technique. *Yale J. Biol. Med.* **28**:650.

Brodish, A., and Long, C. N. H. (1956*c*). Changes in blood ACTH under various experimental conditions studied by means of a cross-circulation technique. *Endocrinology* **59**:666.

Brodish, A., and Long, C. N. H. (1962). ACTH-releasing hypothalamic neurohumor in peripheral blood. *Endocrinology* **71**:298.

D'Angelo, S. A., Snyder, J., and Grodin, J. M. (1964). Electrical stimulation of the hypothalamus: Simultaneous effects on the pituitary-adrenal and thyroid systems of the rat. *Endocrinology* **75**:417.

Egdahl, R. H. (1960). Adrenal cortical and medullary response to trauma in dogs with isolated pituitaries. *Endocrinology* **66**:200.

Egdahl, R. H. (1961). Corticosteroid secretion following caval constriction in dogs with isolated pituitaries. *Endocrinology* **68**:226.

Egdahl, R. H. (1962). Further studies on adrenal cortical function in dogs with isolated pituitaries. *Endocrinology* **71**:926.

Falconi, G. (1966). Maintenance of growth hormone releasing activity in the hypothalamus of long-term hypophysectomized rats. *Experientia* **22**:333.

Hume, D. M., and Wittenstein, G. J. (1950). The relationship of the hypothalamus to pituitary-adrenocortical function. In Mote, J. R. (ed.), *Proceedings of the First Clinical ACTH Conference*, Blakiston, Philadelphia.

Lymangrover, J., and Brodish, A. (1973*a*). Time-course of response to hypothalamic extract and multiple use of lesioned rats for CRF assay. *Neuroendocrinology* **12**:98.

Lymangrover, J., and Brodish, A. (1973*b*). Tissue-CRF: An extrahypothalamic corticotropin releasing factor (CRF) in the peripheral blood of stressed rats. *Neuroendocrinology* **12**:225.

Lymangrover, J. R., and Brodish, A. (1973/4). Physiological regulation of tissue-CRF. *Neuroendocrinology* **12**:234.

McCann, S. M. (1953). Effect of hypothalamic lesions on adrenal cortical response to stress. *Am. J. Physiol.* **175**:13.

Müller, E. E., Arimura, A., Saito, T., and Schally, A. V. (1967). Growth hormone-releasing activity in plasma of normal and hypophysectomized rats. *Endocrinology* **80**:77.

Nallar, R., and McCann, S. M. (1965). Luteinizing hormone-releasing activity in plasma of hypophysectomized rats. *Endocrinology* **76**:276.

Negro-Vilar, A., Dickerman, E., and Meites, J. (1968). FSH-releasing factor activity in plasma of rats after hypophysectomy and continuous light. *Endocrinology* **82**:939.

Porter, J. C., and Jones, J. C. (1956). Effect of plasma from hypophyseal-portal vessel blood on adrenal ascorbic acid. *Endocrinology* **58**:62.

Porter, J. C., and Rumsfeld, H. W., Jr. (1956). Effect of lyophilized plasma and plasma fractions from hypophyseal-portal vessel blood on adrenal ascorbic acid. *Endocrinology* **58**:359.

Porter, J. C., and Rumsfeld, H. W., Jr. (1959). Further study of an ACTH-releasing protein from hypophyseal portal vessel plasma. *Endocrinology* **64**:948.

Redgate, E. S., Fahringer, E. E., and Szechtman, H. (1973). Effects of the nervous system on pituitary adrenal activity. In Brodish, A., and Redgate, E. S. (eds.), *Brain-Pituitary-Adrenal Interrelationships*, Karger, Basel, p. 152.

Saito, T., Sawano, S., Arimura, A., and Schally, A. V. (1967). Follicle-stimulating hormone-releasing activity in peripheral blood. *Endocrinology* **81**:1226.

Sayers, G. (1957). Factors influencing the level of ACTH in the blood. *Ciba Found. Colloq. Endocrinol. Proc.* **11**:138.

Schapiro, S., Marmorston, J., and Sobel, H. (1956). Pituitary stimulating substance in brain blood of hypophysectomized rat following electric shock 'stress.' *Proc. Soc. Exp. Biol.* **91**:382.

Tanaka, A. (1955). A simple method of hypophysectomy on rats. *Shionogi Kenkyusho Nempo* **5**:678.

Weissman, G., and Thomas, L. (1964). The effects of corticosteroids upon connective tissue and lysosomes. *Recent Progr. Horm. Res.* **20**:215.

Witorsch, R., and Brodish, A. (1972*a*). Conditions for reliable use of lesioned rats for assay of CRF in tissue extracts. *Endocrinology* **90**:552.

Witorsch, R., and Brodish, A. (1972*b*). Evidence for acute ACTH release by extra-hypothalamic mechanisms. *Endocrinology* **90**:1160.

6

Barry Cross

Barry Cross was born in Surrey, England, on March 17, 1925. He attended Reigate Grammar School from 1933 to 1943 and graduated from The Royal Veterinary College, London, in 1947. After brief spells in veterinary practice he went to St. Johns College, Cambridge, where he obtained the B.A. hons. in natural sciences in 1949. He was successively Animal Health Trust Scholar and ICI Research Fellow in the Physiological Laboratory, Cambridge, under G. W. Harris. He won the Gedge Prize for physiological research in 1952 and received the Ph.D. degree in 1953. Cambridge University appointed him in 1951 to a demonstratorship in the Zoological Laboratory and he was promoted to a lectureship in 1955 that was later transferred to the Anatomy School. He held a Rockefeller Traveling Fellowship to UCLA in 1957–1958 and was elected a fellow of Corpus Christi College, Cambridge, in 1962. He was awarded the Sc.D. degree in 1963 and 4 years later was appointed professor and head of the Department of Anatomy in Bristol University Medical School. He remained in Bristol until 1974, when he became director of the ARC Institute of Animal Physiology, Babraham, Cambridge.

Dr. Cross has held office in various scientific societies including the Anatomical and Physiological Societies, the Society for the Study of Fertility, the Society for Endocrinology, the Association for the Study of Animal Behaviour, and the Brain Research Association. He was elected a fellow of the Royal Society in 1975 and is currently president of the International Society for Neuroendocrinology.

6

Brain, Breast, and Gonads: A Passage in Neuroendocrinology

BARRY A. CROSS

When I was 14 I resolved to be a veterinarian. It was not a decision lightly made, because my school had placed me on the classics rather than the science side, presumably because my mathematics was even worse than my languages, but I could write tolerable English essays. Nor was the choice based on a sentimental attachment to animals. A school friend had given me *The Science of Life* by H. G. Wells, J. S. Huxley, and G. P. Wells, and reading this from cover to cover convinced me that biology must play a central part in my life. The test tubes, colored precipitates, and sulfurous odors of my bedroom/laboratory were then replaced by anatomized corpses of luckless mice caught in our wartime mousetraps, and my secret cache of potassium cyanide was thereafter used to slaughter Hymenoptera and Diptera captured in the garden. My first thrill of scientific discovery was when, peering down my microscope, I saw a goggle-eyed monster gazing upward from an excavated stronghold in a bee's abdomen. It was a fine specimen of *Stylops*, which parasitizes bees and changes their sexual appearance. Unfortunately, W. Kirby had made the discovery a hundred years before me.

The sections on the ductless glands in *The Science of Life* held a special fascination for me. This was reinforced when, several years later, I reached the Royal Veterinary College and received lectures from E. C. Amoroso and S. J. Folley. Both men had an important influence on the subsequent course of my career, as did H. Rosenberg, who lectured enticingly on the nervous system. Reproduction, lactation, endocrinology, and neurology then

BARRY A. CROSS • Agricultural Research Council, Institute of Animal Physiology, Babraham, Cambridge, England.

became my field of endeavor, and nothing later in the veterinary curriculum supplanted these subjects in my affections. On graduation in 1947 I felt the need for more intensive education in a single discipline, and managed to get myself admitted to St. John's College, Cambridge, to read for the honors degree in physiology. It was indeed a fortunate time to study in the Cambridge Physiological Laboratory, which was then at one of the peaks of its illustrious history. Under E. D. Adrian's leadership, the staff included B. H. C. Matthews, A. L. Hodgkin, A. Huxley, W. Feldberg, W. A. H. Rushton, E. N. Willmer, G. W. Harris, F. G. W. Roughton, and R. D. Keynes. R. A. McCance and E. B. Verney also lectured in the physiology course. Among such a galaxy of talent, it was impossible not to feel a little inspired. The three who undoubtedly had most impact on my developing purpose were Adrian, Verney, and Harris. Adrian so beautifully exemplified the signaling system of the brain, Verney demonstrated the great virtues of surgically prepared unanesthetized trained subjects for detailed endocrinological analysis, and Harris began to piece together the interactions of hormones and the brain.

I needed to persuade Harris to take me on as a research student. Three review articles by S. J. Folley (1947*a*,*b*,*c*) had impressed upon me the extraordinary potency of the suckling stimulus, and I determined to follow up some of the neuroendocrine implications. When Harris was presented with my plan to study the effect of suckling on vasopressin, oxytocin, growth hormone, and ACTH and TSH secretion, he was understandably skeptical of my ability to telescope 10 years work into 3. Probably, too, he was disappointed that I did not ask to work on portal vessels, because his famous review on the neural control of the pituitary gland had just been published (Harris, 1948) and the flow of distinguished endocrinologists to his laboratory to discuss his neurovascular theory had already begun. Also, he was a totally dedicated bench worker and no doubt suspected that I might want Sundays off and an occasional holiday. Nevertheless, he accepted me as a Ph.D. student and from this many blessings followed. Although I worked in his lab for only 2 years, we remained firm friends until his premature death in 1971, and in many of the opportunities that came my way I fancied I could detect his benificent influence.

Another ally at this time was Sir John Hammond, director of the Animal Research Station, Huntingdon Road (subsequently ARC Unit of Reproductive Physiology and Biochemistry). In the summer of 1949 he gave me access to his rabbit colony at the Animal Research Station. In a few weeks there, I was able to evolve the once-daily suckling regime and to show that litter growth rates were the same as on *ad lib* suckling. This daily suckling regime was used in many subsequent researches. The choice of the rabbit as experimental subject was a good example of serendipity, because it soon turned out that this species nurses its young only once or twice a day in a state of nature. Because the litter may double its birth weight in 10 days, the rabbit is endowed with an especially efficient milk-ejection reflex. I could not have chosen a better subject for analyzing the physiological mechanism of this reflex.

Harris taught me how to wrap the rabbit in a towel with only the head protruding, and to pass a stomach tube through a mouth-gag to administer a water load. Then a diuresis curve was obtained by manual expression of the bladder, and the antidiuretic effect of suckling was compared to that of injecting small doses of posterior pituitary hormone intravenously. I became quite adept at the acrobatics involved in these various single-handed manipulations, but there was one stupefying occasion when the rabbit dislodged the gag, bit through the tubing, and promptly swallowed it. The young veterinarian recoiled in dismay, but Harris found the incident extremely amusing and set about a gastrectomy to retrieve the missing tubing. That rabbit subsequently gave some of my best experimental results.

Only about 0.5 mU vasopressin was released by the suckling stimulus in the lactating rabbit (Cross, 1951), and I began to think the quantity of hormone altogether too small to account for the "let-down" or milk ejection as described by Gaines (1915) and Ely and Petersen (1941). At this time, the only rabbit assay for milk ejection potency was that of Turner and Cooper (1941), which depended on a visible swelling of shaved, lactating mammae.

We consulted John Hammond to try to devise a better procedure. He suggested incising the skin over the gland and cannulating a lactiferous duct. One afternoon while trying out various maneuvers with anesthetized lactating rabbits, Harris happened to snip off the end of one of the teats with scissors and saw the half-dozen teat ducts each outlined by a ring of capillary hemorrhage. It was then quite easy to grasp the edge of the teat and insert a fine-tipped glass cannula into a duct and secure it in place with a silk ligature. We replaced saline in the cannula with 1% sodium citrate to prevent milk clots, and this preparation proved a simple and effective means of comparing the milk-ejection activity of various hormonal and pharmacological preparations.

On June 22, 1950, Harris and I performed the first electrical stimulation of the infundibular stem in a lactating rabbit with a teat cannulated and connected to a simple float recorder writing on a smoked drum (Cross and Harris, 1950). Seldom have I experienced such joy as when, after no visible effect had followed stimulation at 1.6, 1.7, and 1.8 cm from the skull surface, lowering the electrode to 1.9 cm and repeating the stimulus at first seemed equally ineffective, then after 40 sec the milk level as if by magic began to rise up the cannula, reaching a peak at about 90 sec and then slowly returned to the starting level. The response was easily reproducible, and in a few more experiments it became clear that the effective site for stimulation was the supraopticohypophyseal (SOH) tract. When the electrode passed 0.5 mm beyond the tract, e.g., into the anterior pituitary, the response disappeared. We had one bad moment when movement of the electrode alone (without passage of current) evoked milk ejection, but this was explained when histological reconstruction of decalcified sections of the hypothalamohypophyseal area showed that the electrode had been pressing on a detached fragment of the bony dorsum sellae protruding into the infundibular process.

In retrospect, this work and the similar studies in the lactating goat (Andersson, 1951*a*,*b*) were too readily accepted as proof of neurohypophyseal mediation of the milk-ejection reflex. As one of our colleagues rather unfeelingly pointed out at the time, we knew that injections of posterior pituitary hormone would evoke milk ejection, and earlier work had shown that infundibular stimulation released posterior pituitary hormone, so we had established nothing new! It was essential therefore to interrupt the reflex pathway, e.g., by SOH tract lesions, and to demonstrate that the sucklings were disadvantaged thereby. The first crucial results from electrolytic lesions in the SOH tract began to emerge while Harris was in Sweden working with Dora Jacobsohn on their famous pituitary autograft experiments in rats. My SOH lesions did not stop milk secretion in the rabbit or alter maternal behavior, but the litters were unable to extract the milk from their mothers' glands, which remained full after the suckling period. Intravenous injection of posterior pituitary extract immediately before suckling enabled the young to empty the glands and obtain the full yield of milk. Harris was pleased to learn of these results and promptly arranged for me to visit Bengt Andersson in Stockholm, who was stimulating milk ejection in conscious goats by the Hess technique applied to the supraoptic nucleus. Moreover, Harris's work with Dora Jacobsohn had also shown that the posterior pituitary gland was necessary to prevent loss of litters in hypophysectomized, lactating rats with pituitary autografts under the median eminence. The rabbit (Cross and Harris, 1952) and the rat (Harris and Jacobsohn, 1952) papers were published in the same year, and the role of the neurohypophysis in milk ejection rapidly became generally accepted.

Looking back at these formative years, I marvel at the cheapness of the experiments. No expensive equipment was used and a single rabbit would be followed through two or three lactations to provide data on milk yield and litter growth, the antidiuretic effects of suckling and other emotional stimuli, the milk-ejection response to electrical stimulation of the hypothalamus, and the effect of SOH or control lesions on milk removal. I liked to compare my favorite does with Verney's "Croonian" bitches, but I did not go so far as to give them all pet names. Rats were much less popular with me, although I mastered Harris's difficult transtemporal approach to pituitary stalk section (Harris, 1950) only to find that it produced equivocal results on lactation, failing to discriminate adequately between deleterious effects on milk secretion and milk ejection. However, during another of Geoffrey Harris's absences abroad, in collaboration with Mary Cotes from F. G. Young's Biochemistry Department in Cambridge, we established that the suckling stimulus (without milk withdrawal) stimulated food intake in rats (Cotes and Cross, 1954).

After 2 years of my Ph.D. work, I was appointed to a university demonstratorship in the Zoological Laboratory to teach veterinary anatomy. Harris and other good friends helped to bring this about, and now I was not only in a position properly to support my young wife and baby daughter, but for the first time I had a laboratory all to myself, in the basement of the Zoological Department, although it had no equipment and there were no facilities for keeping experimental animals. I had to breed my own rabbits at home in Trumpington and transport them by bicycle to the laboratory for experiments. Then John Hammond once more came to my aid and I secured a small research grant from the Agricultural Research Council to purchase equipment and rabbit cages. He also lent me space in the Animal House on the Downing Site, not too far from the Zoological Laboratory. Here I kept my rabbits on the once-daily suckling regime and studied the replacement dose of oxytocin required to restore milk ejection blocked by intravenous Nembutal.

Another piece of luck occurred at about this time. H. B. van Dyke from Columbia had come to England to work at the National Institute of Medical Research at Mill Hill in 1953, and he learned through Geoffrey Harris of our neurohypophyseal lactation work. We all met in the Bun Shop, a pub much frequented by Cambridge researchers, and to my great surprise Harry van Dyke offered to come and spend some time working with me in the Zoological Laboratory on the relative potency of du Vigneaud's purified oxytocin and vasopressin measured by the rabbit milk-ejection method. He was a delightful collaborator and quite unmindful of the great gulf in knowledge and experience that separated us. Our results (Cross and van Dyke, 1953) convinced me that vasopressin could not be the natural milk-ejection hormone because the quantity released by suckling was below the threshold dose. Oxytocin was about 6 times more potent, and must be released in quantities of 20–50 mU in the suckled, lactating doe to account for the ejection of up to 250 g of milk to the litter in the few minutes of the daily nursing period.

The other problem I wanted to settle was the importance of sympathetic activity and adrenomedullary secretion in emotional disturbance of milk ejection which had been postulated by Ely and Petersen (1941). This entailed more acute stimulation experiments to establish the existence of a hypothalamic mechanism for inhibition of milk ejection. Stimuli in the lateral or posterior hypothalamus blocked the milk ejection response to intravenous oxytocin and impaired milk removal by a suckling litter in a manner that could be duplicated by intravenous injection of 1–5 μg epinephrine. Moreover, bilateral adrenalectomy abolished the inhibitory effect of the hypothalamic stimulus on the milk ejection response to a physiological dose of oxytocin given 10 sec later, but only attenuated the response to a dose given *during* stimulation. In this way, the effect of circulating adrenomedullary hormone could be separated from that of direct sympathetic innervation of the mammary gland. The timing of the various intravenous injections in relation to stimulation was critical and, without assistance, these experiments involved considerable agility. Help was near, however, as I was then given a 16-year-old technical assistant, Derek Thurlbourn, who, although one arm was partly disabled by poliomyelitis, proved to be a highly dextrous and resourceful assistant, and he stayed with me for 14 years.

While trying to discover how sympathetico-adrenal activity blocks milk ejection, I found that the myoepithelium of the mammary gland responded not only to circulating oxytocin but also to direct mechanical stimuli (Cross, 1954). Besides accounting for the ability of sucklings to remove some milk from the gland when the milk ejection reflex was inoperative (e.g., anesthesia or SOH lesions), this phenomenon was useful for studying the mechanism of norepinephrine blockade of milk ejection. As the "tap" response was unaffected by norepinephrine and the response to intravenous oxytocin was blocked only if the norepinephrine arrived a few seconds earlier in the gland, I proposed that mammary vasoconstriction, by preventing access of oxytocin to the myoepithelium, was the mode of action (Cross, 1955*a*).

Then began a set of chronic experiments with 15 does to analyze the way an emotional stimulus (forcible supine restraint) blocked milk removal by the litter. Individual does reacted differently, e.g., with no interference, partial block, or complete block to milk removal by the litter. However, since even with complete absence of reflex milk ejection I could seldom demonstrate a significant peripheral blockade of the response to a physiological injection of oxytocin, it was clear that a central inhibition of the reflex preventing release of oxytocin was the principal mechanism for emotional block of milk ejection (Cross, 1955*b*). I was proud of these experiments, because they seemed to me to give a sound physiological basis for a known psychosomatic phenomenon and to extend our understanding of the relationship of the sympathetic system and neurohypophysis. I felt that I had achieved for oxytocin (the milk ejection hormone) what Verney had previously shown, much more elegantly, for antidiuretic hormone.

My hypothalamic stimulation experiments had revealed the intriguing fact that merely moving the electrode a couple of millimeters could change the response from milk ejection to blockade of milk ejection through mammary sympathetic vasoconstriction. This prompted me to look for similar reciprocal effects in the estrous or estrogenized uterus and in the male sexual organs. Some interesting parallels were observed, although the myometrial and myoepithelial responses were qualitatively quite different (Cross, 1958*a*). For example, the estrogenized rabbit uterus was rhythmically active in the absence of circulating oxytocin, which merely enhanced its contractility. However, the myoepithelium could be induced to contract rhythmically, giving a record very similar to that of the uterus if oxytocin was infused intravenously to maintain a continuous high circulating level. Another distinction was that either norepinephrine or stimulation of the hypothalamic sympathetic areas induced a brief tetanic contraction of the uterus before reducing contractile activity. Following up these experiments with studies of the neuroendocrine mechanisms in labor in the rabbit, I was encouraged to find that not only could oxytocin or infundibular stimulation induce parturition in full-term does but also the neurohypophyseal mechanism was adequate to secure delivery of the pups even when other ancillary mechanisms (e.g., voluntary or reflex abdominal contractions) were prevented by Nembutal combined with spinal anesthesia (Cross, 1958*b*). Moreover, the occurrence of milk ejection from cannulated teats during expulsion of young was confirmation of the existence of the Ferguson reflex (Ferguson, 1941). In the male rabbit, T. D. Glover and I (Cross and Glover, 1958) could find no clear effect of oxytocin on the reproductive tract, but stimulation of hypothalamic sympathetic areas elicited a powerful contraction of the vas deferens and seminal vesicle mediated by both nervous and adrenomedullary pathways.

At that time, physiologists were more interested in the location of parasympathetic and sympathetic centers in the hypothalamus, but I felt the dichotomy between oxytocin and norepinephrine mechanisms much more rewarding (Cross, 1958*c*). I remember the excitement of the first occasion I was able to evoke release of oxytocin by stimulation in the paraventricular nucleus after removal of the adrenals. Immediately before, stimulation at the same site had blocked the milk ejection response to intravenous oxytocin. Olivecrona's (1957) work was beginning to convince people, wrongly as we know now, that the paraventricular nucleus was the principal source of oxytocin.

My debt to Harris and to Verney should now be clear. Adrian's influence then began to make itself felt. A nagging thought I had had for some time was that so far only disturbances of hypothalamic function had been exploited (lesions or stimulation); was it not time to monitor the natural workings of the organ? Neither Harris nor Verney had much respect for electrical recording as an analytical tool, but one evening in 1956 at a Royal Society Conversazione I met J. D. Green, who had done the early portal vessel work with Geoffrey Harris and was now at UCLA with Magoun and Sawyer. He fed my outlandish notion of recording hypothalamic activity and suggested I come to the Anatomy Department at UCLA to do it. I was due for a sabbatical year and, to my surprise, Sir James Gray (professor of zoology in Cambridge), whom I mistakenly thought knew me only through my bizarre requirement for lagomorphs in a department devoted to invertebrates and fish, sponsored me for a Rockefeller Fellowship to UCLA in 1957–1958. It was a good year, living in Pasadena with my growing family and commuting on alternate days to the UCLA campus at Westwood for EEG experiments with C. H. Sawyer and Robert Holland, and to the Veterans Administration Hospital at Long Beach for microelectrode recording of single hypothalamic units with J. D. Green. Intracarotid injections of hypertonic saline in lactating rabbits produced dramatic changes in EEG preceding release of oxytocin as indicated by milk-ejection records (Holland *et al.*, 1959*a*). However, a causal relationship between the two parameters was not established, and later work excluded this possibility. A further observation of interest was that the hypertonic stimulus also generated a sympathetico-adrenal discharge capable of diminishing the milk-ejection response to intravenous oxytocin (Holland *et al.*, 1959*b*). In the microelectrode experiments, again with lactating rabbits, we found it unexpectedly straightforward to record from single neurons in the supraoptic and paraventricular nuclei. Their action potentials were unexceptional in waveform and the neurons were responsive to osmotic as well as tactile, nouceptive, and occasionally visual stimuli (Cross and Green, 1959). We did not persevere sufficiently with our attempts to record antidromic potentials in neurosecretory cells, diverted by the rich harvest of hypothalamic single units, and thus we missed an opportunity that delayed progress in this field for another 10 years. Nevertheless, I learned a lot about electrical recording techniques in California that I would have been too awed to inquire about in Cambridge, where it was still the thing to build all your own electronic apparatus. From Tom Sawyer and Johnny Green I found that collaborative friendships could be even more rewarding than making discoveries, although we had an adequate ration of these.

I also acquired expensive tastes in equipment during my Rockefeller year at UCLA. The return to Cambridge in 1958 more or less coincided with the transfer of my lectureship from the Zoological Laboratory to the newly completed Subdepartment of Veterinary Anatomy, attached to the

Anatomy School on the Downing site. My colleague, Ian Silver, and I found ourselves in possession of space and facilities beyond our dreams—a whole floor with operating theater, laboratories, and animal rooms. But there was little research apparatus until Colonel Henry of the USAF appeared, right on cue, and immediately negotiated a sizable contract for myself and Ian Silver to work on hypothalamic unit recording in relation to stress mechanisms and hypoxia. With this and a subsequent NIH grant, we equipped our new laboratories and began a program of research on hypothalamic homeostatic mechanisms, using microelectrode recordings, oxygen cathodes, and a variety of acute and chronic experimental preparations.

The gold-plated needle electrodes used for polarographic recording of oxygen tension proved useful not only for comparing tissue oxygen tensions in different parts of an organ, e.g., gray and white matter of the brain (Cross and Silver, 1962*a*) or testis and epididymis (Cross and Silver, 1962*b*), but also for demonstrating diminution of blood flow, e.g., by arterial occlusion, sympathetic vasoconstriction, or intramammary pressure changes (Cross and Silver, 1962*c*). The vasoconstriction induced in the mammary gland or testis by electrical stimulation of the sympathetic areas of the hypothalamus elevated arterial blood pressure and thus raised oxygen tension in the hypothalamus itself. We showed that severe hypoxic or hypercapnic stimuli activated these hypothalamic sympathetico-adrenal pathways and that the response could be reduced or abolished by hypothalamic lesions (Cross and Silver, 1962*d*). Finally, microelectrode recordings in these hypothalamic regions showed the presence of neurons that responded to the hypoxic and hypercapnic stimuli in a manner consistent with their involvement in the sympathetic discharge (Cross and Silver, 1963). These excursions into hypothalamic homeostatic mechanisms were summarized in a SEB symposium on homeostasis (Cross, 1964). They were fun to do, but they assume a somewhat antique aspect when viewed against the subsequent development of research on biogenic amines.

It was during this period that Charles Barraclough came to work with me and together we obtained the first hypothalamic neuron recordings from rats, which led to attempts to understand ovarian hormone feedback effects at the single-cell level (Barraclough and Cross, 1963). The most dramatic effects were produced by intravenous injection of progesterone, which suppressed unit responses to probing of the vaginal cervix without affecting their response to pain or cold stimuli on the tail. No doubt, as others proposed, this was partly due to the "anesthetic" action of high doses of progesterone, but in later experiments Ian Silver and I obtained evidence that a similar diminution of response in hypothalamic neurons to cervical probing was present in pseudopregnant rats, which could be reversed by ovariectomy. Furthermore, thalamic units did not show any comparable change, which suggested that the influence of progesterone was probably restricted to hypothalamic afferent mechanisms (Cross and Silver, 1965). Graduate students began to make valuable contributions with Alan Findlay recording sensory discharges from mammary glands for the first time, as well as studying estrogen effects on hypothalamic unit activity, and Dennis Lincoln quantifying the neuronal firing characteristics of units in the forebrain of the cyclic female rat. This period of work was reviewed in an article published in the *British Medical Bulletin* (Cross and Silver, 1966).

My last year in the subdepartment was spent with the collaboration of Julian Kitay on a problem that had been gnawing at the consciousness of several neuroendocrinologists for some time, i.e., the functional capacity of the hypothalamus when isolated from all other brain connections. Our approach was to develop a device for manufacturing reproducible hypothalamic islands in rats that were otherwise decerebrate (Cross and Kitay, 1967). It was a quite different preparation from those devised by Bard or by Halász. There were plenty of neurons discharging within the islands, although their pattern of activity differed from that of intact brains, and we planned to use the techniques for studying the action of hormone feedback without the complication of anesthesia or afferent influence from other brain regions. I also intended to present the work at the meeting of the Physiological Society to be held at the Royal Veterinary College, London, in 1967, with my old teacher Amoroso as host for the last time before his retirement. It was not to be, for an outbreak of foot and mouth disease necessitated cancellation of the meeting, and in the meantime I had left Cambridge after a stay of 20 years to succeed J. M. Yoffey as head of the Department of Anatomy at Bristol University. Cambridge had given me far more than I could have possibly expected: to work in three great biological departments, and to experience the social and intellectual stimulation of collegiate life. But with another 25 years of working life ahead, I felt the need for something different, with perhaps a bit more rough and tumble.

The Bristol Medical School was a new building on an elevated site, with views over Bristol to the Mendip Hills. My department was spacious, well equipped for research in microscopic anatomy, and, to my mind, grossly underpopulated, for there were only four permanent members of staff. On the other hand, there were 300 students to teach, far more than I had ever had personal responsibility for in Cambridge. In a couple of years the courses were reorganized and new staff recruited. Research was carried on in a framework of half a dozen semiautonomous groups, with distinct areas of investigation, so that no individual scientist lacked informed colleagues in support, or the stimulus of argument with friends in cognate disciplines. One of these groups was my own neuroendocrine section, which included four recording laboratories and excellent ancillary services. This time, the equipment needs were met by generous grants from research councils in the United Kingdom. Dennis Lincoln joined me from Cambridge, and Richard Dyball, a former veterinary student at Cambridge, was recruited from a postdoctoral fellowship he held with Chandler Brooks. A year later, I appointed a young zoologist, Richard Dyer, as my research assistant to help initially with the program of work on the hypothalamic island preparation.

In a very short time the corridor resounded like a battlefield with the staccato reports of single-neuron discharges from four laboratories. The work with hypothalamic islands showed convincingly that urethane and barbiturates had totally different effects on hypothalamic units (Cross and Dyer, 1971*a*), that oxytocin given intravenously does not change firing rates (Cross and Dyer, 1969), and, most importantly, that the firing rates of anterior hypothalamic units are conditioned by the stage of the estrus cycle, with highest discharge activity in proestrus (Cross and Dyer, 1971*b*). Subsequent work with intact rats revealed that the principal site of accelerated firing was the dorsal portion of the anterior hypothalamic area and contiguous preoptic zone (Dyer *et al.*, 1972). Meanwhile, Dyball, who had done his predoctoral training with Hans Heller in Bristol, pursued the quest for osmoresponsive neurons and control of vasopressin secretion, utilizing the antidromic stimulation technique to identify neurosecretory cells. Two postdoctoral visitors, John Sundsten and Don Novin, joined me to carry out a study of neurosecretory cells in the paraventricular (PV) nucleus of the rabbit. We identified the neurosecretory cells antidromically by stimulation of the neural lobe with an electrode inserted through a foramen conveniently sited in the rabbit basisphenoid bone. Sundsten had been a graduate student of C. H. Sawyer's at UCLA 10 years earlier when J. D. Green and I had first recorded single neurons in the PV nucleus of the rabbit, so it was fitting that he should be the first to see antidromically identified PV cells and help to work out the conduction velocities of their axons and their waveform characteristics (Sundsten *et al.*, 1970; Novin *et al.*, 1970). Another postdoctoral visitor, Bob Moss, followed this opening with a microiontophoretic study of the identified PV neurons, which confirmed the suggestions of various earlier studies that cholinergic afferent synapses were excitatory while aminergic synapses were in the main inhibitory (Moss *et al.*, 1971, 1972*b*). Interestingly, oxytocin itself, delivered microiontophoretically in very small amounts, had excitatory effects on PV neurosecretory cells, although we were careful not to claim a special physiological importance for this phenomenon (Moss *et al.*, 1972*a*). My hope was to study such cells during the milk-ejection reflex, and, because I knew anesthesia blocked the reflex in the rabbit, I encouraged Dennis Lincoln to develop a telemetry system for recording from the conscious animal.

Then occurred one of those events that an investigator is lucky to see once or twice in his career. Jon Wakerley, a young predoctoral student of Dennis Lincoln's, noticed that small intermittent pressure rises occurred in the cannulated gland of an anesthetized, lactating rat with all its litter attached to the teats. Contrary to all expectation from published accounts, my own included, the milk-ejection reflex in the rat can survive anesthesia. Indeed, Wakerley and Lincoln soon found that the periodic ejections in suckled rats occurred at similar intervals in anesthetized and undisturbed conscious rats. This discovery soon led to the identification of oxytocin neurons *in vivo* in both paraventricular and supraoptic nuclei by their twenty- to fortyfold acceleration of firing rate prior to each reflex milk-ejection response (Wakerley and Lincoln, 1973), and scooped other attempts within our group to detect the oxytocin neurons mediating milk ejection in the rabbit (Urban *et al.*, 1971).

During these early years at Bristol, we reached the conclusion that, while much useful information had accrued from recording single-unit activity in the hypothalamus, the future advance of this part of neuroendocrinology depended on a greater knowledge of the connections and chemical specificity of the cells being studied in any particular context. These ideas were developed in several review articles covering the recent contributions of unit recording in neuroendocrinology (Cross, 1973, 1974).

One attraction of the Bristol department for me was the presence of Professor Hans Heller and his MRC Neurosecretion Group in the Pharmacology Department below us. From the outset, our two groups maintained the friendliest links, and when the time came for Hans Heller to retire, the MRC group by common consent moved upstairs and was incorporated into the Anatomy Department under the leadership of Brian Pickering. This, of course, added a much valued biochemical expertise to our neuroendocrine activities, especially with regard to the neurohypophysis. We had also been lucky to interest one of my predecessor's students, John Morris, in the electron microscopy of the paraventricular nucleus. He showed such a flair for careful quantitative EM studes that in a very short time he became widely known for his contribution to the ultrastructure of neurosecretion. It was a new and exciting experience for me to see so many talented individualists cooperate with such enthusiasms to carry forward their mutual research interests in neuroendocrinology. A fitting consummation for the years of electrophysiological, pharmacological, biochemical, and ultrastructural collaboration was the paper given to the 1974 Laurentian Hormone Conference with the simple title "Endocrine Neurons" (Cross *et al.*, 1975).

After a year working in Australia, Hans Heller returned to Bristol and occupied a room in my department where he was able to continue his experiments. I shall always be glad that in this way I could benefit so much from the graceful advice and friendship of this wise old campaigner during his last 2 years. He died in December 1974.

Things were going so well at Bristol that one of the problems I consulted Hans Heller about was whether to go back to Cambridge as Director of the Institute of Animal Physiology at Babraham. I also sought the opinions of Amoroso and T. R. R. Mann, both of whom knew Babraham well and also had a good knowledge of my limitations. Having reached a decision to return, in March 1974, with mixed feelings, I took leave of the department where I had spent 7 happy and productive years.

For the third time, I found myself setting up a neuroendocrine laboratory where none had been before. Fortunately, this time I had the help of three Bristol colleagues who accompanied me to Babraham—R. E. J. Dyball, R. G. Dyer, and J. B. Wakerley. The transformation was remarkably quick, and important new results shortly began to flow from the three newly established laboratories. Dyer and two visiting workers discovered that the number of synaptic inputs from the medial amygdala to neurons in the preoptic or anterior hypothalamic area whose axons pass to the median eminence region is determined by the neonatal "organizational" effects of testosterone, thus providing the first electrophysiological correlate at the single-unit level of sexual differentiation of the brain by androgenic hormone (Dyer *et al.*, 1976). In other studies, Dyball and Wakerley, with collaborators, accumulated persuasive evidence that the previously mysterious "phasic cell" discovered by Lincoln and Wakerley is in fact a particular mode of activity of the endocrine neuron secreting vasopressin. Thus, within 3 years of the discovery of the oxytocin cell (about 50% of recorded cells in PV and SO nuclei), its counterpart, the "vasopressin cell," was identified and the separate characteristics of the two cell types were described (Wakerley *et al.*, 1975). Apart from these two important discoveries, there have been a number of other pleasing advances, some of which echo my own previous preoccupations, e.g., the reversible blockade of milk ejection in the rat by means of radiofrequency currents (Wakerley and Cross, 1975) and the demonstrations that suckling stimulates release of TSH as well as prolactin and oxytocin (Burnet and Wakerley, 1976). The last-mentioned research was made possible by the arrival at Babraham of Keith Brown-Grant and a small MRC research group after the final winding up of Geoffrey Harris's MRC unit at Oxford. This time, the particular expertise added to that of our own group was in the realm of radioimmunoassay. The latest recruit to this group is Barend ter Haar, who worked in the Anatomy Department at Oxford under G. W. Harris but later collaborated with Bernard Donovan at the Institute of Psychiatry in the department begun by Harris in 1953.

Babraham is an excellent place for neuroendocrine research. It has an established reputation for excellence in animal experimentation and has made signal contributions in reproductive and lactational physiology, neuropharmacology, and behavioral studies. There are fine workshops and electronic and general laboratory services, and we enjoy the presence among us of such distinguished senior scientists as E. C. Amoroso and Marthe Vogt. I expect the next 10 years to be highly productive ones. Perhaps the greatest prize will be to attain an understanding of the neurophysiological control of releasing hormones to compare with that which we have finally obtained for oxytocin and vasopressin.

My own passage in neuroendocrinology has been a progression from a single-handed enterprise with minimal facilities through a number of rewarding collaborative endeavors to the leadership of a series of well-supported research groups. I believe the best research is done when a delicate balance is maintained between the natural egoism of the individual scientist and the mutual benefits obtained through cooperative effort. I have never sought to impose a fixed program on my junior colleagues and am pleased to follow the good Cambridge tradition that master and student are equal citizens in the kingdom of ideas. The best hope of the creative scientist is surely that his pupils speedily surpass his own creations.

Note Added in Proof

Readers of this chapter may be interested to know that since it was written Keith Brown-Grant has been appointed Professor in the Faculty of Medicine, Memorial University of Newfoundland, and Barend ter Haar has joined Elsevier Ltd. Richard Dyball has become Reader in the Department of Anatomy once headed by Sir Francis Knowles at King's College Medical School, London. My own former Department of Anatomy at Bristol University is now in the charge of Brian Pickering. John Morris holds a lectureship in Anatomy at Oxford University in the Department Geoffrey Harris made a mecca for neuroendocrinologists. Meanwhile at Babraham good progress has been made in specifying the parameters of the neural signal for release of an ovulatory dose of luteinizing hormone, and in comparing the behavior of magnocellular neurosecretory neurons *in vivo* with that in superfused hypothalamic slices *in vitro*.

REFERENCES

Andersson, B. (1951*a*). The effect and localisation of electrical stimulation of certain parts of the brain stem in sheep and goats. *Acta Physiol. Scand.* **23**:8.

Andersson, B. (1951*b*). Some observations on the neuro-humoral regulation of milk ejection. *Acta Physiol. Scand.* **23**:24.

Barraclough, C., and Cross, B. A. (1963). Unit activity in the hypothalamus of the cyclic female rat: Effect of genital stimuli and progesterone. *J. Endocrinol.* **26**:339.

Burnet, F. R., and Wakerley, J. B. (1976). Plasma concentrations of prolactin and thyrotrophin during suckling in urethane-anaesthetised rats. *J. Endocrinol.* **70**:429.

Cotes, P. M., and Cross, B. A. (1954). The influence of suckling on food intake and growth of adult female rats. *J. Endocrinol.* **10**:363.

Cross, B. A. (1951). Suckling antidiuresis in rabbits. *J. Physiol.* **114**:447.

Cross, B. A. (1954). Milk ejection resulting from mechanical stimulation of mammary myoepithelium in the rabbit. *Nature* (*London*) **173**:450.

Cross, B. A. (1955*a*). The hypothalamus and the mechanism of sympathetico-adrenal inhibition of milk ejection. *J. Endocrinol.* **12**:15.

Cross, B. A. (1955*b*). Neurohormonal mechanisms in emotional inhibition of milk-ejection. *J. Endocrinol.* **12**:29.

Cross, B. A. (1958*a*). The motility and reactivity of the oestrogenised rabbit uterus *in vivo;* with comparative observations on milk ejection. *J. Endocrinol.* **16**:237.

Cross, B. A. (1958*b*). On the mechanism of labour in the rabbit. *J. Endocrinol.* **16**:261.

Cross, B. A. (1958*c*). Hypothalamic control of the secretion of oxytocin and adrenaline. In Curri, S. B. and Martini, L. (eds.), *Pathophysiologia Diencephalica*, Springer, Vienna, pp. 167–181.

Cross, B. A. (1964). The hypothalamus in mammalian homeostasis. *Symp. Soc. Exp. Biol.* **18**:157.

Cross, B. A. (1973). Unit responses in the hypothalamus. In Ganong, W. F. and Martini, L. (eds.), *Frontiers in Neuroendocrinology, 1973*, Oxford University Press, New York, pp. 133–171.

Cross, B. A. (1974). Functional identification of hypothalamic neurones. In Lederis, K., and Cooper, K. E. (eds.), *Recent Studies of Hypothalamic Function*. Karger, Basel, pp. 39–49.

Cross, B. A., and Dyer, R. G. (1969). Does oxytocin influence the activity of hypothalamic neurones? *J. Physiol.* **203**:70P.

Cross, B. A., and Dyer, R. G. (1971*a*). Unit activity in rat diencephalic islands—The effect of anaesthetics. *J. Physiol.* **212**:467.

Cross, B. A., and Dyer, R. G. (1971*b*). Cyclic changes in neurons of the anterior hypothalamus during the rat estrous cycle, and the effect of anesthesia. In Gorski, R., and Sawyer, C. H. (eds.), *Steroid Hormones and Brain Functions*, University of California Press, Los Angeles, pp. 95–102.

Cross, B. A., and van Dyke, H. B. (1953). The effects of highly purified posterior pituitary principles on the lactating mammary gland of the rabbit. *J. Endocrinol.* **9**:232.

Cross, B. A., and Green, J. D. (1959). Activity of single neurons in the hypothalamus: Effect of osmotic and other stimuli. *J. Physiol.* **148**:554.

Cross, B. A., and Harris, G. W. (1950). Milk ejection following electrical stimulation of the pituitary stalk in rabbits. *Nature* (*London*) **166**:994.

Cross, B. A., and Harris, G. W. (1952). The role of the neurohypophysis in the milk-ejection reflex. *J. Endocrinol.* **8**:148.

Cross, B. A., and Kitay, J. I. (1967). Unit activity in diencephalic islands. *Exp. Neurol.* **19**:316.

Cross, B. A., and Silver, I. A. (1962*a*). Some factors affecting oxygen tension in the brain and other organs. *Proc. R. Soc. London Ser. B.* **156**:483.

Cross, B. A., and Silver, I. A. (1962*b*). Neurovascular control of oxygen tension in the testis and epididymis. *J. Reprod. Fert.* **3**:377.

Cross, B. A., and Silver, I. A. (1962*c*). Mammary oxygen tension and the milk-ejection mechanism. *J. Endocrinol.* **23**:375.

Cross, B. A., and Silver, I. A. (1962*d*). Central activation of the sympathetico-adrenal system by hypoxia and hypercapnia. *J. Endocrinol.* **24**:91.

Cross, B. A., and Silver, I. A. (1963). Unit activity in the hypothalamus and the sympathetic responses to hypoxia and hypercapnia. *Exp. Neurol.* **7**:375.

Cross, B. A., and Silver, I. A. (1965). Effect of luteal hormone on the behaviour of hypothalamic neurones in pseudopregnant rats. *J. Endocrinol.* **31**:251.

Cross, B. A., and Silver, I. A. (1966). Electrophysiological studies on the hypothalamus. *Br. Med. Bull.* **22**:254.

Cross, B. A., and Glover, T. D. (1958). The hypothalamus and seminal emission. *J. Endocrinol.* **16**:385.

Cross, B. A., Dyball, R. E. J., Dyer, R. G., Jones, C. W., Lincoln, D. W., Morris, J. F., and Pickering, B. T. (1975). Endocrine neurons. *Recent Progr. Hormone Res.* **31**:243.

Dyer, R. G., Pritchett, C. J., and Cross, B. A. (1972). Unit activity in the diencephalon of female rats during the oestrous cycle. *J. Endocrinol.* **53**:151.

Dyer, R. G., MacLeod, N. K., and Ellendorff, F. (1976). Electrophysiological evidence for sexual dimorphism and synaptic convergence in the preoptic and anterior hypothalamic areas of the rat. *Proc. R. Soc. London Ser. B* **193**:421.

Ely, F., and Petersen, W. E. (1941). Factors involved in the ejection of milk. *J. Dairy Sci.* **24**:211.

Ferguson, J. K. W. (1941). A study of the motility of the intact uterus at term. *Surg. Gynecol. Obstet.* **73**:359.

Folley, S. J. (1947*a*). Endocrine control of the mammary gland. I. Mammary development. *Br. Med. Bull.* **5**:130.

Folley, S. J. (1947*b*). Endocrine control of the mammary gland. II. Lactation. *Br. Med. Bull.* **5**:135.

Folley, S. J. (1947*c*). The nervous system and lactation. *Br. Med. Bull.* **5**:142.

Gaines, W. L. (1915). A contribution to the physiology of lactation. *Am. J. Physiol.* **38**:285.

Harris, G. W. (1948). Neural control of the pituitary gland. *Physiol. Rev.* **28**:139.

Harris, G. W. (1950). Oestrous rhythm, pseudopregnancy and the pituitary stalk in the rat. *J. Physiol.* **111**:347.

Harris, G. W., and Jacobsohn, D. (1952). Functional grafts of the anterior pituitary gland. *Proc. R. Soc. London B.* **139**:263.

Holland, R. C., Cross, B. A., and Sawyer, C. H. (1959*a*). EEG correlates of osmotic activation of the neurohypophyseal milk-ejection mechanism. *Am. J. Physiol.* **196**:796.

Holland, R. C., Cross, B. A., and Sawyer, C. H. (1959*b*). Effects of intracarotid injections of hypertonic solutions on the neurohypophyseal milk-ejection mechanism. *Am. J. Physiol.* **196**:791.

Moss, R. L., Dyball, R. E. J., and Cross, B. A. (1971). Responses of antidromically identified supraoptic and paraventricular units to acetylcholine, noradrenaline and glutamate applied iontophoretically. *Brain Res.* **35**:573.

Moss, R. L., Dyball, R. E. J., and Cross, B. A. (1972*a*). Excitation of antidromically identified neurosecretory cells of the paraventricular nucleus by oxytocin applied iontophoretically. *Exp. Neurol.* **34**:95.

Moss, R. L., Urban, I., and Cross, B. A. (1972*b*). Microelectrophoresis of cholinergic and aminergic drugs on paraventricular neurons. *Am. J. Physiol.* **223**:310.

Novin, D., Sundsten, J. W., and Cross, B. A. (1970). Some properties of antidromically activated units in the paraventricular nucleus of the hypothalamus. *Exp. Neurol.* **26**:330.

Olivecrona, H. (1957). Paraventricular nucleus and pituitary gland. *Acta Physiol. Scand.* **40**:*Suppl.* **136**:1.

Sundsten, J. W., Novin, D., and Cross, B. A. (1970). Identification and distribution of paraventricular units excited by stimulation of the neural lobe of the hypophysis. *Exp. Neurol.* **26**:316.

Turner, C. W., and Cooper, W. D. (1941). Assay of posterior pituitary factors which contract the lactating mammary gland. *Endocrinology* **29**:320.

Urban, I., Moss, R. L., and Cross, B. A. (1971). Problems in electrical stimulation of afferent pathways for oxytocin release. *J. Endocrinol.* **51**:347.

Wakerley, J. B., and Cross, B. A. (1975). Milk ejection in the rat: Reversible blockade of hypothalamic-neurohypophyseal and sympathetico-adrenal pathways by radio-frequency currents. *J. Endocrinol.* **67**:31P.

Wakerley, J. B., and Lincoln. D. W. (1973). The milk-ejection reflex of the rat: A 20- to 40-fold acceleration in the firing of paraventricular neurones during release of oxytocin. *J. Endocrinol.* **57**:477.

Wakerley, J. B., Poulain, D. A., Dyball, R. E. J., and Cross, B. A. (1975). Activity of phasic neurosecretory cells during haemorrhage. *Nature* (*London*) **258**:82.

7

Savino A. D'Angelo

Savino A. D'Angelo, known as Sam, was born in Jersey City, New Jersey, in 1910 and received his secondary school education at Hazelton High School in Hazelton, Pennsylvania. He received all of his advanced education at New York University, where he obtained the A.B. in 1936, the M.A. in 1938, and the Ph.D. in biology in 1940. He put himself through school by playing saxophone in local jazz bands. Except for 4 years in military service working in aviation medicine from 1942 to 1946, he spent nearly his entire research career investigating the control of the thyroid. He developed the stasis tadpole assay for TSH which was sensitive enough to measure the hormone in blood. His work in the hypothalamic control of TSH secretion began in the mid-1950's, and this area continued to interest him. Except for 2 years as an instructor at New York University, his entire academic career was spent at Jefferson Medical College, where he rose from the rank of assistant professor of anatomy in 1949 to that of full professor in 1958. He held a Career Research Award from NIH which was initially granted in 1962. His research was honored by the Cressy-Morrison Award of the New York Academy of Sciences in 1959, and he received the Lindback Award for Distinguished Teaching in 1969. Dr. D'Angelo was a member of many societies, including the American Association of Anatomists, the Americal Physiological Society, the Endocrine Society, the American Thyroid Association, and the Royal Society of Medicine (London). He served on the editorial boards of the *American Journal of Anatomy*, *Endocrinology*, and *Neuroendocrinology*. He published numerous papers in the area of hypothalamic-pituitary interrelationships, control of TSH secretion, and more recently development of TSH control.

Dr. D'Angelo died in the summer of 1976. He will be greatly missed by all of us in the neuroendocrine community.

7

From Tadpole to TRF

SAVINO A. D'ANGELO

My venture into the field of neuroendocrinology can be considered a prime example of phylogenetic "upper mobility." I became a devoted student of the amphibian thyroid gland in my early graduate school days. Later, I climbed to the hypophysis to better understand thyroid function. Eventually, I reached the hypothalamus to gain a better vantage point from which to view the hypophysis. In any event, the experimental efforts of the writer on the mammalian hypothalamic-hypophysial-thyroid system cannot be realistically placed in historical perspective without a few prefatory remarks on the tadpole.

As an outcome of my thesis work (purely morphological) for the doctorate in biology at New York University (1940), I became preoccupied with the belief that the "stasis" tadpole (one whose development was arrested by total inanition) might be a suitable test object for the detection of the thyroid-stimulating hormone (TSH). (Up to this time, at least several dozen different bioassay methods for TSH had been described in the literature.) Following a series of successful trials on these creatures, a preliminary work on the matter of TSH detection was published (D'Angelo *et al.*, 1942), but soon afterward I left the University to enter military service in the aviation physiology program of the USAF. Tadpoles, thyroids, and TSH were soon forgotten. (One would have had to be cretinous to get any endocrinological attention in the early years of the war; later, the adrenal cortex began to attract some attention in connection with the stress of prolonged flight.)

I returned to the University and academic life in 1946, prepared to continue somehow in "altitude" research. That year I attended a symposium, presented by the New York Academy of Sciences, on newer methods for the study of thyroid function. One of the most stimulating topics on the program (to me, at least) was that on TSH, presented by Dr. Alexander

SAVINO A. D'ANGELO • Jefferson Medical College, Thomas Jefferson University, Philadelphia, Pennsylvania 19107.

Albert. In his review of the subject, Albert stressed the urgent need for a method which could accurately define circulatory levels of the thyrotropic hormone, and in this regard made some favorable comments about the stasis tadpole method of assay. Encouraged by his comments, I returned to my laboratory and began to reexamine the problem. With the collaboration of an esteemed colleague, Dr. Albert S. Gordon, and the unstinting cooperation of some graduate students who provided us with tadpoles (they were producing *Rana pipiens* larvae artificially by the thousands for an experimental embryology course given by Professor Roberts Rugh), the tadpole method for the bioassay of TSH was refined and later published (D'Angelo and Gordon, 1950). At about this time, I left the undergraduate Department of Biology at New York University to join the Anatomy Faculty at Jefferson Medical College in Philadelphia. Fruitful collaborations were established there with various members of the Biochemistry and Medical Departments, from which ensued a series of animal and clinical studies, central to which was the assay of circulating TSH using the stasis tadpole method.

The decade of the 1950s was notable for a stream of anatomical and physiological research which began to unravel and clarify the complex neuroendocrine mechanisms underlying the relationship between the brain and the pituitary–target gland systems. By 1955, it had become patently clear (1) that the nerve supply to the pars distalis was too scanty to account for the diverse functional activity of its cells and (2) that the hypofunction of the pituitary gland which ensued after section of its stalk was not solely attributable to a generalized vascular insult. G. W. Harris collated and marshalled the cogent evidence in his now classic monograph (Harris, 1955) and boldly insisted that neural regulation of anterior pituitary secretion was accomplished by special principles from the hypothalamus which were released into the primary capillary plexus of the hypophyseal portal system and conveyed through its venous channels to the pars distalis. Animal experiments designed to test Harris's neurovascular hypothesis proliferated—the placement of electrolytic lesions into different regions of the hypothalamus became a popular and highly useful scientific exercise. It was during this exciting period that I entered the field of hypothalamic investigation. The action was impelled by Dr. Monte Greer's demonstration that electrocautery of the rat's anterior hypothalamus inhibited the hyperplastic response of the thyroid gland to propylthiouracil (PTU). To explain the paradoxical results on thyroid function, Monte (never at an imaginative loss!) postulated the existence of two pituitary thyrotropins, one (for growth of the thyroid) under hypothalamic control and another (a metabolic factor) independent of neural influence. It was this provocative proposition which prompted us to apply our method of TSH assay to the lesioning approach. It was felt that we should be able to measure some change in at least one of these thyrotropins and thereby make some contribution to solution of the problem. To this end, a visit to Greer's laboratory (he was at the NIH in Bethesda at the time) was arranged.

The trip to Bethesda was very useful and Monte was most helpful. He showed us histological sections through the rat brain illustrating the bilateral electrolytic lesions which were effective in inhibiting goitrogenesis. He also demonstrated on a gentle rat (it died on the table) the attachment of earplugs, positioning of the animal in the stereotaxic apparatus, drilling of burr holes in the skull, and insertion of the electrodes using coordinates. Enriched by this stimulating demonstration, we returned to Philadelphia prepared to give it a try in our own laboratory. Working in our unit at this time was a young, eager medical student by the name of Ronald Traum, who had managed to find time in a crowded medical curriculum to engage in a research experience. In due time, a stereotaxic apparatus was purchased, a battery source for electrocautery was rigged up, and, soon after, rats by the dozen came tumbling off Traum's tumbril. By the end of the year we were fully astride the third ventricle (D'Angelo and Traum, 1956).

We presented our first set of results on lesioning of the hypothalamus and TSH secretion at the Spring Meeting of the American Physiological Society in 1957. A detailed work, entitled "An Experimental Analysis of the Hypothalamic-Hypophysial-Thyroid system in the Rat" (D'Angelo and Traum, 1958), was submitted in competition to the New York Academy of Sciences. The laboratory was elated to learn that it had won the Cressy-Morrison Award in the Natural Sciences. The investigation clarified several points of basic and timely interest:

1. Lesions which resulted in diminished thyroid activity were always associated with significant reduction of circulating levels of TSH, whereas hormone stores in the pituitary remained abundant.
2. Adminstration of thyroid hormones to rats with lesions further suppressed pituitary TSH secretion, indicating that the basic negative feedback mechanism between thyroid and hypophysis was direct, and not under neural domination.
3. Destruction of the median eminence did not prevent the marked reaccumulation of TSH stores in the pituitary which characteristically occurred in PTU-fed rats after withdrawal of the goitrogen from the diet.

The last observation led us, later, to a series of new investigations on the dynamics of TSH secretion and the "pituitary TSH rebound phenomenon." The orientation of the laboratory to studies on the hypothalamus provided a source of continual excitement to our technicians and to the succession of medical students who came for research experience. The stereotaxic coordinates routinely employed for electrocautery of the tuberal hypothalamus produced a goodly number of obese rats. The hyperphagia displayed by such beasts constantly amazed workers in the laboratory and wagers were made among them as to the body weight these fat, waddling, docile rats would achieve at autopsy. It was somewhat like running a "weight-watchers" class! Enthusiasm in the laboratory was intensified when our stereotaxic coordinates shifted rostrally, from the median eminence to suprachiasmatic and preoptic regions of the brain. Lesions in these areas usually resulted in mean, lean rats displaying recurrent or persistent estrus. These rats were highly irritable and sensitive to the slightest touch. It became a harrowing experience to take the daily vaginal smears required to validate the constant estrous condition. While their cages were being opened, they would "bound out" on the slightest provocation. These lemuroidean leaps necessitated various precautions: all doors to the animal room had to be closed, and a pair of foot-long, Fischer tongs were always at hand with which to retrieve and return truant rats (by the base of the tail) to their individual cages (we lost an occasional tail this way). Finally, an etherization jar was always available as a last resort to subdue an unmanageable rat.

Despite the travail, our experimental excursions into the rostral brain were quite fruitful. We were able to demonstrate that the preoptic region represents an important neural pathway for regulation of the secretion of several anterior pituitary hormones. In rats rendered persistently estrous by electrocautery of the preoptic area, blood levels of FSH were significantly decreased and unilateral compensatory hypertrophy of the ovary was inhibited (D'Angelo and Kravatz, 1960). Destruction of the preoptic region revealed an incapacity of the pituitary to augment FSH secretion under conditions of enhanced demand (unilateral oophorectomy). Other experiments indicated that the secretion of other pituitary hormones was also potentially limited. When persistent-estrus rats bearing preoptic lesions were chronically exposed to cold (a stress known to activate both TSH and ACTH secretion in the rodent), it was found that

1. Cold exposure further limited gonadotropin function; ballooned uteri and vaginal estrus were replaced by uterine atrophy and anestrus, suggesting withdrawal of FSH support at a time when LH secretion was already deficient.
2. Serum and pituitary TSH levels, normal at animal room temperatures, were decreased in the cold, and the expected acceleration of radioiodine release by the thyroid gland was inhibited.

3. Despite enlarged adrenals, plasma corticosteroid levels in cold-exposed, lesioned rats averaged less than one-third those in intact rats.

The constellation of endocrine changes observed cast serious doubt on the view widely held at the time that discrete anatomical localization existed in the rat's hypothalamus for the neural elements regulating the secretion of the separate anterior pituitary hormones (D'Angelo, 1960).

There was another laboratory at this time in the Philadelphia area which was humming along on the hypothalamus, and at a rate faster than ours. Dr. Sam (Don) McCann had joined the Physiology Department, headed by Dr. John Brobeck (himself a pioneer hypothalamist) in the medical school at the University of Pennsylvania. Dr. Brobeck had occasion to visit our laboratory at Jefferson one afternoon to discuss some aspects of our work on the hypophysis. He casually suggested that a linkup with Dr. McCann would be salutary. Since neither one of us seemed to need ideas from the other, the proposed arrangement was never consummated. We occasionally did communicate, however. I do recall mentioning to Don, one wintry night after a meeting of local endocrinologists, that we were getting persistent estrus in rats with lesions made 7 mm anterior to the earplugs. Imagine my surprise when I heard Don present a paper, not too long after, describing LH concentrations in blood and pituitary of rats lesioned in the preoptic "versus" median eminence regions. He had used a new method of bioassay (devised by Parlow) involving ovarian ascorbic acid depletion. This work was soon followed by another in which LH-releasing effects of crude hypothalamic extracts were first demonstrated. The McCann laboratory has been moving in high gear ever since, notwithstanding a new locale.

Although we and others had clearly established that the diminished thyroid function in hypothalamic-deficient states was clearly referable to reduction in circulating levels of TSH, counterpart studies involving electrical stimulation were then scanty. There were reports indicating that cortical or subcortical excitation induced TSH release from the pituitary and activated the thyroid gland in some species. It had also been demonstrated by Woods in the Harris laboratory of London that direct stimulation of the rabbit's hypothalamus enhanced the secretion of labeled hormone from the thyroid gland. In this connection, we (with Joseph Snyder, a sophomore medical student) began a study in which electrodes were chronically implanted into the hypothalamus of rats (D'Angelo *et al.*, 1964). A series of electrical stimulations delivered to the anterior hypothalamus and the tip of the median eminence induced significant rises in blood levels of TSH, accompanied by histological change in the thyroid indicative of stimulation. These hormonal changes did not occur if these hypothalamic regions were cauterized prior to electrical stimulation. It is gratifying to learn from more recent experiments by others on the precise localization of TRF in the rat brain that anatomical sites showing high concentration of the hypothalamic hormone correspond closely to those which significantly altered TSH secretion in lesion and electrical stimulation studies.

It had become abundantly clear by the early 1960s that the proliferation of articles on neuroendocrinology and new advances in the subject had made the need for a comprehensive symposium a compelling one. A meeting was held, under the auspices of the NIH, to formulate plans for the symposium, and I was invited to review the state of the art as it then existed for the hypothalamic-hypophyseal-thyroid system. The task was accepted, and proved to be a labor of love. In preparing the chapter "Central Nervous Regulation of Secretion and Release of Thyroid Stimulating Hormone," every attempt was made to be as comprehensive as possible from the historical point of view, even to including reports from Iron Curtain countries (provided they contained English summaries). Two incidents occurred later which may or may not have had some bearing on this. I was visited by a Colonel ______ (CIA) who chatted amiably with me about recent European travels and scientific friends. In response to a pointed question, I answered that any discussions I had had with my scientific colleagues abroad were purely scientific and not political. The Colonel never revisited. A few months after submitting an application to NIH for a Research Career Award, I received a telephone call from Washington making gentle inquiries about my Russian "connections." I assured them I had never been in the Soviet Union, and knew none of its scientists personally. This was probably nothing more than a routine check, required by the federal government. In any event, I received the Research Career Award in 1962.[1]

The symposium on neuroendocrinology, held in Miami, Florida, 1961, was a very exciting affair. My presentation (D'Angelo, 1963) was well received and elicited some kind comments from Geoffrey Harris, who agreed with much of what I had presented. There were a few moments of

[1] The awardee wishes to express his personal appreciation to the NIH for this program, which has permitted him to investigate in depth his longstanding interests in the field of endocrine interrelationships. The assurance of a continued, stimulating life-style in the academic environment and the degree of security offered by the Career Award are envied by many scientists. It is regrettable that the program has been curtailed. In these days of changing emphasis from basic research to medical delivery systems, it becomes even more imperative to encourage long-term investigative careers in the basic research process. The continuity of such a program would provide an alternative to other career avenues and would ensure that the well of basic information—so vital to mission-oriented research and technological advance—will not run dry.

trepidation in the ensuing discussion, however. I had emphasized in the talk our failure to influence TSH secretion in rats given vasopressin, and voiced the suspicion that the augmentation of thyroid radioiodine release with the same preparation, as described by himself and others, represented a direct effect on the gland and not one mediated by TSH. Dr. Harris took the floor and began to document, with several data-filled slides, a detailed study on the effects of infusion of lysine-vasopressin on the thyroid gland of the rabbit. Not until the last slide was shown did I heave a sigh of relief! Vasopressin did, indeed, augment thyroidal radioiodine release, but it did so in the hypophysectomized as well as in the intact rabbit. Harris stressed that a sensitive method for measuring TSH in blood was an important prerequisite for testing of the TSH-releasing properties of crude hypothalamic extracts which were likely to be contaminated with vasopressin.

I concluded my symposium presentation by making the obvious prediction that the next decade of neuroendocrinological research would be notable for the isolation, chemical characterization, and synthesis of the family of peptides composing the hypothalamic releasing hormones. [Inwardly, there was the saddening recognition that I would not be able to progress (compete?) in this investigative area with my more versatile colleagues.] In retrospect, it is ironic that the first hypothalamic releasing hormone to be chemically identified and synthesized was TRF (TRH). It was the ACTH-releasing effects of crude hypothalamic extracts which initially occupied the attention of workers in this field, and CRF still remains a hypothetical hormone. Virtually every laboratory investigating the physiology of hypothalamohypophysial mechanisms during the period of 1950 to the mid-1960s managed to concern itself at some time or another with the pituitary-adrenal system. Our investigations into the relationship between the hypothalamus and ACTH secretion were limited and peripheral. We were able to demonstrate, however, that electrical stimulation of the anterior hypothalamus in rats stimulated ACTH release from the pituitary but not if the median eminence was previously destroyed (D'Angelo and Young, 1966). In fact, it appeared that neurogenous stimuli arising in the anterior hypothalamic region activated both TSH and ACTH secretion, but only if the final common path (median eminence) was intact. I would not be surprised if it should be found that localization of CRF in the hypothalamus, as for TRF, is not confined to some single, discrete region.

The investigations which culminated in the isolation and structural elucidation of the thyrotropic releasing hormone from ovine and porcine hypothalami constitute one of the crowning biophysical and chemical achievements in the field of neuroendocrinology. The availability of the synthetic hypothalamic tripeptide hormone fostered a new wave in animal and clinical research which has not yet crested. Meanwhile, the research thrust of our laboratory had veered in other directions. One of these concerned anterior pituitary gland responses in rat pregnancy, specifically in the perinatal period. In a series of studies on maternal-fetal endocrine interrelations, we were able to demonstrate for the first time the release of TSH from the fetal hypophysis after administration of synthetic TRH to pregnant rats on the last day of gestation (D'Angelo *et al.*, 1971). The administration of TRH to mothers resulted in significant changes in levels of radioactivity in blood and thyroid gland, and induced threefold elevations in concentration of circulating TSH in their newborn. *In vitro* experiments revealed that TRH induced TSH release not only from the fetal pituitary but also from the postnatal pituitary (1–3 weeks). The synthetic hypothalamic hormone was found to be equally effective for TSH release in fetus and neonate of the hypophysectomized, pregnant rat as well (D'Angelo and Wall, 1972). The totality of evidence indicated that TRH crosses the placental barrier readily and that the pituitary-thyroid system of the late fetus can be influenced by the hypothalamic hormone. Whether or not endogenous TRH from the mother normally traverses the placenta to influence pituitary-thyroid function in her offspring has not been elucidated. Nor is it known precisely when and where TRH first appears in the rat brain.

Although it was originally postulated that one hypothalamic factor would be found for each specific hormone secreted by the pituitary (unitary hypothesis of Harris), it appears that TRH can cause the release of both TSH and prolactin from the pituitary gland of several species. This duality of function is especially apparent in the hypothyroid individual. We have recently studied acute and chronic effects of TRH administration on TSH and prolactin secretion in the hypothyroid rat (D'Angelo *et al.*, 1975). As in other species, it was noted that intravenous injection of TRH caused prompt release of both TSH and prolactin. The intensity of the acute response could be enhanced in goitrous rats by single injections of T_4. Chronic treatment with TRH (in the drinking water) resulted in diminished release of the pituitary hormones. There was some suggestion from electron microscopy that the thyrotroph and lactotroph cells of the adenohypophysis may have been structurally damaged. An analysis of TRH receptors in these pituitary cell types after acute and prolonged administration of the hypothalamic hormone is urgently needed. Recent reports of psychobiological effects of TRH in normal and endocrinopathic individuals raise a new dimension with respect to clinical usage of the hypothalamic hormone. The antidepressant effects elicited in individuals after intravenous injection of TRH are believed to be independent of the pituitary and thyroid and may represent a pharmacological rather than a physiological response. Although the mechanism by which TRH produces subjective effects in the human subject is still unknown, the possible deleterious effects of chronic treatment with TRH on the pituitary (and on the brain as well) certainly warrant further investigation.

REFERENCES

D'Angelo, S. A. (1960). Hypothalamic and endocrine function in persistent estrous rats at low environmental temperature. *Am. J. Physiol.* **199**:701.

D'Angelo, S. A. (1963). Central nervous regulation of the secretion and release of thyroid stimulating hormone. In *Advances in Neuroendocrinology*, University of Illinois Press, Urbana, (1964). pp. 158–210.

D'Angelo, S. A., and Gordon, A. S. (1950). The simultaneous detection of thyroid and thyrotrophic hormones in vertebrate sera. *Endocrinology* **46**:39.

D'Angelo, S. A., and Kravatz, A. S. (1960). Gonadotrophic hormone function in persistent estrous rats with hypothalamic lesions. *Proc. Soc. Exp. Biol. Med.* **104**:130.

D'Angelo, S. A., and Traum, R. E. (1956). Pituitary-thyroid function in rats with hypothalamic lesions. *Endocrinology* **59**:593.

D'Angelo, S. A., and Traum, R. E. (1958). An experimental analysis of the hypothalamic-hypophysial-thyroid relationship in the rat. *Ann. N.Y. Acad. Sci.* **72**:239.

D'Angelo, S. A., and Wall, N. R. (1972). Maternal-fetal endocrine interrelations: Effects of synthetic thyrotrophic releasing hormone (TRH) on the fetal pituitary-thyroid system of the rat. *Neuroendocrinology* **9**:197.

D'Angelo, S. A., and Young, R. (1966). Chronic lesions and ACTH: Effects of thyroid hormones and electrical stimulation. *Am. J. Physiol.* **210**:795.

D'Angelo, S. A., Gordon, A. S., and Charipper, H. A. (1942). Thyrotropic hormone assay in the tadpole. *Endocrinology* **31**:217.

D'Angelo, S. A., Snyder, J., and Grodin, J. M. (1964). Electrical stimulation of the hypothalamus: Simultaneous effects on the pituitary-adrenal and thyroid systems of the rat. *Endocrinology* **75**:417.

D'Angelo, S. A., Wall, N. R., and Bowers, C. Y. (1971). Demonstration of TSH release from the fetal hypophysis in pregnant rats administered synthetic TRH. *Proc. Soc. Exp. Biol. Med.* **137**:175.

D'Angelo, S. A., Wall, N. R., Bowers, C. Y., and Rosa, C. G. (1975). Effects of acute and chronic administration of TRH on TSH and prolactin secretion in normal and hypothyroid rats. *Neuroendocrinology* **18**:161.

Harris, G. W. (1955). *Neural Control of the Pituitary Gland*, Arnold, London.

8

Bernard T. Donovan

Bernard T. Donovan was born in London in 1927 and has spent most of his scientific life at the Institute of Psychiatry. After a spell as an assistant chemist at the Wellcome Laboratories of Tropical Medicine, and another as a radio electrician in the Royal Navy, he was a science student at Chelsea Polytechnic and University College, London, and graduated in 1952. He held a Medical Research Council Scholarship (1952–1954) and a Beit Memorial Fellowship (1954–1958) before joining the teaching staff of the Institute. His work in London has been punctuated by periods as visiting professor in the Institute of Physiology, University of Lund, in the Department of Anatomy, School of Medicine, University of California, San Francisco, and in the Department of Physiology-Anatomy, University of California, Berkeley. His joint studies with J. J. van der Werff ten Bosch on hypothalamic lesions and puberty in rats won them the Ciba Foundation Prize for Basic Research Relevant to the Problems of Ageing in 1957.

Dr. Donovan has long been associated with the *Journal of Endocrinology* and became its editor in 1974. Together with G. W. Harris he edited a three-volume treatise on *The Pituitary Gland* (Butterworths) which appeared in 1966, and with J. J. van der Werff ten Bosch he wrote the *Physiology of Puberty* (Edward Arnold), which was published in 1965; he also wrote *Mammalian Neuroendocrinology* (McGraw-Hill, 1970). He delivered the Laqueur Memorial Lecture to the Dutch Society for Endocrinology in 1970, and was G. W. Harris Memorial Lecturer at the International Congress of Physiological Sciences in Paris in 1977.

8

Portal Vessels and Puberty

BERNARD T. DONOVAN

As the 1940s gave way to the 1950s, the major preoccupation of those concerned with the relationship between the nervous and endocrine systems was to tag the neuro- on to endocrinology. Major contributions had been made by men such as G. Roussy and M. Mosinger, who published their *Traité de Neuro-endocrinologie* in 1946, but the nature of the functional link between the brain and pituitary gland was still highly controversial. Indeed, the role of the brain in influencing the secretion of several hormones of the pars distalis remained to be established. That the brain exercised some trophic role over hypophysial activity, as implied by the embryological development of a close association, was tacitly accepted, but little more. The reputation of my tutor in this field, G. W. Harris, was established by his classical studies at Cambridge of the effects of stimulation of the hypothalamus on hypophysial activity, and by his work, alone and in collaboration with Dora Jacobsohn, on the consequences of complete transection of the pituitary stalk. His *Physiological Review* on the neural control of the pituitary gland, published in 1948, was also of seminal importance. In 1952 Harris became the first Fitzmary Professor of Physiology at the Institute of Psychiatry, which is associated with the Maudsley Hospital in London. Since a pioneer Department of Neuroendocrinology was then founded, it is worth describing the events that led up to his appointment.

Interest in the psychiatric implications of endocrinology had been manifest at the Maudsley Hospital long before World War II, when Edward Mapother recognized that endocrinological studies were highly relevant to psychiatry and approached several people, among whom was Solly Zuckerman, in the hope of promoting research. These inquiries lapsed upon the outbreak of war, but were resumed relatively soon after the Institute of Psychiatry was established at the Maudsley Hospital in 1948. Now the

---

BERNARD T. DONOVAN • Department of Physiology, Institute of Psychiatry, De Crespigny Park, London SE5 8AF, England.

prime mover was Aubrey Lewis, the professor of psychiatry, who told me that at dinner one evening at a Ciba Conference in April 1951, "I sat next to Geoffrey Harris, I sat on his right. I hadn't met him before and found him very congenial and very interesting, and in fact the notion was set in my mind that he might be the man we were looking for. I enjoyed my contact with him and formed a high opinion of his, well, what shall I say, his superficial qualities. Frank Beach was sitting opposite." At that time the prospects of a senior post in Cambridge for Harris were not very bright and he was amenable to the overtures from London. While on a visit to Cambridge, Aubrey Lewis decided to look in on Geoffrey. "I wasn't sure that he would be in, but I thought that it would be very nice to go and see him, and there he was sitting on a high stool in the lab. He told me what he was doing and talked to me in a rather desultory way for an hour or two and then he said he would drive me to the station. On the way I put to him bluntly the question of whether he would be interested in coming to the Maudsley. As far as I can recall now, he was cautious and I pictured him saying to himself, 'Those looney-bin doctors! I don't know whether they can live with the goods they promise so light-heartedly,' and I pictured him going around inquiring of a few people just what the status of the place was and what facilities he might reasonably expect in an institution so different from what he was accustomed to in Cambridge. The upshot was that on the strength of his tentative willingness to be considered for the post, I spoke to the Lord Mayor of London for that year, Sir T. Brown, who was chairman, I think, of our Finance Committee or General Purposes Committee of the Hospital, and asked whether it would be possible to deflect an appropriate sum from the endowments to serve a chair that would run for 10 years on a pattern set by the Nuffield Foundation, who had endowed a number of chairs." It was indeed possible, and the Institute Report for 1951–1952 records that "First amongst events of high importance is the establishment of the Department of Experimental Neuroendocrinology under the direction of Dr. G. W. Harris. Much progress had been made toward the creation of a Chair of Physiology tenable at this Institute, and it was intended that the occupant of this Chair will be head of the newly-established department." While the negotiations were proceeding, Harris was first appointed as a senior lecturer, but 1 year later Professor Lewis could write in the Institute Report (for 1952–1953): "A noteworthy event in the year under review was the appointment of Dr. G. W. Harris to the Fitzmary Chair of Physiology. The wisdom of our choice was emphasized shortly afterwards by the election of Professor Harris to Fellowship of the Royal Society."

The first home of the new Department of Neuroendocrinology (Experimental was never included in the title) was a prefabricated, one-story building which contained an experimental room, an operating theater, and several animal rooms. It cost £8000. This space was never wholly allocated to the department, and during the first few months office work was carried out in an animal room.

I joined Geoffrey Harris in July 1952, on a temporary basis before taking up a Medical Research Council Scholarship in September of that year. My home was the experimental room, then completely lacking in equipment, but possessing bare shelves and empty cupboards. It was my task, and that of Michael Cymbalist, an equally raw graduate, to scour catalogues in search of the equipment we needed. This was not arduous, because there were then few suppliers of scientific apparatus and our requirements were not particularly sophisticated. Essentially, we were planning for animal surgery and pharmacological experiments utilizing organ baths and smoked drums.

My first collaborative work with G. W. H. was the collection of hypophyseal portal vessel blood from cats and rabbits subjected to electrical stimulation of the hypothalamus, and we got down to serious work in late September. In both species the aim was to expose the pituitary stalk, after removing the tongue and soft palate, and then to collect the blood oozing from transected portal vessels. Heparinization was necessary before incising the portal vessels and this meant that all bleeding had to be stopped with certainty before injection of the anticoagulant. Not surprisingly, this provided excellent training for the novice surgeon. Because of the broad skull floor and easier approach to the pituitary stalk, the cat was soon preferred to the rabbit, but the collection of portal vessel blood was painfully slow. At that time, attention was directed toward the estimation of adrenergic activity in the blood samples, because Markee *et al.* (1952) had indicated that the neurohumoral agent affecting luteinizing hormone secretion in the rabbit and rat was adrenergic in character. To our chagrin, adrenergic or noradrenergic activity was never convincingly demonstrable on the rat preparations of Michael Cymbalist, and we gradually became discouraged.

Alongside the collection of portal vessel blood, more productive experiments concerned with the slow injection of epinephrine, norepinephrine, or control solutions into the hypothalamus or pituitary gland of rabbits were in hand. As before, these studies were prompted by the work of Markee *et al.* (1952), who had concluded that the hypophysis released luteinizing hormone upon exposure to epinephrine. A needle attached to a micrometer syringe filled with the test solution was fixed to a stereotaxic machine and lowered vertically between the cerebral hemispheres to reach the hypothalamus or to pass through the pituitary stalk to the hypophysis. Every minute for 100 min, the micrometer was advanced by hand, my hand, to expel a minute quantity of fluid from the needle into the surrounding tissue, and 2 days later the animals were killed to check for the occurrence of ovulation. Many rabbits were used in this way, and, while it was confirmed that the injection of epinephrine or norepinephrine into the pituitary gland could cause ovulation, it was also found that the acidity of the solution injected was responsible for the reaction. Ovulation could be induced by the injection of acidic solutions, and did not ensue when the epinephrine solution was brought to a neutral pH before administration. Besides their intrinsic interest, these experiments taught much about the need for rigorous control experiments in physiological research, despite the seeming retardation of progress that is entailed.

In the spring of 1953 our experimental menagerie was enlarged by the addition of ferrets. Like rabbits, these animals conveniently ovulate after coitus, indicating the operation of a neuroendocrine control mechanism, and it was relatively easy to approach the pituitary gland and pituitary stalk surgically by a parapharyngeal route. We wished to determine whether the direct application of a neutral solution of epinephrine to the pituitary gland, in conjunction with iontophoresis of the amine, would release luteinizing hormone and cause ovulation. Follicular rupture did not ensue, although the rather crude technique of bathing the subtemporally approached hypothalamus with an epinephrine solution was more promising. However, these exploratory studies were interrupted by the publication on May 30, 1953, of a provocative letter to *Nature* by A. P. D. Thomson and S. Zuckerman of the Department of Anatomy, University of Birmingham. Under the heading, "Functional Relations of the Adenohypophysis and Hypothalamus," the letter gave a preliminary account of a study of the effect of pituitary stalk section on light-induced estrus in the ferret, where animals exposed to the stimulus of extra illumination during the winter months failed to come into heat, although "the distal end of the proximal stump of the stalk had become either adherent to the superior surface of the anterior part of the pituitary, or closely connected to it by reaction tissue. In both cases there was a clear anastomosis between the vessels in the two structures." In one critical case, however, the pars tuberalis had been so severely damaged that it was not possible to recognize either cells which could be designated as tuberalis cells or the primary loops of the portal vessels. In another ferret, paper, new bone, and fibrous tissue intervened between the pituitary gland and the part of the brain anterior to the base of the stalk, there being no trace of any vascular connexion between the two. Both of these animals reacted to light by becoming estrous. Naturally, to Thomson and Zuckerman, these experiments showed that "the gonads of the ferret will respond to additional light during the winter, when there appears to be no possibility of impulses emanating in the retina being transmitted via the hypothalamus to the pars distalis of the pituitary either directly by a nervous pathway or indirectly, through the pituitary-portal system of vessels."

Zuckerman had long been a critic of the portal vessel hypothesis and had argued that "the critical proof that the anatomical integrity of the portal circulation is essential for the normal functioning of the anterior pituitary, or for that matter, any other anatomical link (e.g., neural) between the hypothalamus and pars distalis, would be the demonstration that pituitary grafts do not maintain normal bodily function, or alternatively that normal function is impossible after the pituitary is separated from the hypothalamus" (Zuckerman, 1952). Here was that demonstration, or so it seemed. Clearly the work of Thomson and Zuckerman could be of major significance and required confirmation. And it was natural for us to set about the task at once.

A major point of criticism of the Birmingham workers was that their parapharyngeal surgical approach to the pituitary stalk allowed herniation of the hypothalamus through the necessary hole in the floor of the skull and so disrupted the normal anatomical relationships. Displacement of the paper plates inserted between the hypothalamus and hypophysis to block regeneration of the portal vessels also frequently occurred. We aimed to avoid this difficulty by approaching the pituitary stalk subtemporally, using the technique developed by Harris (1950) for work on the rat and already applied to ferrets. It was not easy to manipulate pieces of waxed paper between the optic nerves and internal carotid arteries and then backward to lie between the carotids and between the cut ends of the pituitary stalk, but it was possible. The first stalk-section operation was attempted on June 18, but failed because the right internal carotid artery was accidentally severed. The next operations were successful and a series of stalk-sectioned ferrets, with and without plate insertion between the hypothalamus and hypophysis, and blank-operated ferrets was prepared during the summer. On October 26, a total of 24 ferrets were transferred from natural daylight conditions to racks provided with fluorescent lights which were switched on for 16 hr a day. The windows of that animal room (which had been the first departmental office) were blacked out to avoid the complication of natural variations in daylight. All animals were carefully examined at weekly intervals. To our horror, one animal started to show the vulval swelling indicative of the onset of heat 1 week after transfer to the new conditions, and that was a stalk-sectioned animal in which a barrier had been inserted to prevent portal vessel regeneration. Four animals (three stalk-sectioned and one blank-operated) had followed this example after a further 3 weeks, so that doubts were being expressed concerning the validity of the experiment. Fortunately, as we saw later, portal vessels were apparent in all five animals, with the paper plates being misplaced in the two animals provided with them.

The study was completed in February 1954, when all the animals were killed and the vital histological work was begun. It was the very good practice of Geoffrey Harris to require that his research students carry out *all* of the procedures involved in an experiment, and, while we grumbled at the labor involved and the limitations imposed on the pace of experiments, we appreciated the desirability of this training. Our acceptance of the regime was reinforced by observation of the fumblings of visiting research workers, who had previously enjoyed the ministrations of skilled laboratory technicians. Nevertheless, there was some urgency about the pace of the anatomical studies, for G. W. H. was to depart for a spell of study leave in Berkeley, California, in the autumn, and I had to complete and write up sufficient work for the award of a Ph.D. degree before then.

In the course of evaluating the ferret results, Geoffrey Harris and I were shown the slides of the hypothalamohypophysial region of the remarkable Thomson and Zuckerman ferrets during a visit to the Department of Anatomy. I would have needed hours to extract anything useful from a study of the slides, but my senior colleague was much more perceptive and considered that a significant proportion of the blood vessels in the paraffin sections did not contain india ink, and that conclusions based on incompletely injected specimens were unreliable. This point well illustrates the scientific problems involved in proving a negative. The most rigorous tests were needed before concluding that no blood vessels linked the hypothalamus to the pars distalis in those ferrets that came into estrus after exposure to long days, although some animals with such connections failed to respond to light. The situation is reminiscent of the controversy over the existence of a functional innervation of the pars distalis, a controversy resolved by the application of electron microscopic techniques to the gland.

We aimed for conclusive observations in our ferrets by thorough perfusion of the heads with indian ink and by cutting thick sections (130–200 μm) of the nitrocellulose-embedded pituitary region, which is an excellent way of demonstrating a vascular network. The preliminary findings were summarized in a letter to *Nature*, which was submitted on June 15, 1954, and published 3 months later, on September 11. Essentially, we found that the appearance of estrus could be correlated with the presence of vascular connections between the median eminence of the tuber cinereum and the anterior pituitary gland. These connections were clearly visible in all eight stalk-sectioned ferrets that became estrous. Study of the three animals in this group in which plates had been inserted showed that the plates had been misplaced (two were situated too anteriorly and one obliquely) and that vascular connections had been reestablished around them. No case of estrus was observed in the absence of vascular connections. The anestrous animals had few, if any, vascular connections between the median eminence and anterior pituitary gland, although the animals in this group (one simple section and nine with plates) showed a well-marked primary plexus and a well-vascularized adenohypophysis.

The controversy over the functional significance of the hypophysial portal vessels was publicly aired at a meeting on the human adrenal cortex, held at the Ciba Foundation June 22–25, 1954. Zuckerman was to speak on this topic, and it was clear that his remarks could not go unchallenged. In the course of a long and fascinating critique of the evidence for the view that the portal vessels are a critical component in the mechanism through which the hypothalamus controls the pars distalis (Zuckerman, 1955), he described the Birmingham study on the ferret, and a clash became inevitable. Harris opened by remarking that "I should like to say how much I enjoyed Professor Zuckerman's philosophy, but if I may say so, I don't think his attitude to the literature was very critical," and went on to give reasons. Then he proceeded to summarize our observations. In his turn, Zuckerman replied at length, and the flavor of the occasion is conveyed by the following extract:

> I do not propose taking up with Professor Harris the validity of reports of experiments that have appeared in the literature. I, and I imagine he, have no direct knowledge of the work, and I can only hope that the investigators to whose studies I referred were both competent and concerned not to mislead their readers in what they wrote. . . . Professor Harris derives the conclusion that the pituitary portal vessels are a necessary part of the pathways whereby light induces oestrus in the ferret, from the fact that oestrus did not occur in 12 of 20 ferrets in which the stalk was divided, and that in these animals few, if any, vascular connections had re-established themselves between the median eminence and anterior pituitary gland. We can match these negative observations with a far larger number of cases where ferrets have failed to respond to light after some surgical intervention in the region of the pituitary stalk or the pituitary. On the other hand, I regard these negative observations as being irrelevant to the primary problem. The propositions which we have set out to test derive directly from Professor Harris's thesis that the anterior pituitary gland ceases to function normally when it is effectively separated from the hypothalamus, and further, that when the anterior pituitary gland does function normally—in the particular experiments which we are now considering, when oestrus occurs in a ferret after exposure to light—the pituitary portal vessels drain directly into the anterior pituitary. In order to test the first proposition, Dr. Thomson and I have tried, over a period of more than six years, to obtain one or more preparations in which there is an absolutely effective barrier between the pars distalis and the overlying hypothalamus, and where the animal concerned responded to light. This we have finally achieved. If we had achieved it only once, it would effectively have disposed of the first proposition which we set out to test. The second proposition, that the observation of normal pituitary function implies that the pituitary portal vessels drain directly into the anterior pituitary, has also been shown to be invalid, since we now have several ferrets—additional to those in which there was a barrier between the hypothalamus and pituitary—which have responded to light when no direct vascular connection can be demonstrated between the pars distalis and the remains of the proximal stump of the pituitary stalk containing the primary capillary net.
>
> Professor Harris's third point concerns the thickness of our sections. Here our experience simply does not accord with Professor Harris's. 10 μ sections adequately

> demonstrate such vascular connections as exist between the pars distalis and the tissues by which it is surrounded. It just so happens that in our critical preparations these connections are present in all places except where Professor Harris's thesis demands that they should be. . . . In what he calls our critical preparations we have been concerned to show whether the primary capillary net of the pituitary portal vessels is present, and if it is, whether there is an effective barrier to the passage of capillaries or any other small vessels from the hypothalamic region to the pars distalis. We now have two three-dimensional reconstructions which show precisely the position of the barrier of cellulose surrounded by reaction tissue. In both this impervious barrier happens to satisfy, completely, the anatomical requirements of our operation—and the two animals in question responded to light. . . . Had Professor Harris spent more than a minute glancing at our sections he would have realized that whether or not any vessels remained uninjected, others which were properly injected were present everywhere, as I have already emphasized, except where they should be according to his thesis. Dr. Thomson and I have spent not one minute looking at our many series of serial sections, but weeks. Other anatomists have also examined our preparations for far longer than did Professor Harris or Mr. Donovan. Even if one could ever tell whether every minute vessel was full of ink—which I doubt—the suggestion that a significant proportion of visible blood vessels did not contain Indian ink is quite incorrect. And to repeat, it is also irrelevant, since what matters is whether there was a barrier impervious to the direct passage of blood vessels from the hypothalamus to the pars distalis—which it was in the specimen looked at by Professor Harris and Mr. Donovan.

The rejoinder of Zuckerman has been quoted extensively because it well illustrates the importance attached in this particular controversy to the aberrant findings in a very few individuals, and the neglect of a much greater mass of concordant observations. This is underlined by the efforts made "over a period of more than six years, to obtain one or more preparations in which there is an absolutely effective barrier between the pars distalis and overlying hypothalamus, and where the animal concerned responded to light." It is undeniable that logic argues against the portal vessels being effective in such an animal, but logic also requires that the highly exceptional nature of such individuals be taken into account and that the results be examined with the utmost rigor. We had found that vascular connections between the hypothalamus and hypophysis could be made around the edges of the plate, in the connective tissue investment. Nevertheless, Zuckerman has remained skeptical about the significance of the portal vessels and returned to the subject in a book of essays published in 1970, when he wrote "But even if I am alone in my disbelief, what I do know is that the evidence against the speculation that the pituitary portal vessels constitute the critical functional link between the hypothalamus and the pars distalis, which remains the text book story, is even more convincing now than what it was when I first expressed my contrary view several years ago."

My involvement in this argument, albeit in a silent capacity, was not allowed to interfere with the examination of adrenergic substances as a putative neurohumoral agent controlling LH secretion in the rabbit, because the deadline of thesis completion loomed ever nearer. By virtue of much use of the small hours, the thesis was completed and submitted to the University of London on the earliest date allowed by the regulations, exactly 2 years after registration for the degree. The oral examination was held in Cambridge, in the rooms of my External Examiner, John Hammond, within a few days, and by September 11 the Academic Registrar could acknowledge receipt of the examiners' report. The degree was formally conferred a month later.

Bigger things were afoot in the laboratory. Additional, temporary, accommodation had now been built for us and several other departments of the flourishing Institute of Psychiatry, and we could now breathe a little more expansively. Up until then the experimental room had been crammed with experimenters, with Michael Cymbalist being preoccupied with his organ bath in one corner, thyroid activity in rabbits being assessed by Keith Brown-Grant and Si Reichlin in another, a histological technician trying to process slides on a table by the door, and animal surgery being undertaken by Geoffrey Harris or myself on an operating table extending peninsularwise into the center of the room. It made for a happy family atmosphere but did not exclude family squabbles, particularly when the temperature inside the laboratory soared above the 80s in the summer. The squabbles faded when the new space came into use, and I could spread my wings a little.

Of even greater importance was the arrival of J. J. van der Werff ten Bosch (Koos), in September 1954, from the Department of Endocrinology of the University of Leiden, for a 2-year spell as a British Council Scholar, and a lunchtime diet of soup and sandwiches. Our collaborative work, which then began, has been always most enjoyable, although he learned more English than I Dutch. Since it was easy to transmit my fascination with ferret physiology to Koos, we decided to explore several facets together. On the probably mistaken premise that an increase in the secretion of follicle-stimulating hormone precipitated the onset of estrus, we tried to hasten ovarian development in the sexually quiescent ferret by prolonged electrical stimulation of the hypothalamus. This was no easy task, for extended periods of brain stimulation had to be applied to animals kept in the dark or under environmental conditions that would not advance the onset of estrus. We resorted to keeping ferrets equipped with hypothalamic electrodes and a leather harness in lightproof but ventilated boxes, or in light-tight compartments under the laboratory bench. The experiments extended over two winters, but were unsuccessful in that estrus was not advanced; however, the results took on a new significance when the effect of hypothalamic lesions in ferrets later became apparent.

Lesions had been made in the anterior hypothalamus, just behind the optic chiasma, in an effort to interrupt any nerve fibers that might leave the optic nerves to convey information directly from the eyes to the hypothalamus in mediating the endocrine response to photic stimulation. The existence of such fibers was controversial, but we felt that nothing was to be lost by an experimental test of their possible physiological significance and, late in 1954, duly made our lesions and placed the animals under the stimulating conditions of long days. We paid little attention to the animals that came quickly into heat, being especially concerned with the all-too-few animals in which the response was blocked. Nevertheless, the results were sufficiently promising to prompt an extension of the study during the following year. However, during the winter of 1955–1956, the long-day accommodation was preempted for another study, so that we had to return our five ferrets to short-day conditions after placement of the lesions and look on this occasion for interference with the onset of the natural breeding season. We did not have to wait that long, because, unexpectedly, the lesioned animals came into heat. By this time, Geoffrey Harris was back with us and he was quick to appreciate the likely significance of this finding, so that with his encouragement, and financial backing for the purchase of further animals, my notebook for January 19, 1955, laconically records: "In view of surprising onset of oestrus in 3 ferrets with hypothalamic lesions, and in view of possible importance of this observation a larger series was prepared." The larger series consisted of ten further animals with lesions, and 12 blank-operated females, because in work of this kind an adequate number of controls was vital. Reassuringly, five of the lesioned animals rapidly came into heat long before the blank-operated controls, so that the reproducibility of the response was assured and the reality of the effect established. Now we could ponder the meaning.

At that time, speculation was free of the constraints now placed on it by present knowledge of plasma gonadotropin concentrations, and it was reasonable to conclude that lesions in the anterior hypothalamus produced estrus by destruction of or interference with an area normally inhibiting the secretion of pituitary gonadotropin, thus permitting the discharge of hormone. During sexual quiescence, the area would be tonically restraining gonadotropin secretion, while stimuli known to cause pituitary activation, such as exposure to prolonged illumination, would depress the anterior hypothalamus and allow discharge of gonadotropin in increased amount. In this light, our failure to alter ovarian function in the ferret by prolonged hypothalamic stimulation was readily explained: we were simply activating an inhibition.

Our excitement at the discovery of this novel effect had little chance to subside, as another implication began to be appreciated. Among the meager collection of clinical observations that could be used as evidence for an active role of the brain in influencing human sexual function were those associating precocious puberty with the occurrence of hypothalamic tumors. The parallel between diencephalic damage and hastened sexual development in children and ferrets was too striking to be ignored, so we decided to see whether puberty could be advanced in rats by hypothalamic lesions. Here again, we owed everything to the imagination, stimulating presence, and encouragement of Geoffrey Harris. The first lesions in 14-day-old rats were made on February 28, 1956, and a series of some 29 males and females was prepared over the next few days. Two weeks later our hopes were realized, when it became clear that puberty occurred earlier in rats suffering damage to the anterior hypothalamus. After determining the location of the lesions, we tried to sift the animals in terms of their response in order to define the region of the hypothalamus concerned with sexual development, but never fully achieved this aim. We could only state that a rather well-defined area behind the optic chiasma seemed to be concerned in the response, and it is little consolation that others have since been unable to be more precise.

The ferret observations were reported to the Physiological Society at a meeting in Manchester on April 20, 1956, and were soon published in the *Journal of Physiology* (Donovan and van der Werff ten Bosch, 1956). They did not pass without challenge, because later in that year Herbert and Zuckerman placed lesions in the thalamus, caudate nucleus, or fornix, or performed control operations, and found that estrus occurred much earlier in both sets of operated females than in unoperated animals. To them (Herbert and Zuckerman, 1957), the results suggested that the "acceleration of oestrus following operations on the brain provides no justification for assigning any specific sexual or reproductive function to any part of the brain of the ferret. In particular, they show that the fact that oestrus may be accelerated when lesions are made in the anterior hypothalamus provides no more basis for positing the existence of a sexual centre in that part of the brain than the present experiments do for the view that such a centre exists in the thalamus or caudate nucleus." We had, of course, included blank operations in our investigation and found no nonspecific acceleration of estrus, and in the winter of 1956, coincidentally with the work of Herbert and Zuckerman, had made lesions in the thalamus and amygdala, alongside further blank operations, without advancing estrus. The results from a subsequent series of animals prepared earlier than usual, at the end of October 1957, agreed with those of Herbert and Zuckerman in that ferrets with lesions in the thalamus or cerebral cortex came into estrus early. In reporting this work (Donovan, 1960), I commented that perhaps the neural mechanisms governing gonadotropin secretion are peculiarly susceptible to disturbance at the very beginning of the winter anestrous period and still feel that there is merit in the idea.

By now, the production of contradictory findings in London and in Birmingham was becoming almost routine. Abrams *et al.* (1954) had reported that removal of the cervical sympathetic ganglia in anestrous female ferrets prevented the onset of estrus that normally followed exposure to prolonged illumination. Koos van der Werff ten Bosch and I repeated these experiments in the winter of 1954, found that the onset of estrus was delayed, not prevented, and attributed the modified response to light to the ocular effects of sympathectomy, which reduced the amount of light reaching the retina. We were probably wrong about the mode of action of sympathectomy, but our experiments remain valid.

The basis of the discrepant findings of the London and Birmingham laboratories remains to be determined. Koos (van der Werff ten Bosch, 1960) suggested that the protracted sojourn of ferrets under laboratory conditions may modify ovarian activity in an unexpected way, and it is indeed likely that the conditions under which ferrets are kept before inclusion in an experiment affect the results obtained. It seems no longer safe to choose animals for long term experiments solely on the basis of the apparent level of ovarian activity at the time of selection. The previous sexual history needs to be taken into account, as was found in experiments involving the transfer of ferrets to long days at different times of year, for under some circumstances long days inhibited rather than promoted gonadotropin secretion in anestrous females (Donovan, 1967). Ideally, we should have resolved the contradictions in collaborative experiments with our Birmingham colleagues, but these were never undertaken—not because of personal antagonism, for we agreed to differ in the most friendly way and mutually remained puzzled, but mainly through pressure of other work.

Koos van der Werff ten Bosch had to return to the Netherlands in the spring of 1956 and so our period of close collaboration ended. We wanted to foster international collaboration and tried to carry on with joint experiments in London and in Leiden, but these never got very far. There is a great difference between collaborative work in the same or a neighboring laboratory, when co-workers can press one another for results, and collaboration at a distance, when local problems and affairs always seem the more demanding. Complete separation proved to be a long drawn-out affair, because we decided to write a monograph, *Physiology of Puberty*. The preparation of the book took much more time than we had imagined, and 9 years elapsed before it appeared (Donovan and van der Werff ten Bosch, 1965). Nevertheless, with the departure of Koos, a new phase in my career began, but that is another story.

REFERENCES

Abrams, M. E., Marshall, W. A., and Thomson, A. P. D. (1954). Effect of cervical sympathectomy on the onset of oestrus in ferrets. *Nature (London)* **174**:311.

Donovan, B. T. (1960). The inhibitory action of the hypothalamus upon gonadotrophin secretion. *Mem. Soc. Endocrinol.* **9**:1.

Donovan, B. T. (1967). Light and control of the oestrous cycle in the ferret. *J. Endocrinol.* **39**:105.

Donovan, B. T., and Harris, G. W. (1954). Effect of pituitary stalk section on light-induced oestrus in the ferret. *Nature (London)* **174**:503.

Donovan, B. T., and van der Werff ten Bosch, J. J. (1956). Oestrus in winter following hypothalamic lesions in the ferret. *J. Physiol.* **132**:57P.

Donovan, B. T., and van der Werff ten Bosch, J. J. (1965). *Physiology of Puberty*, Edward Arnold, London.

Harris, G. W. (1948). Neural control of the pituitary gland. *Physiol. Rev.* **28**:139.

Harris, G. W. (1950). Oestrous rhythm, pseudopregnancy and the pituitary stalk in the rat. *J. Physiol.* **111**:347.

Herbert, J., and Zuckerman, S. (1957). Effect of cerebral lesions upon oestrus in ferrets. *Nature (London)* **180**:547.

Markee, J. E., Everett, J. W., and Sawyer, C. H. (1952). The relationship of the nervous system to the release of gonadotrophin and the regulation of the sex cycle. *Recent Progr. Horm. Res.* **7**:139.

Roussy, G., and Mosinger, M. (1946). *Traité de Neuro-endocrinologie*, Masson, Paris.

Thomson, A. P. D., and Zuckerman, S. (1953). Functional relations of the adenohypophysis and hypothalamus. *Nature (London)* **171**:970.

van der Werff ten Bosch, J. J. (1960). Discussion. *Mem. Soc. Endocrinol.* **9**:15.

Zuckerman, S. (1952). The influence of environmental changes on the pituitary. *Ciba Found. Colloq. Endocrinol. Proc.* **4**:213.

Zuckerman, S. (1955). The possible functional significance of the pituitary portal vessels. *Ciba Found. Colloq. Endocrinol. Proc.* **8**:551.

Zuckerman, S. (1970). *Beyond the Ivory Tower*, Weidenfeld and Nicolson, London, p. 59.

9

Elemér Endröczi

Elemér Endröczi was born in 1927 and was graduated from the Medical School of the University of Pécs, Hungary. He began to study neural transmission at the Institute of Physiology, University of Pécs, and since 1950 has been concerned with the interactions between the neuronal and endocrine systems in controlling adaptive and reproductive behavior. In collaboration with K. Lissák, he has published books on *The Neuroendocrine Control of Adaptation* and *The Role of Limbic Structures in Neuroendocrine Regulation and Learning Processes*. As a fellow of the Ford Foundation and of the Canadian Medical Research Council, he spent long periods in the Department of Anatomy at the University of California in Los Angeles and in the Department of Physiology at Laval University, Quebec.

After 23 years in Pécs, he was appointed head of the Central Research Division of the Postgraduate Medical School and became professor at the Institute of Experimental and Clinical Investigations in Budapest. Currently, his main research interest is the study of the molecular bases of hormonal action on the brain and behavior, and his staff is working on the role of the monoaminergic neuronal systems in hormone-induced behavioral changes and with hormone receptors in the central nervous system.

Development of Neuroendocrine Research in the Institute of Physiology at the Medical School of the University of Pécs, Hungary

ELEMÉR ENDRÖCZI

There was no neurophysiological or endocrinological research tradition in the Institute before World War II. After the death of the former director in 1943, Kálmán Lissák, then associate professor of the Institute of Physiology, University of Debrecen, was appointed to this position. Lissák had studied the cholinergic nature of neural transmission in Otto Loewi's laboratory in Graz and had been trained in classical neurophysiological techniques in Trendelenburg's Institute in Berlin between 1933 and 1936. Then he went to Boston and spent nearly 2 years as a Rockefeller Fellow in Walter B. Cannon's laboratory at the Harvard Medical School. This period was very productive: he studied acetylcholine and epinephrine content in the peripheral vegetative nerves and in the sensory nervous system, and was deeply influenced by Cannon's concept of the nervous integration of bodily changes and emotions, as well as by the concept of emergency function of the autonomic nervous system.

At the end of World War II, the staff of the Institute, inherited from the former director, consisted of several assistants. It was then hard

ELEMÉR ENDRÖCZI • Central Research Division, Postgraduate Medical School, Budapest, Hungary.

to maintain the teaching of medical students, and the Institute had to cope with physiology, biology, and biochemistry. Although there was no remarkable damage to the Institute, the devastation and ruin throughout the country hindered daily life and resulted in a shortage of elementary supplies. To aid in the supply of vital medicines, insulin and oxytocic extracts were successfully produced in the Institute. Only home-purified sodium chloride could be used for Ringer solution, and the most important laboratory animal was the frog. Major journals and books were received as gifts from friends and international organizations. In 1947 several medical students of a new generation entered the Institute. After the examinations in physiology and biochemistry, they were full of enthusiasm, but none guessed that they would spend more than two decades in the Institute, or that they would participate in the development of a new physiological science.

The establishment of the new Academy of Sciences was a landmark in the development of scientific life in Hungary, because a planned organization of research and increasing support made it possible to begin work throughout the country. The first postwar Annual Meeting of the Hungarian Physiological Society in 1948 and the appearance of the first issue of *Acta Physiologica Hungarica* were early signs of these activities in physiology, biochemistry, and pharmacology.

The initiation of neuroendocrine research at the Institute about 1950 was brought about by at least three different factors. There was an increasing interest in the influence of the hormonal system on the adaptation of the organism, inherited from Cannon's concept, while the earlier studies of Professor Lissák prompted continuation of investigations in the field of neural transmission. In fact, during the first 2 years our main interest was to study cholinergic and adrenergic transmission under different experimental conditions in both neural and "nonneuronal" (e.g., placental) tissues. Highly sensitive bioassays of the transmitters in the frog heart and in different *in situ* and *in vitro* preparations were routine procedures in the laboratory. The other influence which determined our thinking on the nature of the nervous organization of somatic and vegetative processes originated from Pavlov's teaching. As a result of the closer political and cultural contacts with Soviet science, and upon the suggestion of the Professor, we took the first steps in learning the concepts and methodology of the conditioned reflex. The first Pavlov chamber for dogs was built in 1951, despite considerable reluctance on our part during these ideologically hectic years. However, not too much time was needed to understand the importance of the objective methodology of the conditioned reflex in the study of brain and behavior relations, and during the next few years a second chamber was constructed at our request.

A third influence which certainly affected our interest in the neuroendocrine adaptation of higher mammals originated with H. Selye's adaptation theory. Our "brain-oriented" mind in the 1950s accepted the stress concept with considerable reluctance, because it did not involve the integrative and discriminatory role of the central nervous system in the response of the organism to noxious stimuli. We supposed that the central neuronal organization played a decisive role in the endocrine response and that both facilitatory and inhibitory influences determined the activation of the endocrine system. Nevertheless, the characteristic reactions of the endocrine system to nonspecific noxious stimuli attracted our attention and led us to begin investigations of the pituitary-adrenal axis. Our first papers, published in 1950, were related to the influence of the frontoorbital cortex to responsiveness of pituitary-adrenal function in dogs. An augmented response of the pituitary-adrenal axis to stressors in frontal lobectomized dogs led us to assume that the rostral forebrain exerts an inhibitory influence on endocrine adaptation.

We realized very early that special instruments were necessary to study the role of subcortical structures in controlling the endocrine system. With the kind assistance and suggestions of J. Szentágothai, who became the director of the Institute of Anatomy and Histology in 1945, the first stereotaxic apparatus for cats, and later dogs and rats, was constructed in the workshop of the Institute. The implantation of chronic electrodes for stimulation and for recording of electrical activity from subcortical areas greatly contributed to the extension of our methodology in both neurophysiological and neuroendocrinological research. This kind of approach to brain and behavior relations was revolutionary for classical students of neurophysiology and was accepted with reluctance in the early 1950s. During these years, a loose organization of two groups had developed in the Institute: "electrophysiology" was the special interest of E. Grastyán's group and they began to study the electrical correlates of the development of temporary linkages and motivated behavioral reactions. The "neuroendocrine group," under my leadership, differed both conceptually and methodologically. Both groups, with five to six assistants, received continuous support from the Academy of Sciences.

In the early 1950s, the lack of sensitive and specific techniques for studying pituitary-adrenal function greatly stimulated our efforts to develop appropriate biochemical methods for the assay of corticosteroids in peripheral blood. The paper chromatographic separation of corticosteroids from adrenal venous blood collected through chronically implanted cannulas in dogs and acutely in barbiturate-anesthesized rats, cats, and rabbits led to a better understanding of the composition of the adrenocortical secretory products in different species and under different environmental conditions. We found that dogs with marked "internal inhibition" (resistance to the extinction of a conditioned response by nonreinforced conditional stimuli) showed a higher ratio of hydrocortisone to corticosterone in the adrenal venous blood than those with a lesser degree of inhibition during the course of reversed conditioning or that of the extinction of a conditioned response. These new observations stimulated other laboratories interested in the study of the influence of hormones on brain function, and this type of approach attracted the attention of both Eastern and Western colleagues and led to the development of both scientific and friendly personal connections.

In the early 1950s the pioneer discoveries of G. W. Harris on the neurohumoral control of the anterior pituitary gland greatly influenced the development of our interest in the hypothalamic control of the pituitary-adrenal axis. We were among the first to demonstrate blockade of the activation of pituitary ACTH secretion by electrolytic lesions of the tubero-infundibular area in rats. This work was performed in collaboration with Szentágothai's school, who helped us to control the size of the lesions. The influence of the Harris laboratory became more personal when I was able to spend several weeks with Professor Harris and B. T. Donovan (this was my first trip to Western countries, in 1959). Subsequently, it was always a pleasure to meet G. W. Harris and spend days in discussion of recent developments in neuroendocrine research with him and B. T. Donovan, both in London and in Pécs.

The influence of the servomechanistic and cybernetic approach to behavioral research, and the neuronal and hormonal mechanisms underlying these processes, also influenced our thinking about the central organization of neuroendocrine regulation. This prompted us to implant minute amounts of corticosteroids into different parts of the hypothalamus, and we found that hydrocortisone implants in the median eminence area could block pituitary ACTH secretion in rats. This study was the first in the field of the feedback regulation of the pituitary-adrenal secretion, although the use of ovarian implants and the resultant change in pituitary gonadotropin secretion (B. Flerkó and V. Bárdos from the Szentágothai school) had already shown the existence of a steroid-sensitive region in the hypothalamus.

In the late 1950s, using chronically implanted cannulas, we studied the possible role of known neurotransmitters on the control of pituitary ACTH secretion. It was found that cholinergic stimulation of the basal and medial hypothalamus and the ascending activation system of the brainstem led to marked ACTH release in cats. Moreover, we reported that the local application of epinephrine to the basal and medial hypothalamus inhibited stress-induced activation of pituitary ACTH secretion. This paper is frequently cited in the literature in connection with cholinergic activation, but another conclusion, namely, the adrenergic nature of the inhibitory control, has been neglected. In recent years, adrenergic inhibition of pituitary ACTH secretion has been proved in different laboratories, although the involvement of other monoamines in this process cannot be excluded.

Increasing international efforts for peaceful coexistence—especially in science—greatly influenced the development of global scientific connections. The Moscow Colloquium on Higher Nervous Activity in 1958 provided a gathering of leading scientists in neurophysiology to discuss the nature of the conditioned reflex and its electrophysiological correlates. This meeting was a landmark in the development of neurosciences, and the participants decided to establish the International Brain Research Organization (IBRO), which was followed by a rapid development of international scientific connections. Professor Lissák, with his personal and friendly contacts with both Eastern and Western brain research scientists, greatly contributed to this development. In 1960, as a fellow of the Ford Foundation, I was able to spend a long period in the UCLA Department of Anatomy, and this visit was repeated in 1963. Personal meetings with scientists like Professors Magoun, Lindsley, and French, who greatly influenced neurophysiological research all over the world, and the work in Professor Sawyer's laboratory in collaboration with J. Hilliard, were highlights. During this visit, we studied progestin secretion in rabbits in response to gonadotropins and copulation. By the use of the LH-induced increase of ovarian progestin secretion, we could study the distribution of LHRF activity within the central nervous system of dogs and rabbits. In addition to the considerable amount of LHRF activity in the basal and medial hypothalamus, appreciable activity could be detected in extrahypothalamic structures, especially in the rostral and basal limbic regions and the midbrain. These observations did not fit into the customary picture of hypophysiotropic regulation of the anterior pituitary function, and these findings were mostly neglected and treated skeptically in the literature. More recently, immunohistochemical methods and experiments involving deafferentation of the hypophysiotropic area strongly suggest an extrahypophysiotropic and extrahypothalamic origin of LHRF neurons.

By the early 1960s our interest had become strongly focused on steroid production of the adrenals and gonads. The chromatographic separation of the steroids from the venous blood of the target organs and their sensitive measurement with different physicochemical methods facilitated the work. These advances in methodology led to a number of investigations of the metabolism of steroid hormones. More papers were published and a new research line was developed. In order to study steroid metabolism and gonadal hormone secretion, G. Telegdy spent a long period in Miami, and later in Diczfalusy's laboratory in Stockholm. Despite the loose link between investigations of the steroid metabolism of the fetoplacental unit, or of the steroid hormone production of gonads, the original neuroendocrine research theme of the group, these training visits were important for Telegdy's scientific career and for learning up-to-date biochemical methods.

The involvement of endocrine factors in learning processes and motivated behavioral reactions attracted the attention of many investigators long before development of our own research group. As part of a "brain-research-oriented" institute, we were influenced by a variety of neurophysiological and experimental psychological concepts, and we wished to understand the mechanism of hormonally and humorally induced "drives" and of the hormonal "conditioning" of behavioral reactions. These interests led to adventures in experimental psychology and neurophysiology, and such studies yielded important information and led us to glance at the real nature of the brain–behavior relationship. The registration of electrical activity in freely moving animals and observations of conditioned reflex behavior simultaneously with measurement of endocrine parameters made it possible to analyze the correlations between the activation and inhibition of pituitary–target organ function and the excitability states of the central nervous sytem. Administration of a polar corticosteroid such as hydrocortisone or cortisone produced marked changes in acquisition or extinction of the conditioned response and led to changes in the development of a "neurotic" state in conflicting situations in dogs and cats. Striking differences in the response to corticosteroid treatment could be observed, and we concluded that corticosteroid-induced changes in behavioral reactions were manifested by alterations of newly acquired temporary linkages, and that the steroids influenced both the acquisition and extinction of a conditioned response by facilitation or inhibition of internal inhibitory processes.

The influence of ACTH on behavior was regarded as mediated through the adrenal cortex, although there were some observations which indicated an extraadrenal action of this trophic hormone. Thus it was found that ACTH administration suppressed the copulatory activity of male rabbits and resulted in an inhibition of polysynaptic reflexes. These effects of ACTH could not be duplicated by administration of corticosteroids and so indicated an extraadrenal influence of the peptide. The observations of David de Wied of Utrecht in relation to the influence of ACTH on the extinction of avoidance response in rats, which initiated many studies in this field of neuroendocrinology, alongside our observations on behavioral changes as related to pituitary-adrenocortical hormones, led to a close and friendly contact between the two groups. Observations in human subjects, in collaboration with de Wied and with T. Fekete of the Clinic of Neurology and Psychiatry, University of Pécs, revealed that administration of ACTH 1–24 and ACTH 1–10 produced a disinhibition of the electrocorticographic response to habituation and relieved anxious and depressive states. These observations are of historical interest, for they prompted many laboratories to study the influence of peptide hormones on behavior. B. Bohus, who decided to prolong his visit to Utrecht indefinitely, and L. Korányi, who is still in Pécs, have contributed much to the understanding of the effect of pituitary-adrenocortical hormones on conditioned-reflex behavior.

Numerous observations accumulated during the first decades of our neuroendocrine research group were reviewed in several books (*Neuroendocrine Control of Adaptation*, K. Lissák and E. Endröczi, eds., Akadémia Kiadó, Budapest, 1960, and Pergamon Press, 1962; revised and translated editions in Hungarian, Russian, and Slovakian in 1965). This period was extremely active, with many guests in the laboratory and with each member of the group spending some time abroad. Alongside recognition of the scientific activity of the Institute, individual reputations have been enhanced, with Professor Lissák being elected vice-president of the International Union of the Physiological Sciences and the elder members of the group participating more in international scientific life. Thus we have been involved in the organization of a number of international symposia and in the establishment of the International Society of Psychoneuroendocrinology and the European Brain and Behaviour Society. Active collaboration and friendly connections have been developed between our group and other laboratories in the socialist countries; mutual exchanges of co-workers with N. A. Yudaev's laboratory in Moscow and with L. Mikulaj's group in Bratislava also have contributed to the development of our scientific activities.

In the late 1960s, there were six or seven staff members in the neuroendocrine group, but disintegration began when I was appointed to the Postgraduate Medical School in Budapest. Just before I left Pécs, I spent 8 months in collaboration with the group of Professor Fortier in the Department of Physiology at Laval University, Quebec. This period was highly productive and it was a pleasure to work in the very friendly atmosphere of St. Foy. Returning to Budapest from Canada, and after 23 years in Pécs, I was charged with the leadership of experimental research in the Postgraduate Medical School, although the Research Division was soon fused with the Institute of Experimental and Clinical Investigations. On the basis of this Institute, the Ministry of Health later established the National Institute of Laboratory Investigations, which generated more work and responsibilities than ever before. Nevertheless, a new neuroendocrine group with young co-workers had been developed and research could continue. The oldest member of the new staff, Cs. Nyakas, began work in Pécs.

Generally speaking, our approach to neuroendocrine research has not changed, although the methodology has been widely extended. The possibility of the use of a broad spectrum of isotope techniques prompted us to look at molecular aspects of hormonal actions on brain functions. Using labeled steroids, we have studied the characteristics of steroid receptors both in brain and peripheral target organs. We have seen that the transfer of labeled corticosterone from cytosol into the nuclear compartment in the hippocampus is positively correlated with the intensity of internal inhibition in rats. Animals showing a rapid extinction of the conditioned avoidance response displayed a greater transfer of tritiated corticosterone. In extensive ontogenetic studies we investigated the permanent changes of behavioral reactions as a result of early postnatal corticosterone treatment. We found an impairment of avoidance conditioned reflex behavior, increased exploration, and an impaired transfer of corticosterone from cytosol into the nuclear compartment in the hippocampus of adult rats after early corticosterone administration. Moreover, we have also found permanent changes in the norepinephrine content and turnover in the brain stem and midbrain in early corticosterone-treated rats.

Our attention became engaged with the effects of hormones on the brain monoaminergic system, and efforts were made to develop techniques to study the catecholamine and serotonin content and turnover in rats. From the behavioral point of view, the use of labeled catecholamines and precursors has facilitated research into the molecular bases of learning. An increased norepinephrine turnover in the hippocampus and midbrain of rats, with a faster acquisition of the conditioned response in shuttlebox experiments and the opposite changes in catecholamine metabolism in rats with a faster extinction of the avoidance response, led us to assume that the brain norepinephrine neuronal system plays no essential role in the fear-induced motivational state but is involved in the organization of goal-directed motor patterns. This assumption is supported by the positive correlation between norepinephrine turnover and the number of avoidance responses in both acquisition and extinction sessions, although these rats received less painful electrical shocks as unconditioned stimuli. There was no difference in the norepinephrine turnover of rats showing good or poor performance during avoidance conditioning when they were tested under resting conditions.

The monoaminergic control of FSH secretion in prepubertal rats also is being traced. Using local implantation of different monoamines and blocking agents in the basal and medial hypothalamic region, we have found that α-adrenergic receptors are involved in the activation of LHRH/FSHRH release. In other work the feedback influence of ovarian hormone production on the maturation of the FSH secretion during the first postnatal 3 weeks is being examined. At all times during this period, ovarian hormones seem to exert a negative-feedback influence on pituitary FSH release. The mechanism underlying the high FSH peak at or about 14 days of age is still obscure.

In studying the possible role of monoamines in ACTH-induced changes in behavioral reactions in rats, we have found that intraventricular injection of ACTH 1-24 and ACTH 4-10 resulted in marked facilitation of norepinephrine turnover in different brain regions. An amphetaminelike influence of ACTH peptides on extinction of avoidance behavior, and observations showing increased norepinephrine release in response to ACTH peptides, may be considered as a new approach to the mechanism of ACTH action on behavioral reactions. In addition to the monoaminergic action of the ACTH peptides other direct effects on neuronal system cannot be excluded.

Brain receptors for steroid hormones have attracted the attention of many investigators recently. While studying corticosteroid receptors in relation to learning behavior, we were prompted to investigate the interaction of some psychopharmacological agents on both corticosteroid and sex steroid receptors in the brain and the anterior pituitary gland. A drastic displacement of estradiol and corticosterone occurred from receptor sites following administration of chlorpromazine and *Rauwolfia*, but not of dehydrobenzperidol, to adrenalectomized and ovariectomized rats. This led us to assume that certain tranquilizers producing activation of pituitary ACTH and prolactin secretion and blockade of gonadotropin release exert their influence by competing for specific steroid receptors. The mechanism of this action is not yet fully understood, because the major tranquillizers do not displace steroid hormones from specific binding sites under *in vitro* conditions.

The disintegration of the parent group in Pécs and the establishment of new teams continued in 1975 when Gy. Telegdy was appointed to the Institute of Pathophysiology, University of Szeged. After ending his studies on steroid metabolism in the fetoplacental unit, he went on to investigate the monoaminergic control of pituitary ACTH secretion.

Very soon after the end of World War II the development of two independent schools of neuroendocrine research in Hungary began in Pécs. One was launched by J. Szentágothai in the Department of Anatomy, with such followers as B. Flerkó, B. Halász, and B. Mess. The other school, under the leadership of K. Lissák, employed more neurophysiologically oriented concepts and was directed to the study of the interactions of neuronal and humoral factors in the organization of behavioral reactions. Personal friendship and mutual appreciation, as well as a readiness to help, have always characterized the contacts between the two schools and been a feature of Pécs science.

Professor Lissák, now 70, is still active, and neuroendocrine research continues despite the loss of the elder members of the Institute. The influence of the Lissák school on the development of neurophysiology and neuroendocrinology has been great, for it initiated research in a field of medical science that had no tradition in Hungary. Neuroendocrinology is a new branch of endocrine research and its acceptance as an independent discipline has hardly been longer than two to three decades. The new generation that has grown up during the past two decades can continue work in a more peaceful and comfortable atmosphere than during the pioneering days.

SELECTED PUBLICATIONS OF THE NEUROENDOCRINE RESEARCH GROUP AT THE INSTITUTE OF PHYSIOLOGY IN PÉCS

Endröczi, E. (1972). *Limbic System, Learning and Pituitary-Adrenal Function*, Akadémiai Kiadó, Budapest, 154 pp.

Endröczi, E., Lissák, K., and Hartmann, G. (1967). Meso-diencephalic inhibitory and activatory mechanisms. *Acta Physiol. Acad. Sci. Hung.* **31**:117.

Endröczi, E., Lissák, K., Fekete, T., and de Wied, D. (1970). Effects of ACTH on EEG habituation in human subjects. In de Wied, D., and Weijnen, W. M. J. H. (eds.), *Pituitary, Adrenal and the Brain*, Vol. 32 of *Progress in Brain Research*, Elsevier, Amsterdam, pp. 254–263.

Lissák, K. (1969–1976). *Recent Developments of Neurobiology in Hungary*, Vols. I–V, Akadémiai Kiadó, Budapest.

Lissák, K., and Endröczi, E. (1960). *Die neuroendokrine Steuerung der Adaptationstätigkeit*, Akadémiai Kiadó, Budapest, 175 pp.

Lissák, K., Endröczi, E. (1962). Some aspects of the effect of hippocampal stimulation on the endocrine system. In *Physiology de l'Hippocampe*, CNRS Syposium, Montpellier, pp. 463–473.

Lissák, K., and Endröczi, E. (1965). *The Neuroendocrine Control of Adaptation*, Pergamon Press, Oxford, 186 pp.

Lissák, K., Endröczi, E., and Medgyesi, P. (1957). Somatische Verhalten und Nebennierenrindentätigkeit. *Pflugers Arch. Ges. Physiol.* **117**:265.

10

William Etkin

William Etkin was born in New York City on December 10, 1906. He received the B.S. degree at City College of New York in 1928, the M.A. at Cornell University in 1930, and the Ph.D. at the University of Chicago in 1934. He progressed from instructor to professor of biology at the City College of New York during 1934–1966, served as lecturer at Columbia University during 1950–1952, and was associate professor of anatomy and later professor of anatomy and biology at Albert Einstein College of Medicine and Yeshiva University from 1955 to 1972. He was a research associate at the American Museum of Natural History from 1934 to 1950, a consultant to the Graduate Record Office of the Carnegie Foundation from 1940 to 1944, and a member of the Columbia Seminar on Human Genetics and Evolution from 1954 to 1958.

Dr. Etkin is a member and chairman of several committees of the American Society of Zoologists, a fellow of the Animal Behavior Society, and a member of the American Association of Anatomists, the Endocrine Society, the Society for Experimental Biology and Medicine, and the Society for Development and Growth. He served as president of the AAUP chapter at City College of New York and as a panel member of the Regulatory Biology Section, National Science Foundation, and is a member of the editorial board of *Neuroendocrinology*. (Photo by Walter Sussman.)

Searching for the Clocks of Metamorphosis

WILLIAM ETKIN

At the beginning of my scientific career, I became caught up in the problem of tadpole metamorphosis: how does the animal "know" when to start and how to time the sequence of events involved? This search led, of course, into endocrinology, then into neuroendocrinology and into numerous side paths. The call by the Editors for my contribution to this series I interpret in terms of an informal accounting of the interplay of ideas, personalities, and events that propelled and buffetted me in this journey.

THYROID AND THE TIMETABLE OF METAMORPHOSIS

I lit upon the metamorphosis problem as an undergraduate biology major at City College of New York ambitious to become a teacher of biology and to do research in animal development. Professor A. J. Goldforb, the long-time secretary–editor of the *Proceedings of the Society for Experimental Biology and Medicine*, was my professor, and a most marvelous teacher he was. All his courses were conducted in the Socratic manner, especially his seminar on growth and development. He tried to get us to ask analytical questions of the laboratory material or seminar papers rather than to seek confirmation of textbook facts. We all tried to carry through an experiment for our own term report, and mine on the temperature effect on *Drosophila* locomotion was more elaborate than most because I had the

WILLIAM ETKIN • Emeritus Professor of Anatomy, Albert Einstein College of Medicine, Bronx, New York, 10461.

advantage of a cellar "laboratory" in my parents' house. But though the professor gave me an "A" for effort, I was not satisfied with it. That spring (1928) I discovered a wonderful forest pond and became fascinated by the changing biota, including tadpoles, through the season. The wonder of how animal development fits into the annual weather cycle caught my imagination, perhaps all the more because of my urban background. After graduation, I had a job as tutor in the Biology Department at City College and took graduate courses at Columbia and NYU. At NYU I took Professor Gudernatch's seminar and read some of the literature on thyroid and metamorphosis. But I was dissatisfied with the course work at both places; it was too citified for my new enthusiasm for ecology! I spent a summer at Cornell mostly nature-studying in the Ithaca countryside. In the following fall I was admitted to Cornell Graduate School in Anatomy, at the Medical School in New York and at Ithaca for the summers. Here I came into contact with Drs. Stockard and Nonidez in New York and Drs. Kingsbury and Adelmann at Ithaca. Dr. Stockard was my official mentor and sponsor for my research topic in tadpole metamorphosis. Actually, I saw little of him. He told me at the outset that he would not recommend me for a job in anatomy because I was Jewish. Despite that remark, delivered in a most matter-of-fact tone, he was always friendly, loved to tell me about his dog colony and endocrine work, and gladly recommended me for doctoral work at Chicago when I finished my Master's degree. Actually, I worked with Dr. Nonidez, who was a gifted teacher and a helpful guide. I saw his original slides demonstrating the parafollicular (C) cells in the dog thyroid and some of Dr. Papanicolaou's early work on the menstrual cycle. I felt quite at home in the non-WASPish side of the anatomy floor! It was all fascinating, but I stuck to my tadpoles.

In my home laboratory I did the descriptive study of metamorphosis upon which all my later work was founded. I followed the changes in tadpole metamorphosis by measuring growth and morphological changes, with daily observations on individual animals. I came to recognize that the larval period was one of growth without significant morphological change. Its length varied greatly in different species in correlation with the animal's ecology. Species breeding in temporary ponds had periods appropriate to the life expectancy of the ponds; the larger-growing species bred in permanent waters. This larval growth period was followed by one of gradual morphological change, notably the rapid growth of the legs relative to the body. Since this was known then to depend on the presence of the thyroid, I considered this period to be early metamorphosis or prometamorphosis. With the emergence of the forelimbs from the gill chamber in which they had been hidden, there began a rapid series of startling morphological changes involving the resorption of the tail, gills, larval mouthparts, etc., which transformed the tadpole into the frog. This I called "metamorphic climax." Whereas the larval period varied greatly among species, the periods of prometamorphosis and climax were remarkably similar in duration in large and small species, requiring only slightly longer time in bullfrogs than in the small wood frogs (Etkin, 1932). Clearly, this pattern suggested that thyroid activity was minimal and without noticeable effect during the growth period, and increased gradually during prometamorphosis to reach a high at climax. This study focused the problem for me. What determines the initiation of prometamorphic change? What regulates the timing of the sequence of events of prometamorphosis which require 3–5 weeks at 23°C?

It was an obvious suggestion that the pattern of metamorphosis is controlled by the level of thyroid activity. Bennet Allen and others had already explored this idea in qualitative terms and had found that the size and histological evidences of thyroid activity were consistent with this notion. But their studies did not examine the *rate* of change involved. For my Master's thesis, I did a quantitative correlational analysis of the rate of thyroid growth in relation to body growth and to metamorphic changes. I found that thyroid growth rate and histological signs of activity accelerated at the beginning of prometamorphosis, reached a maximum at the beginning of climax, and before the end of climax went back into inactivity (Etkin, 1930, 1936). Was the timetable of metamorphosis the consequence of the rising level of thyroid homone tripping a sequence of tissue thresholds?

I wanted to get away from New York and see more of the world. Against everyone's advice to finish my doctoral studies at Cornell, I gave up my tutorship at City College (a mad idea in the depth of the depression in 1931) to go to Chicago to work under Carl Moore. The choice of Chicago came about partly through an accidental meeting with Libbie Hyman during a vacation visit to Woods Hole and partly because I was impressed by Frank Lillie's paper on twinning in cattle and Carl Moore's elegantly simple work on the function of the scrotum. I went with the understanding that Dr. Moore would allow me to continue my quantitative analysis of thyroid action in metamorphosis, on the persuasive argument of the analogy between control of metamorphosis and that of puberty in mammals.

Thinking to determine the level of thyroid activity in relation to metamorphosis by measuring metabolic rate, I studied the oxygen consumption of tadpoles through metamorphosis. To my astonishment, I found that the rate did not go up at all. When consideration is given to the loss of water from the tissues and of intestinal contents during climax, the rate remained stable and even decreased slightly. Later, working with Dr. R. Root at City College, I extended the study of thyroid and metabolism to fish and found here, too, no accelerating action of thyroid (Etkin *et al.*, 1940). Despite much confirmatory evidence, the notion that thyroid action on cold-blooded animals differs in this respect from that in homeotherms still has not penetrated far into endocrinological thinking.

My doctoral thesis, based on a quantitative study of the effects of different concentrations of thyroxine on the rate of metamorphic change, substantiated the concept that the pattern (that is, the temporal spacing out) of metamorphic events depends on the pattern of increasing activity of the thyroid gland (Etkin, 1935*a*). I found that changes in tissue sensitivity with age play no significant role. The effect in each tissue varies quantitatively with the concentration of hormone in an ordinary stoichiometric manner, and no true physiological thresholds for hormone action appear to exist. The idea of a succession of threshold effects ticking off the events of metamorphosis had no basic validity. Rather, each metamorphic change depends on concentration of hormone times the time of application. Low concentrations induced late events if given sufficient time. A normal timetable of metamorphic change can be induced in thyroidectomized tadpoles only by a regulated application of increasing concentrations of hormone, and only in this way can a viable froglet be produced.

The pattern of activation of the thyroid was thus the proximal "clock" of metamorphosis. The problem of the patterning of metamorphosis now moved back to that of the control of pituitary thyrotropic activity. Was this an autonomous developmental phenomenon or was the pituitary controlled by another endocrine gland in the same way as it influences the thyroid? Or perhaps the pituitary is regulated by nerves from the brain?

A CLOCK FOR THE MASTER GLAND

At a session of Moore's endocrinology seminar, the question came up for discussion of what controls the pituitary, the "master gland" of the endocrine system, in its activation of the sex system. I remember vividly thinking out in a single flash my postdoctoral program: the transplantation of the pituitary primordium. By such a procedure I would be able to settle, I thought, the question of whether the production of thyrotrophic hormone by the pituitary is controlled by the brain, by some other endocrine organ, or by its own autonomous clock mechanism. I remember the moment all the more clearly because it came just before Dr. Moore asked me a question in seminar and I was caught daydreaming!

At any rate, I was reappointed to City College as an instructor in February 1934, a stroke of luck, because jobs were desperately scarce at the time. City College offered no research facilities to its staff, but luck was still with me. Dr. G. K. Noble had persuaded the Trustees of the American Museum of Natural History to establish an experimental laboratory for the study of animal behavior and evolution. He had become enthused by the work of the European ethologists and set about developing a laboratory to apply their ideas to the evolution of behavior. Unfortunately, in 1934 when the laboratory was completed, little money was available for staff. He was therefore glad to welcome me to work on tadpole endocrinology, using his facilities. I was made a research associate and shared an office with Dr. Libbie Hyman, who had begun writing her famous treatise on the invertebrates.

My situation was ideal. I had complete freedom to develop as a teacher and as a research worker. I there perfected my secret research weapon. At City College I could not be called upon for administrative chores because I had to go to the Museum for my research, and at the Museum I had to be let alone because I was not on the payroll. Later, at Einstein Medical School I developed the same relationship, and my chairmen all around went along enthusiastically with my game. Even the federal government cooperated. At the Museum the Works Progress Administration provided me with part-time technical assistance and at Einstein NSF was even more generous.

THE SWEETS OF SERENDIPITY

The transplantation of the pituitary primordium was done at the tailbud stage of the embryo. The initial results were surprising. First out, I discovered that I could tell whether a pituitary graft had taken successfully because the animals, under the influence of the developing pars intermedia, assumed a dark color due to the hormone of the pars intermedia of the gland. Thus if an animal with its own pituitary primordium transplanted to the tail developed with a silvery appearance, I knew that the graft had failed to survive. If the animal was normal in pigmentation, I probably had failed to take out the entire primordium at operation (generally such animals metamorphosed normally). The successfully grafted animals were those which grew rapidly and became much darker than normal, often becoming black giants. I had assumed that sectioning the tail to examine the histology of the graft would be laborious. The graft was made into the tailbud and thus could end up most anywhere in the tail. But again I was pleasantly surprised. At the site of implantation the characteristic lips and other mouthparts of the tadpole also developed; thus the area of the graft was readily found.

The excess pigmentation of the graft hosts immediately suggested that the pars intermedia of the gland is normally controlled by inhibitory innervation from the brain. The histological sections of the grafts in the black giants showed the pars intermedia to be enormously hypertrophied, with the cells much enlarged. I confirmed this interpretation by developing an operation for cutting the infundibular lobe of the brain in fully grown tadpoles and showing that this led to hypertrophy and hyperactivity of the pars intermedia (Etkin and Rosenberg, 1938; Etkin, 1941). Later, I found that if the epithelial primordium is transplanted with adjacent brain tissue the pars intermedia that develops is often of normal size and activity (Etkin, 1943). Since the only other instance of overgrowth in a gland after deprivation of its nerve supply then known to me was that of the denervated corpus allatum in the roach, shown by Berta Scharrer, I wrote to her about it at the time. Little did I think then that the fates would bring us together in research again many years later. Working with adult frogs, I confirmed the inhibitory control by innervation here, too, but because of the intervention of World War II did not complete my study or publish until going to Einstein (Etkin, 1962*a*).

It would be convenient to complete the account of our work on pars intermedia here before returning to trace the work on metamorphosis. Subsequent studies at the light-microscope level showed the presence of some typical neurosecretory fibers in the pars intermedia and I drew what now appears to be the premature conclusion that these must be the controlling fibers (Etkin, 1962*b*). However, Dr. Linda Saland in her doctoral thesis found that at the EM level the only fibers to be seen permeating the gland were aminergic fibers. The peptidergic fibers, presumably the ones we had seen in the light microscope, are confined to the area of contact with the pars nervosa (Saland, 1968). A subsequent study by Dr. Mona Castel displayed the extraordinary activation of the protein-synthesizing mechanism of the cells after section of the hypothalamus (Castel, 1972). Both studies indicated that the initial effect of removal of inhibition is on release of the hormone from storage granules, with subsequent activation of synthesis.

HOW I FOUND THE ANSWER (THE WRONG ONE)

In the mid-1930s the question of the innervation of the epithelial pituitary was unsettled. In Nonidez's laboratory I had acquired experience with various current silver methods and the difficulties attendant upon their interpretation. I could not satisfy myself in 1936 whether the tadpole showed nerve fibers in the anterior lobe or not. I expected the transplantation to clarify this, as indeed it did for the pars intermedia; however, the results for the thyrotropic function of the anterior lobe were perplexing. The black animals with their own primordia transplanted to the tail grew as well as or better than the controls, but when the latter transformed, the grafted animals remained larval. I fixed many of my first group of experimentals after the controls had all metamorphosed to check the graft and the hypothalamic area for pituitary tissue. I found fine-looking anterior lobes with clear, well-granulated acidophils in the grafts. If I had fixed all the animals at that time, I would have had clear evidence (i.e., the "statistically significant" results beloved of journal editors) that without contact with the brain the anterior lobe cannot activate the thyroid. The conclusion to be drawn would have been that the anterior lobe is controlled through its connection to the brain. Had I drawn this conclusion I would have been well ahead of my times! But I was rescued from this happy fate by a few "aberrant" animals.

I had noticed that some of the grafted animals showed hind leg growth slightly in excess of that to be expected from hypophysectomized animals of comparable size. I had a feeling that, given a further chance, they might show progressive metamorphic development. I had absorbed Carl Moore's dictum that "one good case is worth all the statistical figuring in the world."

In any event, I kept the most likely looking of the animals and was rewarded in the course of the next few weeks in seeing several of them show definitive activation of hind leg growth, a clear indication of thyroid stimulation. Finally, one of these animals entered metamorphic climax, with the emergence of its forelegs and shedding of the larval beaks. Now I faced a critical decision. If I kept this animal longer until perhaps its tail would begin to show resorption, the evidence for thyroid activation would be complete. On the other hand, the giant grafted tadpoles were subject to sudden death brought on, it would appear, by the splanchnomegaly they developed which seemed to interfere with respiration. Animals whose tissues were fixed even a few hours after death were useless for histological analysis. Yet it was essential in each case to examine the graft and the normal site of the pituitary and the thyroid to make that "one good case." I therefore fixed all my advanced animals then and there. Histological study showed, indeed, that the animals were all as intended, with no normal-site pituitary tissue, a well-developed grafted gland, and a large thyroid.

Only one conclusion seemed possible. The pituitary has its own genetically determined developmental clock which sets off the metamorphic mechanism at the appropriate time for the species and controls the rate of thyroid activation. As with many other embryonic primordia, the developmental rate was somewhat retarded in a transplant. Clearly, it could not be controlled by another endocrine gland in the way that the pituitary controls the thyroid, for such a hormone would reach it wherever it was placed. Animals with transplanted thyroids had been shown to metamorphose at the normal time, and I had confirmed that result. It was equally apparent that its normal innervation, if it had such, played no role, because it could never reach it in the tail. I never suspected that there could be any other alternative. (I had seen Ernst Scharrer's beautiful slides of neurosecretory neurons in the fish hypothalamus at Woods Hole in 1938, but, of course, never dreamed that this had anything to do with my problem.) The conclusion that the anterior lobe was governed by an autonomous developmental clock in its thyrotropic hormone production which I drew in my 1938 paper (in somewhat different phraseology since biological "clocks" had not yet come into fashion) was somewhat disappointing (Etkin, 1938). It did not offer much hope of further analysis because the nature of such clocks was, and remains, one of the fundamental mysteries of developmental biology.

HOW I CONFIRMED THE WRONG ANSWER (ALMOST)

One way to test the hypothesis of a self-differentiating clock in the pituitary would be with multiple transplants whose additive effects might produce normal or even precocious metamorphosis. After some preliminary work, I planned an experiment as follows. The experimental hosts would be hypophysectomized, and their own pituitary primordium would be implanted into one orbit; three other primoridia would be implanted into the other orbit or placed in the lower jaw region. By keeping all glands in the head region, I could serially section the head later and ascertain the status of the normal pituitary site, the four transplanted pituitaries, and the thyroid glands in one series of sections. Of course, besides normals, control animals with one pituitary in either the orbit or jaw region would be required.

The results were astonishing. Almost as soon as they completed embryonic development, many of the animals with four pituitaries showed rapid leg growth, and, in little more than a week, metamorphic climax set in. It was the picture of extreme precocity I had often seen when subjecting embryos to high concentrations of thyroxine. Histological study revealed everything as expected in the four-pituitary animals, including highly activated thyroid glands (Etkin, 1935*b*).

On the face of it, this experiment supported the concept of self-differentiation of pituitary function. Yet I could not but wonder at the extraordinary precocity displayed. In any case, I had not very long to wait to discover that I had stumbled again into luck (or serendipity) for better or worse. When another clutch of eggs became available, I repeated the experiment on a more extensive scale. Several days after setting up the experiment, I looked the animals over and noted with satisfaction that several of the four-pituitary animals already showed the first signs of precocity in enlargement of their hind leg buds. But then I spotted one of the single-pituitary-graft controls, with the graft in the jaw region, which also had begun to sprout hind leg buds! Could I have mixed up the dish labels? I felt certain that this was not so, for I had been especially careful in setting out the animals. Another explanation flashed through my mind. Could it be that when the embryonic pituitary is brought close to the thyroid the latter is activated! Could the real reason the four-pituitary animals were precociously activated be that in the insertion of two or three pituitaries under the jaw the first was pushed deeply into the jaw region where it contacted the thyroid primordium?

It did not take me long to prove that this bizarre but exciting idea was indeed correct. With the next batch of eggs, I was able to show that one pituitary carefully placed next to the thyroid primordium would produce precocity. Indeed, the procedure could be reversed and the thyroid transplanted to the pituitary region to produce the same effect. The precocity is so extraordinary that I never tired, in later years, of repeating the experiment with variations just to see it happen again. Could it be that the primordium produces a pulse of thyrotropin when it first differentiates, and, since the circulation is not yet established, the area around the gland remains as a thyrotropic field (Etkin, 1939)? Considering the latent periods of thyrotropin and thyroid hormone effects, this seems a possible explanation, but I have never been able to explore the problem to my satisfaction. However, what was clear in 1939 was that I had no supporting evidence that the metamorphic clock was built into the developmental physiology of the pituitary gland. Soon thereafter, Dr. Noble died suddenly, war clouds gathered, and at least until 1955 I was diverted to other things.

GOODBYE TO ENDOCRINOLOGY

The end of the war found me immersed in teaching responsibilities at City College and unable to continue work at the American Museum. In developing a biology course for nonscience students, I found that students who struggled to keep awake while I explained the intricacies of the life histories of plants or other esoteric subjects erupted with excitement when I talked about social behavior of animals from the ethological viewpoint, a subject I had gotten into through my association with the Museum laboratory. I decided to go into behavior research and in 1952 began a program at the New York Zoological Society's Bronx Zoo. I will not discuss this work further except for a neuroendocrine aspect. When eventually I got to Scharrer's laboratory in the Einstein Medical School, I found the neurology and psychiatry people interested in ethology, which was just then coming into fashion in America. I organized a series of seminars in the area, leading eventually to the publication of a textbook (Etkin, 1964*a*). There was so much new neuroendocrinology in the experimental chapters, especially those by Daniel Lehrman, Frank Beach, and David Davis, that I felt it necessary to write an introduction to the area for the beginning of the book. I had worked with Frank Beach at the American Museum and had helped shepherd Danny through City College while he was spending much too much of his time on bird observation work with G. K. Noble. (As a professor at City College from 1934 to 1966 I like to think I helped to interest several students in research in neuroendocrinology. Besides Danny, names likely to be familiar to neuroendocrinologists include Solomon Berson, Milton Diamond, Clifford Grobstein, Joseph Jailer, and Barry Komisaruk.)

HELLO TO NEUROENDOCRINOLOGY

Man proposes, fate disposes. I had moved my family to a residential area near the edge of the city. Here, in the early 1950s, the city bought a plot of farmland for a new hospital center. Yeshiva University established its new Albert Einstein Medical School in association with the center a half mile from my home. When Ernst and Berta Scharrer were invited to lead the Anatomy Department, I recognized that this was indeed the hand of fate. I wrote to Ernst and received an immediate and enthusiastic welcome to the department. The medical school buildings went up with amazing speed; the new faculty was bubbling with ideas and bursting with energy. The Scharrers turned out to be the storybook scientists of my youthful dreams, completely immersed in their research and teaching, always available for help, and a boundless source of information. Ernst was interested in everything I did, and, because he was always there in the evenings when I was, we had opportunity for lively discussions not only about neuroendocrinology but also about my interests in social behavior and human evolution, the philosophy of science, and, of course, politics. Himself a rationalist and idealist who had given up the Catholicism he had been raised in, he never ceased to wonder at the paradoxes of human behavior. The freedom of science in an institution established by a traditional Jewish religious organization intrigued him no end and raised innumerable questions he loved to explore. He gave of himself generously in teaching as well as in research, often spending an evening developing a set of blackboard diagrams for a single lecture. He was ever ready to spend an hour discussing my work or a manuscript, always playing down his own contribution with an impish sense of humor (see accompanying sketch, Fig. 1). For me, these were indeed

FIGURE 1. Sketch by Ernst Scharrer accompanying his comments on a manuscript of the author's and signed, "your humble servant, E. Scharrer."

halcyon days, brought to a tragic end by his accidental death while on a collecting trip in 1965.

As I became familiar with the newest work in neuroendocrinology, particularly with Harris's concept of hypothalamic influence exerted through the portal system, it became obvious that a new interpretation of the results of the 1938 pituitary transplant paper was possible. The hypothalamic substances could have reached the transplanted pituitary in diluted form through the general circulation and thus account for the late and slow prometamorphic changes induced in the hosts. Although my best grafted hosts had reached the beginning of climax, they had not been allowed (for the reasons given above) to run long enough to demonstrate the full activation of the thyroid necessary to carry through the normal climax. If this concept of neurosecretory control was valid, the transplanted gland should fail to achieve the high level of thyrotropin necessary for climax. Clearly, my first priority at Einstein was a large-scale repetition of the earlier experiment to produce a sufficient number of grafts carrying their hosts through prometamorphosis to permit testing for completion of climax. This experiment yielded clear-cut results. All animals, after going through a slow and delayed prometamorphosis, remained in metamorphic stasis. The growth effects, which were published (Etkin and Lehrer, 1960), clearly showed that the grafted pituitary released more than normal amounts of a growth factor, presumably prolactin, and thus added a new factor to the metamorphic equation.

Our interpretation that the transplanted pituitary may be secreting prolactin was suggested by Everett's work on the rat. Sandra Masur, an undergraduate honors student, tested this idea for amphibians by transplanting the gland in the red eft (land phase of the newt). She showed that it induced water drive in the animal, an effect previously traced by Grant specifically to prolactin (Masur, 1962).

With regard to metamorphosis, a more elegant demonstration of hypothalamic control could be made with salamanders, in which the metamorphic process corresponds to the climax phase of anuran metamorphosis. With the help of a medical student working on a summer project, we were able to carry through a "Harris-type" demonstration. Animals close to metamorphosis were collected, and the student, Walter Sussman, cut the connection between hypothalamus and pituitary and implanted a tiny barrier, less than a millimeter on a side, between them. Histological analysis showed that metamorphosis occurred only in those specimens in which blood vessels were able to surmount the barrier. Those lacking vascular connection between hypothalamus and pituitary failed to metamorphose (Etkin and Sussman, 1961).

The dreamlike quality of research at Einstein in those early years is illustrated by my experience with this experiment. Walter applied for a summer project. I asked him what his area of specialization would be. He replied, "Ophthalmology." "This requires very fine surgery," I said. "That's what I like," he answered. We went into the field and in one glorious afternoon collected over 50 *Ambystoma* larvae at just the right stage, a stroke of luck in itself. I spent a day with Walter developing an operative approach. I did not see him for the next week. He then came to say that he had set up the entire experiment as planned. Astonished, I said, "Let's see how you operate." He demonstrated a different procedure, much simplified but requiring skillful manipulation under the dissecting scope. My pride as a teacher of microsurgery dropped several notches! I was so delighted with the clarity of this experiment and so excited about going on to test the prolactin idea that I forgot to write up the metamorphic stasis results of the main experiment!

HOW MANY WRONGS MAKE IT RIGHT?

By 1960 I was convinced that the clock mechanism regulating metamorphosis lay in the development of the hypothalamic neurosecretory system and its associated portal vessels in the median eminence. I knew from earlier work that there were no evident portal vessels in the tadpole comparable to the system described by Green in the adult frog. We made a study of the development of this circulatory system in relation to metamorphosis and found that indeed the vessels develop with the progress of prometamorphosis, so that by the beginning of climax the system is substantially like that of the adult (Etkin, 1965). The indications were, therefore, that the hypothalamic influence comes into play at the beginning of prometamorphosis, and as it matures the activation of the pituitary increases to a maximum at climax. The metamorphic clock at the next level after thyroid, therefore, lay in the development of the hypothalamus.

In search of an elegantly simple experiment to illustrate this idea, I thought, "How neat it would be to thyroidectomize some tadpoles, raise them beyond the metamorphosis of controls, and then study their hypothalamic-pituitary regions. Of course, the median eminence would stand out clearly matured as in control froglets, whereas the rest of the animal's body would still be larval." When the first slides of this series came through, I was so astonished to find this was not so that I searched frantically for some error in the identification of the initial tissue. But there was none, as was made clear when subsequent specimens came through the histology mill. The hypothalami of the giant thyroidectomized animals were completely larval in form, with no median eminence thickening. The elusive clock of metamorphosis failed to develop in the absence of the thyroid.

Could it be itself dependent on the metamorphosis-inducing action of the thyroid?

Another simple experiment showed that this was indeed the case. When giant thyroidectomized tadpoles were sent through artificial metamorphosis by treatment with a series of increasingly concentrated solutions of thyroxine, their median eminences matured as expected. What we faced, therefore, was a positive feedback mechanism whereby a very low level of thyroid activity before metamorphosis slowly stimulated development of the hypothalamic-thyrotropin mechanism, which in turn activated the pituitary to further activate the thyroid to increase the maturation of the hypothalamus, etc., etc. Such a circular feedback mechanism would start off very slowly indeed but once well started would lead to a period of rapid buildup (prometamorphosis) followed by an explosive release of thyroid hormones at climax. Such is the nature of positive feedback mechanisms. The matured hypothalamus, like other matured tadpole tissue (i.e., the legs), would be expected to lose its sensitivity to thyroid, and negative feedback control would then take over. This concept nicely accounts for the patterning of metamorphic changes. In terms of clock mechanisms, we must think that the timing of events depends on the parameters of the equation of positive feedback. Neat, I thought! The difference between bullfrog and wood frog, according to this theory, resides in the differing parameters of feedback, very slow in the bullfrog, faster in the wood frog. Once the feedback reaches the rapid levels of prometamorphosis, the differences between species become less marked in terms of time, and all species take more nearly the same pattern of metamorphosis as I had found in 1932 (Etkin, 1963, 1964*b*, 1966*a*).

If this hypothesis was true, subjecting tadpoles to very low concentrations of thyroxine during early life should "prime the pump" of feedback and the tadpoles should early go through a normal rate of prometamorphosis followed by climax. This would be a tricky experiment to carry out because the thyroxine intended to prime the pump would at the same time activate a slow rate of prometamorphic change in the general tissues. I had to run the experiment three times, each time with a variety of concentrations, before I was able to satisfy myself that it definitely did not work. There was no priming of the pump. The thyroxine-treated animals grew more slowly, as was expected from other studies, but they metamorphosed at the same time and in the same pattern as the controls (unpublished results).

The only explanation I could think of for this result was that the hypothalamus does not become sensitive to positive feedback from the thyroid until the normal age for prometamorphosis to begin in the species. I did not find this concept congenial, because earlier studies (Etkin, 1950) had indicated that all externally visible metamorphic tissues acquire thyroxine sensitivity at about the same time in a late embryonic stage. In any case, it is possible to test the concept of late acquisition of metamorphic competence in the hypothalamus by treating thyroidectomized or normal tadpoles of various ages with the requisite sequence of thyroxine concentrations to bring them through climax. The examination of the state of the median eminence development in them should settle the question. This experiment did indeed show that tadpoles artificially brought through climax before the normal time failed to develop their median eminences, whereas those that paralleled or trailed the controls showed complete maturation of this structure (Etkin, 1966*b*).

CURRENT GROPINGS

The concept of the metamorphic clock we are now left with seems to be slightly more complex. The initial clock mechanism depends on genetically determined developmental factors that bring the hypothalamus into responsiveness to thyroxine at a particular time in the animal's life. This activates prometamorphosis. The positive feedback mechanism then takes over and determines the timetable of metamorphic events. Further probing of the initiating clock requires elucidating the nature of thyroxine sensitivity and of how tissues acquire it.

As described above, the growth effects of single pituitary grafts had alerted us to a possible role of prolactin in metamorphosis. Amos Gona undertook to explore this lead. In his doctoral thesis he was able to show that large doses of prolactin could stop the metamorphic process even in the midst of climax (Gona, 1967; Etkin and Gona, 1967). Using the *in vitro* tail fin disk method developed by Derby (1968) in our laboratory, we were able to show that, at least in part, this antagonism was expressed at the peripheral level (Derby and Etkin, 1968; Etkin *et al.*, 1969). Nicoll, working in Bern's laboratory, was finding similar effects at this time. We were thus led to the theory that prolactin production predominates in the larval phase of tadpole development and is replaced by TSH activity as the hypothalamus matures under positive feedback stimulation from the thyroid (Etkin, 1970; Etkin and Gona, 1973).

In 1967 Dr. Guillemin kindly offered us some of the TRF material he was preparing from mammalian hypothalami for testing on tadpoles. However, we were unable to detect any metamorphosis-stimulating effects of this material (Etkin and Gona, 1968). Subsequently, we confirmed this negative finding with larger quantities of the synthetic TRF when that became available (unpublished). The recent reports that this material releases prolactin as well as TSH in the rat raise further intriguing questions as to how the hypothalamus and pituitary interact in the life history of the tadpole. Our tadpole work suggests that before metamorphosis the pituitary is largely but not completely independent of the hypothalamus and produces mostly prolactin. When the positive feedback system is activated at the beginning of prometamorphosis, the pituitary shifts to TSH production as detailed above. In view of the evidence cited earlier that the pituitary in embryonic stages produces a thyrotropic field around it, the nature of the interactions of hypothalamic factors, pituitary, and thyroid before metamorphosis remains to be explored.

I retired from active research at Einstein in 1972 but have continued some aspects of the tadpole work in consultation with Dr. Max Hamburgh, Dr. Joseph Osinchak, and Mr. Young So Kim at City College. With Dr. Hamburgh we are using the *in vitro* tail disk system to study the nature of tissue response to thyroid action (Kim *et al.*, 1976). With Dr. Osinchak we are using EM techniques to explore the problem of the development of hormone production in the differentiating pituitary primordium.

Of course, there is no end to the problems of amphibian metamorphosis and no doubt another generation of workers will find this a fruitful field to work in. Perhaps someone will eventually realize the ambition I went to Chicago with—to find the key to puberty in the study of metamorphosis. Our own "shot in the dark" in applying thyroid to the differentiation of the rat median eminence proved abortive (Kikuyama and Etkin, 1964). Wherever they go, may these searchers find as much fun and personal satisfaction as I had in seeking the elusive clocks of animal development.

REFERENCES

Only work done in our laboratory is given specific citation here. Other references can be found from our papers.

Castel, M. (1972). Ultrastructure of the anuran pars intermedia following severance of hypothalamic connection. *Z. Zellforsch. Mikrosk. Anat.* **131**:545.

Derby, A. (1968). An *in vitro* quantitative analysis of the response of tadpole tissue to thyroxine. *J. Exp. Zool.* **168**:147.

Derby, A., and Etkin, W. (1968). Thyroxine induced tail resorption *in vitro* as affected by anterior pituitary hormones. *J. Exp. Zool.* **169**:1.

Etkin, W. (1930). Growth of the thyroid gland of *Rana pipiens* in relation to metamorphosis. *Biol. Bull.* **59**:285.

Etkin, W. (1932). Growth and resorption phenomena in anuran metamorphosis. *Physiol. Zool.* **5**:275.
Etkin, W. (1935*a*). The mechanisms of Anuran metamorphosis—Thyroxin concentration and the metamorphic pattern. *J. Exp. Zool.* **71**:317.
Etkin, W. (1935*b*). Effect of multiple pituitary primordia in the tadpole. *Proc. Soc. Exp. Biol. Med.* **32**:1653.
Etkin, W. (1936). The phenomena of Anuran metamorphosis—The development of the thyroid gland. *J. Morphol.* **59**:69.
Etkin, W. (1938). The development of thyrotropic function in pituitary grafts in the tadpole. *J. Exp. Zool.* **77**:347.
Etkin, W. (1939). A thyrotropic field effect in the tadpole. *J. Exp. Zool.* **82**:463.
Etkin, W. (1941). On the control of growth and activity of the pars intermedia of the pituitary by the hypothalamus in the tadpole. *J. Exp. Zool.* **86**:113.
Etkin, W. (1943). The development control of pars intermedia by brain. *J. Exp. Zool.* **92**:31.
Etkin, W. (1950). The acquisition of thyroxine-sensitivity by tadpole tissues. *Anat. Rec.* **108**:541 (abstr.)
Etkin, W. (1962*a*). Hypothalamic inhibition of pars intermedia activity in the frog. *Gen. Comp. Endocrinol. Suppl.* **1**:148.
Etkin, W. (1962*b*). Neurosecretory control of the pars intermedia. *Gen. Comp. Endocrinol.* **2**:161.
Etkin, W. (1963). Metamorphosis activating system of the frog. *Science* **139**:810.
Etkin, W. (1964*a*). *Social Behavior and Organization among Vertebrates*, University of Chicago Press, Chicago (paperback version as *Social Behavior from Fish to Man.*)
Etkin, W. (1964*b*). Metamorphosis. In Moore, J. (ed.), *Physiology of the Amphibia*, Academic Press, New York, pp. 427–468.
Etkin, W. (1965). The phenomena of amphibian metamorphosis. IV. The development of the median eminence. *J. Morphol.* **116**:371.
Etkin, W. (1966*a*). How a tadpole becomes a frog. *Sci. Am.* **214**:76.
Etkin, W. (1966*b*). Hypothalamic sensitivity to thyroid feedback in the tadpole. *Neuroendocrinology* **1**:293.
Etkin, W. (1970). Endocrine mechanism of amphibian metamorphosis, an evolutionary achievement. *Mem. Soc. Endocrinol.* **18**:137.
Etkin, W. (1978). The thyroid, a gland in search of a function. *Perspect. in Biol. and Med.* (in press).
Etkin, W., and Gilbert, L. (eds.) (1968). *Metamorphosis, A Problem in Developmental Biology*. Appleton-Century-Crofts, New York.
Etkin, W., and Gona, A. G. (1967). Antagonism between prolactin and thyroid hormone in amphibian development. *J. Exp. Zool.* **165**:249.
Etkin, W., and Gona, A. G. (1968). Failure of a mammalian TRF preparation to elicit metamorphic response in tadpoles. *Endocrinology* **82**:1067.
Etkin, W., and Gona, A. G. (1974). The evolution of thyroid function in poikilothermic vertebrates. In *Handbook of Physiology: The Thyroid*, American Physiological Society, Washington, D.C.
Etkin, W., and Kim, Y. S. (1978). An interpretation of the roles of the thyroid in vertebrate evolution, In *Evolution of Vertebrate Hormones*. (A. Epple and P. K. T. Pang, eds.), Texas Tech University Press, Lubbock.
Etkin, W., and Lehrer, R. (1960). Excess growth in tadpoles after transplantation of the adenohypophysis. *Endocrinology* **67**:457.
Etkin, W., and Rosenberg, L. (1938). Infundibular lesion and pars intermedia activity in the tadpole. *Proc. Soc. Exp. Biol. Med.* **39**:332.

Etkin, W., and Sussman, W. (1961). Hypothalamo-pituitary relations in metamorphosis of *Ambystoma. Gen. Comp. Endocrinol.* **1**:70.

Etkin, W., Root, R., and Mofshin, B. (1940). The effect of thyroid feeding on oxygen consumption of the goldfish. *Physiol. Zool.* **13**:415.

Etkin, W., Derby, A., and Gona, A. (1969). Prolactin-like antithyroid action of pituitary grafts in tadpoles. *Gen. Comp. Endocrinol. Suppl.* **2**:253.

Gona, A. (1967). Prolactin as a goitrogenic agent in amphibian. *Endocrinology* **81**:748.

Kikuyama, S., and Etkin, W. (1964). Thyroid influence on development of hypothalamic-hypophyseal system in rat. *Am. Zool.* **4**:228 (abstr.)

Kim, Y., Hamburgh, M., Frankfort, H., and Etkin, W. (1977). Reduction in the latent period of the response to thyroxin by tadpole tail discs fused to discs pretreated with thyroxin. *Devel. Biol.* **55**:387.

Masur, S. K. (1962). Autotransplantation of the pituitary in the red eft. *Am. Zool.* **2**(**4**):299 (abstr.)

Saland, L. C. (1968). Ultrastructure of the frog pars intermedia in relation to hypothalamic control of hormone release. *Neuroendocrinology* **3**:72.

11

Béla Flerkó

Béla Flerkó was born in Pécs, Hungary, in 1924, and received his M.D. from the University Medical School of Pécs. He entered the Anatomy Department at Pécs in 1943, and became its head in 1964. His interest in and research work on neural control of secretion of gonadotropic hormones began in the early 1950s. He is a corresponding member of the Hungarian Academy of Sciences, president of the Hungarian Society of Endocrinology and Metabolism, a member of the Executive Committee of the International Brain Research Organization and of the Central Council of the International Society for Research in Reproduction, founding member of the European Society for Comparative Endocrinology, and member of the IUPS Commission on Endocrinology. He is a member of the editorial boards of *Acta Biologica Academiae Scientiarum Hungaricae*, *Endocrinologica Experimentalis*, and *Endokrinologie*, and former member of the editorial boards of *Experimental Brain Research* and *Neuroendocrinology*. He was vice-president of the International Society of Neuroendocrinology from 1972 to 1976.

Neurohormonal Feedback Control of Gonadotropin Secretion

BÉLA FLERKÓ

It was in 1970 that I paid a visit to Dr. Meites's laboratory in Michigan, and he invited me to give a seminar. He introduced me by mentioning, among other things, that I received the M.D. degree from the university at which I am now professor of anatomy and that I still live in the same old family house in Pécs where I was born in 1924. Pécs is located on the southern slope of the 1500-ft Mecsek Hill in the south of Hungary. It was in Pécs, and as early as 1367, that the first Hungarian university was founded by King Louis the Great. As a consequence of destruction by the Turks in 1543, and 150 years of occupation, the town had few inhabitants when the Turks were expelled in 1686. A High School of Theology and later a High School of Law were established here, but the university did not function until 1923 when the University of Pozsony was transferred to Pécs.

Brain research began with the appointment of J. Szentágothai as head of the Department of Anatomy in 1946, and gradually expanded in the following years along three main lines: functional anatomy of synapses and elementary reflex arcs, functional anatomy of neuroendocrine regulation, and experimental neuroembryology. When Szentágothai took the chair of anatomy in Pécs, the department had no graduates on its teaching staff, and only third- and fourth-year medical students were available as tutors. Like my colleagues, I did not do any research at that time but helped Szentágothai run the dissections and histological courses. He reorganized teaching and research in an admirably short time, and succeeded in transplanting his enthusiasm for teaching and research to all of us in spite of

BELA FLERKÓ • Department of Anatomy, University Medical School, Pécs, Hungary, Szigeti ut 12.

postwar limitations in equipment and chemicals. He has been a unique teacher—not only in science but also in humanities, art, and philosophy. During his stay in Pécs, we walked at least 15 km on the Mecsek Hill nearly every Sunday, and talked about science, books, nature, art, philosophy, problems, the department, people, colleagues. He always has been perceptive in human relations and a good judge of promise and ability. He sought the greatest output in research and teaching that could be comfortably achieved, not only for the sake of the department but also mainly for the credit of the worker. My time with him was most happy and it was a great privilege to work in this way for 17 years in Pécs.

Szentágothai (1975) described very vividly in his chapter in the first volume of this series how research in neuroendocrinology began in the department. As far as I was concerned, I remember that Szentágothai asked me one day in which field of medicine I wished to work after I received the M.D. degree. My answer was gynecology. "Then, you should study the development and life cycle of ciliary epithelium in the female genital tract," he said. So for 3 years I did tedious and time-consuming qualitative and quantitative histological work, and published my first paper on "Development and Decay of the Cilia" in Hungarian in 1950. In the meantime, I became interested in the hormonal background of the cyclic epithelial changes in the female genital tract of rats and guinea pigs. Simultaneously, from the brilliant work of G. W. Harris and J. D. Green on the one hand and of J. W. Everett and C. H. Sawyer on the other I realized the significance of the hypothalamus and the hypophysial portal system in the control of anterior pituitary hormone function. By the end of the 1940s, the hypothalamus had been explored to an extent sufficient to indicate that it was a rich field for experiments. Thus I began to study the effects of electrolytic lesions in various parts of the hypothalamus of rats and rabbits on the histology of the female reproductive system, especially on the ciliary apparatus of the oviduct and uterus.

Using the stereotaxic instrument designed by Szentágothai, I quickly realized the importance of symmetrical bilateral lesions in producing changes in ovarian, uterine, and tubal histology. One of the most striking histological manifestations was that 4–6 weeks after lesions had been made in the tuberal part of the hypothalamus of rabbits, about 30% of the total epithelial cells of the oviduct became so-called pin cells, whereas maximally 0.1% were found in the intact animal (Flerkó, 1951). The pin cells are the degenerate forms of nonciliated epithelial cells in the process of being squeezed out from the epithelial surface, and their appearance in such an enormous abundance was associated with ovarian atrophy and with decreased secretory activity of the uterus (Flerkó, 1953). A few days after estrogen administration, but not after that of progesterone, the pin cells and decreased uterine secretory activity always present in spayed, hypophysectomized or tuberal-lesioned rabbits disappeared, indicating that these striking epithelial changes were due to an insufficient estrogen supply following ovariectomy, hypophysectomy or tuberal hypothalamic lesions (Flerkó, 1953, 1954*a*). This agreed with the findings of other investigators showing that destruction near the origin of the pituitary stalk, or stalk section, totally prevented or at least greatly reduced the production and/or release of the gonadotropic hormones (GTH).

Effects of bilateral electrolytic lesions located more orally, i.e., in the preoptic-anterior hypothalamic area were, however, completely different. As in guinea pigs and rats, anterior hypothalamic lesions in rabbits produced ovarian, tubal, and uterine changes characteristic of a permanent estrogen action evoked by continuous FSH release (Flerkó, 1953). My attention was especially attracted by the polyfollicular ovaries and permanent vaginal cornification associated with anterior hypothalamic lesions. Hillarp's (1949) explanation of the polyfollicular ovaries and associated phenomena failed to account for the absence of a negative estrogen feedback on FSH secretion, and I assumed, therefore, that a nervous mechanism in the anterior hypothalamus was necessary for the inhibitory action of estrogen on FSH secretion (Flerkó, 1954*b*). Although Hohlweg and Junkmann (1932) had raised the possibility more than two decades earlier that the gonadal hormones might influence gonadotropic functions through a hypothetical "sexual center" located somewhere in the brain, their suggestion remained unheeded, and it was generally believed until the mid 1950s that the peripheral hormonal milieu influenced synthesis and release of GTH exclusively at the level of the anterior pituitary. Some of my early experimental findings (Flerkó, 1954*b*) already seemed to support the assumption that bilateral anterior hypothalamic lesions eliminated, or at least reduced, the inhibitory action of estrogen on GTH secretion, but the evidence was rather circumstantial. Therefore, I looked for more evidence. For a start, I found that in parabiotic rats the inhibitory action of estrogen (1 μg per day) on the castration-induced rise of GTH secretion could be greatly diminished by anterior hypothalamic lesions (Flerkó, 1956, 1957*a*). The results of this and of some other experiments suggested, furthermore, the presence of estrogen-sensitive nerve cells in the hypothalamus.

At this time, Szentágothai also became very much interested in neuroendocrinology, especially in the neurohumoral feedback control of anterior pituitary function, and suggested that I graft ovarian tissue fragments into the hypothalamus to substantiate the idea of estrogen-sensitive neurons. Szentágothai (1975) is right in mentioning in his chapter of this series that I first considered this idea to be too absurd to test experimentally; however, his picturesque description of this story is somewhat exaggerated. Indeed, when my wife, Dr. Vera Bárdos, a most unselfish and untiring collaborator, completed the histological processing of this material, I was much too preoccupied with the evaluation of some other experiments, and placed the slides of the intrahypothalamic ovarian grafts aside. These I began to study only a few weeks later, when Szentágothai became impatient, and found the first clear evidence (Flerkó and Szentágothai, 1957) for the existence of estrogen-sensitive neurons in the brain. In fairness, I should add that Szentágothai, in spite of his impulsive temperament, was pleasant to work with. He established a smooth-running organization for maximum effective use of the facilities available, suggested excellent ideas to pursue, and offered continuing encouragement as well as sharp criticism. He created a friendly atmosphere for the staff that grew gradually during the 1950s. It was the right environment for creative life, self-expression, and development. As the number of associates increased, Szentágothai was able to secure additional space. In spite of this, however, from the mid-1950s we were always crowded until we moved into a new building in 1971.

With my wife and a bright enthusiastic medical student, Gy. Illei, now chief gynecologist of the teaching hospital in Szombathely, I continued to collect evidence for the role of the indirect, negative neurohumoral estrogen (Flerkó and Bárdos, 1959, 1960, 1961) and testosterone (Flerkó and Illei, 1957) feedback in the control of gonadotropin secretion. Apart from Hohlweg and Daume (1959), only Donovan and van der Werff ten Bosch (1956*a*,*b*) and Assenmacher (1957) appeared to agree with these ideas in the beginning. In the United States, E. M. Bogdanove was the first to support the hypothesis by confirming (Bogdanove and Crabill, 1959) our finding that estrogen-induced inhibition of follicle development and consequent formation of corpora lutea in infantile rats were diminished by anterior hypothalamic lesions (Flerkó, 1957*b*). Later, he interpreted this experiment differently without assuming a feedback via the hypothalamus, and undertook experiments to confirm the concept of Moore and Price (1932), i.e., that the adenohypophysis and its target glands are linked together in a self-contained system of purely hormonal interactions. Other American scientists, such as Lisk (1960), Davidson and Sawyer (1961), and McCann (1962), furnished evidence in favor of the role of an indirect, neurohormonal feedback in the control of the anterior pituitary function. These findings stimulated us to publish our concept of the hypothalamic control of the anterior pituitary in a monograph (Szentágothai *et al.*, 1962) which has been one of Szentágothai's major contributions to neuroendocrinology. At that time, he (Szentágothai *et al.*, 1962) described the tuberoinfundibular tract as the main source of nerve fibers terminating on or in the immediate vicinity of the capillary loops penetrating the median eminence.

On the basis of Szentágothai's neurohistological findings and of Halász's first experimental results concerning the hypophysiotrophic area (Halász *et al.*, 1962), the concept of the "parvicellular neurosecretory system" as the lower level of the "double neural mechanism" controlling the function of the anterior pituitary gland could be formulated. B. Mess's findings on the habenular mechanism of thyrotropin release, together with my and several other laboratories' findings on gonadotropin control, provided the concept of the "release regulating system," localized "into different parts of the hypothalamus and of the related rhinencephalic structures" which were assumed to supervise the "parvicellular neurosecretory system" (Szentágothai *et al.*, 1962). Although we were four in formulating the above hypothesis, it was solely due to Szentágothai that it was formulated in a monograph in due time and in an acceptable manner.

My first joint work abroad occurred in Prague in 1956, when I worked with J. Bureš in the Physiology Department of the Czechoslovak Academy of Sciences. We implanted electrodes into the hypothalamus and other brain areas of rats and studied the bioelectrical activity of these brain regions after having injected estradiol into the animals. We recorded bioelectrical activities of increased frequency and amplitude from the anterior hypothalamus after treatment with estradiol and found no change in the bioelectrical activity of other brain regions. Because of the small number of our experimental animals, we wanted to repeat this experiment during the following year, but were prevented from doing so by circumstances. Thus I never published this finding, which furnished the first evidence for the existence of hypothalamic neurons whose firing rate could be influenced by increasing the level of estrogen in the blood. I spent about 2 months in the Bucharest Endocrinological Institute in 1958, and in the laboratory of J. J. van der Werff ten Bosch in A. Querido's clinic in Leiden in 1959. In 1961 I was invited to give a paper at the Symposium on Current Concepts of Neuroendocrinology in Miami, Florida, in December, and then spent 6 months with C. H. Sawyer and R. A. Gorski at UCLA. It was with special pleasure and great expectation that I accepted Sawyer's invitation and support, for I considered him and Everett my teachers in neuroendocrinology, being greatly influenced by their ideas and brilliant work at Duke University. My expectations were surpassed at UCLA, and the 6 months spent there certainly belong among my most cherished memories.

When I returned to Pécs in the middle of 1962, I found the usual intense scientific atmosphere in the department, although the prospect of Szentágothai's leaving for Budapest was already having an effect. He left in

1963, and I took over his chair with pride but also with trepidation. The research program of the department has since been confined to the fields of neuroendocrine regulation and experimental neuroembryology. By then, we had good histological equipment, including a cryostat, histo- and spectrophotometers, and stereotaxic apparatus for dogs, cats, rabbits, and small rodents. Later, the Biomedical Division of the Population Council very generously supported our research in reproductive neuroendocrinology, and we had important financial support from the Hungarian Academy of Sciences as well as from the so-called Small Supplies Program of the World Health Organization. This made it possible to raise the level of laboratory facilities to an almost up-to-date level (electron microscope, fluorometer, liquid scintillation counter, equipment for column chromatography, etc.), and we have recently introduced radioimmunoassay (RIA) and immunohistological techniques.

In the early 1960s, together with P. Petrusz, my interest became gradually focused on the mechanism of sexual differentiation of the brain apparatus controlling gonadotropin secretion. Barraclough and Gorski (1961) were the first to show that the so-called preoptic LH trigger was deranged in female rats which received early postnatal androgen treatment. On the basis of the concept of the anterior hypothalamic FSH control apparatus, launched by Donovan and van der Werff ten Bosch (1956*a*,*b*) and myself (Flerkó, 1956, 1957*a*), it could be expected that not only the preoptic LH trigger but also the FSH mechanism might be affected by such treatment. This supposition was supported by our results showing that the sensitivity or responsiveness to estradiol of the anterior hypothalamic FSH control apparatus was considerably reduced in adult male (Petrusz and Flerkó, 1965) as well as in androgen-sterilized female rats (Petrusz and Nagy, 1967).

Simultaneously, many investigators, including us, reported that the responsiveness to estradiol of peripheral tissues (uterus, vagina, anterior pituitary) was reduced in rats injected with a single dose of testosterone in the first few postnatal days. These findings indicated that neural and non-neural target tissues of estrogens are similarly affected by early postnatal androgen action. A preliminary experiment of ours (Flerkó and Mess, 1968) revealed that the estradiol-binding capacity of pituitaries and uteri of androgen-sterilized rats was significantly reduced as compared to that of intact controls. This raised the possibility that early postnatal androgen action might interfere with the development of estrogen-binding proteins, i.e., with estradiol uptake and/or retention by estrogen-responsive tissues. Many findings, gained in the 1960s with isotope techniques, indicated that the middle and anterior hypothalamus, including the preoptic area, is direct target tissue for estradiol, and the estrogen sensitivity of these brain areas is also reduced in androgenized rats (Petrusz and Flerkó, 1965; Petrusz and Nagy, 1967). We postulated, therefore, that early postnatal androgen administration might damage the specific trapping mechanism in estrogen-sensitive hypothalamic neurons. This has been confirmed first by our findings (Flerkó *et al.*, 1969) showing that the estradiol-binding capacity of the preoptic–anterior hypothalamic area and of the middle hypothalamus, as well as that of the nonneural target tissues (uterus and anterior pituitary), was significantly reduced in rats which received 1.25 mg testosterone 2 days after birth. No change occurred in the posterior hypothalamus and parietal cortex of the same animals. This finding, corroborated by many authors, proved that perinatal androgen action interfered with the normal development of estrogen-receptor proteins and hence with the normal uptake or retention of estradiol by the limbic-hypothalamic estrogen-responsive neurons. Accordingly, these neurons became functionally inactive in mediating neurohormonal estrogen feedback controlling cyclic gonadotropin release and ovulation. Thus the severe impairment of the specific estradiol-trapping mechanism of certain neurons may account for the loss of ovulation and cyclic gonadotropin release in androgen- and probably also in light-sterilized (Illei-Donhoffer *et al.*, 1974) rats.

Experimental findings indicated the involvement of norepinephrine (NE) in the release of GTH. Since hypothalamic NE neurons appear to be sensitive to estrogen, it appeared reasonable to assume that the action of estrogen on the neurons triggering the ovulatory gonadotropin release might involve changes in NE biosynthesis. In other words, the preovulatory positive feedback action of estrogen would enhance NE synthesis in the neurons triggering LHRH release. With a reduction in the number of estrogen-binding sites in the trigger neurons in androgen- and light-sterilized rats, the preovulatory estrogen surge cannot enhance NE synthesis and hence these neurons cannot stimulate the LHRH-synthesizing neurons to release sufficient LHRH for the discharge of an ovulatory dose of GTH (Flerkó, 1975). The situation is similar in anterior hypothalamic-lesioned rats, where the trigger neurons are destroyed or separated from the LHRH-synthesizing neurons. When stimulated, however, by NE injected intraventricularly, the rats made anovulatory by anterior hypothalamic lesions, by exposure to neonatal androgen action, or by continuous illumination released sufficient LHRH to induce ovulation (Tima and Flerkó, 1974, 1975).

From 1962, B. Halász and his group explored the functional capacity of the hypophysiotropic area (HTA), using intrahypothalamic pituitary grafts and electrolytic lesion techniques. Although these techniques were essential for elucidation of the problem, a surgical method was necessary for the complete isolation of the HTA. This gave the impetus to Halász to develop in 1963 his simple but ingenious knife to cut all or a part of the nervous pathways entering the HTA. He tested the effect of total or partial deafferentation of the HTA mainly at UCLA in 1964, as described in his chapter of this book, and I wish only to underscore the significance of his contribution to the development of neuroendocrinology, as acknowledged by the von Euler award at the Fifth International Congress of Endocrinology in 1976. I regretted very much his departure for Budapest in 1971, when he took the chair in the Second Anatomy Department of Semmelweis Medical University.

For the time being, my research interests focus mainly on the immunohistological localization of the neural elements synthesizing the releasing and inhibiting hormones. The necessary peroxidase-labeled antibody technique was acquired by G. Sétáló in the laboratory of P. Nakane in Denver (Colorado), where he spent 1 year supported by a Population Council Fellowship, and we are collaborating successfully with A. V. Schally and A. Arimura in the field of immunohistological localization of LHRH- and somatostatin-synthesizing neural elements (Sétáló *et al.*, 1975*a,b*, 1976). The immunohistological group, directed by G. Sétáló, is presently studying the development of these elements as well as the functional differentiation of the anterior pituitary cells and the presumed role of the various releasing hormones in this process. Research aimed at elucidating the way in which various experimental procedures induce anovulatory sterility associated with polyfollicular ovaries is still in progress in my laboratory in collaboration with L. Tima, who studied protein hormone RIA in the laboratory of C. A. Barraclough at Maryland University, Baltimore, where I also spent a happy and very useful 6 months in 1973. Mess and his group are studying the neural control of TSH secretion and the role of serotoninergic mechanisms in pinealectomy-induced ovulation in rats made anovulatory experimentally. Since his return from UCLA, I. Lengvári has continued to explore the neural afferents necessary for the maintenance of the diurnal ACTH rhythm. Unfortunately, our research output is not as large as I should like it to be because of the heavy teaching load and the continuous experimentation with teaching "reforms." Fortunately, there is a good side to everything: we can only slightly contribute to the ever-increasing avalanche of scientific papers, which constitutes a real danger, and we do not have time for mere fact hunting, which—I think—is far from being the essence of research. A research worker might spend her or his whole life recording all possible information about a research subject, and yet nothing whatsoever of any interest might emerge from such efforts if appropriate concepts were lacking. Research finds its own rewards, and, fancying myself as a *Homo ludens*, I believe that science should proceed at a more leisurely pace and give more priority to ideas.

REFERENCES

Assenmacher, I. (1957). Nouvelles données sur le rôle de l'hypothalamus dans les régulations hypophysaires gonadotropes chez le canard domestique. *C. R. Acad. Sci. (Paris)* **245**:2388.

Barraclough, C. A., and Gorski, R. A. (1961). Evidence that the hypothalamus is responsible for androgen-induced sterility in the female rat. *Endocrinology* **68**:68.

Bogdanove, E. M., and Crabill, E. (1959). unpublished. Quoted from Bogdanove, E. M, and Schoen, H. C. (1959). Precocious sexual development in female rats with hypothalamic lesions. *Proc. Soc. Exp. Biol.* **100**:664.

Davidson, J. M., and Sawyer, C. H. (1961). Effects of localized intracerebral implantation of oestrogen on reproductive function in the female rabbit. *Acta Endocrinol. (Kbh.)* **37**:385.

Donovan, B. T., and van der Werff ten Bosch, J. J. (1956*a*). Oestrus in winter following hypothalamic lesions in the ferret. *J. Physiol. (London)* **132**:57.

Donovan, B. T., and van der Werff ten Bosch, J. J. (1956*b*). Precocious puberty in rats with hypothalamic lesions. *Nature (London)* **178**:745.

Flerkó, B. (1951). Einfluss experimenteller Hypothalamuslaesionen auf das Eileiterepithel. *Acta Morphol. Acad. Sci. Hung.* **1**:5.

Flerkó, B. (1953). Einfluss experimenteller Hypothalamuslaesionen auf die Funktion des Sekretionsapparates im weiblichen Genitaltrakt. *Acta Morphol. Acad. Sci. Hung.* **3**:65.

Flerkó, B. (1954*a*). Die Epithelien des Eileiters und ihre hormonalen Reaktionen. *Z. Mikrosk. Anat. Forsch.* **61**:99.

Flerkó, B. (1954*b*). Zur hypothalamischen Steuerung der gonadotrophen Funktion der Hypophyse. *Acta Morphol. Acad. Sci. Hung.* **4**:475.

Flerkó, B. (1956). Die Rolle hypothalamischer Strukturen bei der Hemmungswirkung des erhöhten Östrogenblutspiegels auf die Gonadotrophinsekretion. *Acta Physiol. Acad. Sci. Hung. Suppl.* **9**:17.

Flerkó, B. (1957*a*). Le rôle des structures hypothalamiques dans l'action inhibitrice de la folliculine sur la sécrétion de l'hormone folliculostimulante. *Arch. Anat. Microsc. Morphol. Exp.* **46**:159.

Flerkó, B. (1957*b*). Einfluss experimenteller Hypothalamusläsion auf die durch Follikelhormon indirekt hervorgerufene Hemmung der Luteinisation. *Endokrinologie* **34**:202.

Flerkó, B. (1975). Perinatal androgen action and the differentiation of the hypothalamus. In Brazier, M. A. B. (ed.), *Growth and Development of the Brain*, Raven Press, New York.

Flerkó, B., and Bárdos, V. (1959). Zwei verschiedene Effekte experimenteller Läsion des Hypothalamus auf die Gonaden. *Acta Neuroveg. (Wien)* **20**:248.

Flerkó, B., and Bárdos, V. (1961). Absence of compensatory ovarian hypertrophy in rats with anterior hypothalamic lesions. *Acta Endocrinol. (Kbh.)* **36**:180.

Flerkó, B., and Illei, G. (1957). Zur Frage der Spezifität des Einflusses von Sexualsteroiden auf hypothalamische Nervenstrukturen. *Endokrinologie* **35**:123.

Flerkó, B., and Mess, B. (1968). Reduced oestradiol-binding capacity of androgen sterilized rats. *Acta Physiol. Acad. Sci. Hung.* **33**:111.

Flerkó, B., and Szentágothai, J. (1957). Oestrogen sensitive nervous structures in the hypothalamus. *Acta Endocrinol. (Kbh.)* **26**:121.

Flerkó, B., Mess, B., and Illei-Donhoffer, A. (1969). On the mechanism of androgen sterilization. *Neuroendocrinology* **4**:164.

Halász, B., Pupp, L., and Uhlarik, S. (1962). Hypophysiotrophic area in the hypothalamus. *J. Endocrinol.* **25**:147.

Hillarp, N. Å. (1949). Studies on the localization of hypothalamic centres controlling the gonadotrophic function of the hypophysis. *Acta Endocrinol.* (*Kbh.*) **2**:11.

Hohlweg, W., and Daume, E. (1959). Über die Wirkung intrazerebral verebreichten Dienoestroldiacetats Bei Rattan. *Endokrinologie* **38**:46.

Hohlweg, W., and Junkmann, K. (1932). Die hormonal-nervöse Regulierung der Funktion des Hypophysenvorderlappens. *Klin. Wochenschr.* **11**:321.

Illei-Donhoffer, A., Flerkó, B., and Mess, B. (1974). Reduction of estradiol-binding capacity of neural target tissues in light-sterilized rats. *Neuroendocrinology* **14**:187.

Lisk, R. D. (1960). Estrogen-sensitive centers in the hypothalamus of the rat. *J. Exp. Zool.* **145**:197.

McCann, S. M. (1962). A hypothalamic luteinizing hormone-releasing factor. *Am. J. Physiol.* **202**:395.

Moore, C. R., and Price, D. (1932). Gonad hormone functions, and the reciprocal influence between gonads and hypophysis with its bearing on the problem of sex hormone antagonism. *Am. J. Anat.* **50**:13.

Petrusz, P., and Flerkó, B. (1965). On the mechanism of sexual differentiation of the hypothalamus. *Acta Biol. Acad. Sci. Hung.* **16**:169.

Petrusz, P., and Nagy, É. (1967). On the mechanism of sexual differentiation of the hypothalamus; decreased hypothalamic estrogen sensitivity in androgen-sterilized female rats. *Acta Biol. Acad. Sci. Hung.* **18**:21.

Sétáló, G., Vigh, S., Schally, A. V., Arimura, A., and Flerkó, B. (1975*a*). LH-RH-containing neural elements in the rat hypothalamus. *Endocrinology* **96**:135.

Sétálo, G., Vigh, S., Schally, A. V., Arimura, A., and Flerkó, B. (1975*b*). GH-RIH-containing neural elements in the rat hypothalamus. *Brain Res.* **90**:352.

Sétálo, G., Vigh, S., Schally, A. V., Arimura, A., and Flerkó, B. (1976). Immunohistological study of the origin of LH-RH-containing nerve fibers of the rat hypothalamus. *Brain Res.* **103**:597.

Szentágothai, J. (1975). Under the spell of hypothalamic feedback. In Meites, J., Donovan, B. T., and McCann, S. M. (eds.), *Pioneers in Neuroendocrinology*, Vol. 1, Plenum, New York.

Szentágothai, J., Flerkó, B., Mess, B., and Halász, B. (1962). *Hypothalamic Control of the Anterior Pituitary*, Akadémiai Kiadó, Budapest.

Tima, L., and Flerkó, B. (1974). Ovulation induced by norepinephrine in rats made anovulatory by various experimental procedures. *Neuroendocrinology* **15**:346.

Tima, L., and Flerkó, B. (1975). Ovulation induced by the intraventricular infusion of norepinephrine in rats made anovulatory by neonatal administration of various doses of testosterone. *Endokrinologie* **66**:218.

12

William F. Ganong

William F. Ganong was born on July 6, 1924, in Northampton, Massachusetts. He attended Harvard College, Georgetown College, the University of Virginia Medical School, and Harvard Medical School. He received his A.B. cum laude from Harvard in 1945 and his M.D. magna cum laude from Harvard in 1949. After internship and residency training at the Peter Bent Brigham Hospital in Boston, a tour in the U.S. Army, and 3 years of postdoctoral training at Harvard, he became assistant professor of physiology at the University of California, Berkeley, in 1955. He moved with the department to San Francisco in 1958, becoming associate professor in 1960 and professor in 1964. He assumed his present position as chairman of the Department of Physiology, University of California, San Francisco, in 1970.

Dr. Ganong has served as a member of the council and president of the American Physiological Society. He has served as a member of the council and president of the Association of Chairmen of Departments of Physiology, a member of the council and vice-president of the International Society of Neuroendocrinology, and a member of the council of the Endocrine Society. He is a member of the American Society for Pharmacology and Experimental Therapeutics, the Council for High Blood Pressure Research of the American Heart Association, the Society for Experimental Biology and Medicine, the International Brain Research Organization, and the Society for Neuroscience. He has served on various advisory committees for the National Institutes of Health, including Neurology A Study Section, a special NIH study section on training, and a task force on hypertension. He has served on the editorial boards of *American Journal of Physiology*, *Endocrinology*, *Proceedings of the Society of Experimental Biology and Medicine*, *Neuroendocrinology*, *Journal of Pharmacology and Experimental Therapeutics*, and *Neuroscience*. He was elected corresponding member of the Chilean Endocrine Society of 1966, selected as Faculty Research Lecturer, University of California, San Francisco, in 1968, and awarded the Italian Instituto Farmacoterapico Italiano Golden Hippocrates Award in 1970.

12

The Brain and the Endocrine System: A Memoir

WILLIAM F. GANONG

I suppose my career first began to turn toward neuroendocrinology on December 7, 1941. At that time, I was majoring in government at Harvard College and planning to enter public administration. After the Japanese attack on Pearl Harbor, I was drafted and initially assigned to the infantry. However, the Army subsequently decided to send me back to school, first for pre-engineering studies and then for a premedical and medical education. The change from public administration to medical science is one I have never regretted, but it is one that occurred because of world events rather than personal planning.

I first became interested in the brain at the University of Virginia Medical School, the school to which the Army sent me for my medical education. Laboratory assignments at Virginia were made alphabetically, as they are in most medical schools, so my laboratory partner was Donald Ferguson. Ferguson was a blithe spirit with considerable intellectual curiosity. He and I became intrigued by the way the brain is put together, and we decided to ask the Anatomy Department for an additional brain to dissect on our own time. I remember this created considerable consternation in the department, there being barely enough brains to go around, and we were referred instead to the relevant literature. However, this was the start of a fascination I have had ever since with what is certainly the most remarkable organ in the body.

I subsequently transferred to Harvard Medical School, where, in 1949, I took a 1-month elective course in George Thorn's laboratory. That month was a memorable one for several reasons. First, the day-to-day operation of the laboratory was in the hands of Peter H. Forsham, who became and remains to this day a valued friend, advisor, and colleague. Second, Thorn's

WILLIAM F. GANONG • Department of Physiology, University of California, San Francisco, California 94143.

laboratory was investigating the use of the number of circulating eosinophils as an index of the concentration of circulating glucocorticoids, and I became interested in using this index to study the regulation of adrenocortical secretion. Dr. Forsham's wife, Connie, taught me to count eosinophils, and we studied the eosinopenic response to repeated doses of epinephrine. A third important event was an introduction to Gregory Pincus. I was greatly impressed with Pincus's comment at that time that the central problem in endocrinology was the regulation of pituitary secretion. The fourth and perhaps most important event was that I heard about the work of David Hume, and, through Dr. Forsham, I met Dr. Hume.

In 1949 David Hume was a junior member of the surgical faculty at Harvard. He was interested in the body's response to injury and the increase in ACTH secretion produced by surgical stress. There were a number of reports of gonadal atrophy in patients with hypothalamic disease, and Harris and Green's work on the hypophyseal portal vessels was just becoming known. Hume hypothesized that the multiple stimuli which increase ACTH secretion funnel through the rest of the brain to the hypothalamus and affect ACTH secretion by way of this organ. To test his hypothesis, he produced lesions of the hypothalamus in dogs, and tested their eosinopenic response to surgical stress. He found that appropriate lesions blocked the stress response. He and his associates also prepared an extract of hypothalamic tissue obtained from a local slaughterhouse and found that the extract caused eosinopenia when injected in dogs with the hypothalamic lesions that blocked the eosinopenic response to stress. An abstract reporting this work and postulating the existence of a corticotropin-releasing factor was published in 1949 (Hume, 1949), 6 years before the paper by Saffran *et al.* (1955) which is usually cited as the first demonstration of CRF. Hume subsequently published a more detailed report of the lesion experiments (Hume and Wittenstein, 1950) at about the same time that de Groot and Harris (1950) reported that hypothalamic lesions blocked the lymphopenic response to stress in rabbits. As far as I can tell, there were two reasons a full report on the hypothalamic extracts did not appear. First, it was not Hume's style to rush into print. He was busy with his surgery, his teaching, and his research, and, at least in those days, publishing was something he did when he got around to it. Secondly, the collection of hypothalamic tissue for preparation of the new hypothalamic pituitary-regulating factor was turned over to a commercial company, and the results became erratic. Hume once told me that he learned to his consternation some years later the probable reason for the erratic results; some of the slaughterhouses supplying the brain tissue to the company making the extract became confused about the part of the brain desired and began to send pieces of medulla oblongata instead of hypothalamus. Incidentally,

Hume was trying to destroy the paraventricular nuclei, and, on the basis of his lesions, he thought these nuclei controlled ACTH secretion. Subsequent histological analysis demonstrated that he had missed the paraventricular nuclei and lesioned the median eminence instead.

Hume's work led me to explore the literature on the subject of the control of ACTH secretion in detail, and I submitted a paper on the subject for the annual prize offered by the Boylston Medical Society at Harvard Medical School. Dr. Hume served as my faculty sponsor, and my paper won the prize for 1949. I also did a few experiments with Hume in the spring of 1949, but most of my time was devoted to completing my senior year in medical school.

I subsequently began my postdoctoral training as an intern and junior assistant resident at the Peter Bent Brigham Hospital. In 1951, I was drafted again and spent 18 months as a medical officer in Korea and Japan. This was largely a period of clinical activity, but there were three events of neuroendocrine interest. ACTH and corticoids were being used extensively by Thorn and Forsham to treat patients with various inflammatory and allergic diseases, and, as a house officer, I helped care for their patients. Impressed by the description of the Guillain-Barre syndrome as a "hive of the nervous system," I convinced J. S. Stillman to treat a patient with this syndrome with ACTH. The resulting improvement was great, and we published the case history in the *New England Journal of Medicine* (Stillman and Ganong, 1952). Some of my colleagues and students like to point out that my first published paper reports an experiment in which the *n* was one, but I must say in self-defense that ACTH has subsequently been proved to be of considerable benefit in the treatment of the Guillain-Barre syndrome and related neurological diseases.

Another neuroendocrine event stemmed from a project suggested by the late Samuel Levine during my residency. Dr. Levine, who was one of the most talented clinical investigators I have ever known, thought that patients with a short PR interval had a greater-than-normal incidence of paroxysmal tachycardia even if they did not have the prolonged QRS complexes of the Wolff-Parkinson-White syndrome. While Bernard Lown and I were doing the record-room research that proved him correct (Lown *et al.*, 1952), we noticed that patients with Cushing's syndrome had short PR intervals and patients with Addison's disease had long PR intervals. This chance observation suggested that adrenal glucocorticoids can accelerate transmission in the neural conducting system of the heart. The publication that resulted (Lown *et al.*, 1955) has been cited by Henkin and others as part of the early evidence that glucocorticoids can affect the function of peripheral nerves.

The third event of neuroendocrine interest occurred in Korea. My Army unit, the First Cavalry Division, was holding the line north of Seoul when we were hit by a major outbreak of hemorrhagic fever, a disease new to Western physicians. There is evidence that the Chinese troops opposite us were also struck by the disease, and there was even talk about mysterious and nefarious influences causing the outbreak. We thought that it was probably a rickettsial disease, although at the time we could not prove it (Ganong *et al.*, 1953). However, the disease turned out to be of considerable endocrine interest because a significant number of those surviving it developed pituitary infarction similar to that occurring in women with post-partum uterine hemorrhage. This served to focus attention once again on the vulnerability of the pituitary circulation and demonstrated that there was nothing special about the female in this regard; the male can also infarct his pituitary if he has an episode of hypotension in the presence of diffuse vascular disease.

I had planned to continue my training in internal medicine upon my return from Korea, but for a number of reasons I decided to join Dr. Hume and work as a postdoctoral fellow instead. One problem was that Hume was in the Department of Surgery at Harvard, and I wanted to be able to count my research time for board credit in internal medicine. This problem was solved by a joint appointment in Dr. Thorn's laboratory and the unique title of Fellow in Surgery and Medicine—probably one of the few times that title has ever been awarded.

As with most new ideas, there was considerable criticism of Hume's hypothesis of hypothalamic control of the secretion of ACTH and other anterior pituitary hormones when it first appeared. Some argued, for example, that it was absurd to think that the brain, with so few cell types, could secrete enough hormones to regulate all the hormones of the anterior pituitary. Others argued that the eosinophil count was an inaccurate index of adrenocorticoid secretion. Searching for better indices, I began to study compensatory adrenal hypertrophy. My first published report of laboratory research (Ganong, 1954) was a study of whether or not compensatory adrenal hypertrophy could be explained by the fact that unilaterally adrenalectomized animals had one-half as much adrenal tissue to react to a given amount of circulating ACTH. The experiment was an important preliminary if we were to use compensatory hypertrophy as an index of ACTH secretion in dogs with hypothalamic lesions. I found that compensatory hypertrophy could not be explained on the basis of less adrenal tissue for the same amount of ACTH, and I concluded that increased ACTH secretion must be involved. This work was done in Hume's laboratory and in full consultation with him, but, to help my career, he refused to put his name on the paper. Two other aspects of this paper are of interest. The first is that Hans Selye visited the laboratory when I was about to publish it, and I was immensely crestfallen when, after I had eagerly described my experiment to him, he told me that he had done the same experiment (with the same result) 10 years previously. This is why the paper was submitted and published in the Notes and Comments section rather than the regular portion of *Endocrinology*. The second point is that the results stood, as far as I know, until Mary Dallman, formerly a postdoctoral fellow in my laboratory and currently a member of the faculty in my department, challenged them (Dallman *et al.*, 1976). Dr. Dallman's challenge is still a matter of some debate, but I think it is fair to say that she at least demonstrated in 1976 that what I thought was simple in 1964 is really quite complex and is in need of additional study.

Hume and I were subsequently able to show that lesions of the median eminence abolished compensatory adrenal hypertrophy (Ganong and Hume, 1954). We also found that median eminence lesions had no effect on the compensatory adrenal atrophy produced by corticoids (Ganong and Hume, 1955). This suggests that the two responses are mediated in different ways.

The years with Hume were active and productive in other ways. We demonstrated with Cowie and Steenburg (Cowie *et al.*, 1954; Steenburg and Ganong, 1955) that corticoids were not "used up" at a more rapid rate during stress, as Sayers had hypothesized. With Donald Fredrickson, now director of the National Institutes of Health, we showed that the dog was like the rat in having its thyroid under neural control via hypothalamic control of TSH secretion (Ganong *et al.*, 1955). We developed stereotaxic techniques for use in dogs (Hume and Ganong, 1956), and, in an effort to come as close as possible to collecting pituitary effluent blood, we developed a technique for obtaining blood from the cavernous sinus (Ganong and Hume, 1956*a*). Finally, a study that started out as an effort to transplant the pituitary became instead a study of the effect of partial hypophysectomy when the transplants failed, and we analyzed the sequence in which peripheral endocrine function is compromised with loss of increasingly greater amounts of anterior pituitary tissue (Ganong and Hume, 1956*b*). I suppose the moral of this story is that almost every study is good for something, even if it is not the purpose for which it was originally intended. The study was of additional fundamental interest, however, because at the time there was some question about the effect of hypothalamic lesions being due to damage to the pituitary secondary to disruption of the portal vessels. By demonstrating that the sequence of loss with increasing pituitary disruption was gonads–thyroid–adrenals, whereas selective thyroid or adrenal inhibition could be produced by appropriate hypothalamic lesions, we were able to show that the effects of hypothalamic lesions were specific.

It was in 1955 that I finally decided to give up internal medicine and become a basic scientist. Leslie L. Bennett, whom I had met when he was on sabbatical leave in Dr. Thorn's laboratory in 1949, offered me a position as a beginning assistant professor of physiology at the University of California, Berkeley. After some debate with myself as well as with my wife and family, I turned down offers in the East and moved to Berkeley. I moved with the Department of Physiology from Berkeley to San Francisco in 1958, and have remained at the University of California ever since. Except possibly on payday, I have never regretted the decision to go into basic science. In the hands of a talented few, clinical research provides basic answers, but I need to have better control of the variables and pursue the answers to fundamental questions in experimental animals.

I suppose it is worth noting that I had never had any formal training in physiology beyond my freshman medical course when I accepted Dr. Bennett's offer, although I had been a volunteer teaching assistant in the student laboratories at Harvard. While I was working in Hume's laboratory, I asked Eugene Landis, chairman of the department at Harvard, if he thought I should get a Ph.D. in physiology in addition to my M.D. He advised against it. The lack of a Ph.D. has never been a problem from the professional point of view, but from a scholarly point of view I have regretted the decision not to work for the degree; I believe I would have more depth in areas in which I feel deficient and more range in my research if I had obtained the Ph.D.

At the University of California, I decided to set as my research goal the elucidation of the factors regulating aldosterone secretion. At that time, some argued that the pineal gland and diencephalon were involved in the regulatory process (Farrell, 1958), but an increasing number of questions were cropping up about this hypothesis. John Luetscher of Stanford Medical School had a method in operation at the time for measuring aldosterone, and we carried out a collaborative investigation of the effects of brain lesions on aldosterone secretion. We found that the median eminence was involved in the process via control of ACTH secretion (Ganong *et al.*, 1959), but we could find no evidence that the rest of the diencephalon and neighboring orbital frontal cortex were involved. These results led us to look outside the nervous system for other aldosterone control mechanisms.

It was at this point that I first began to work with Patrick Mulrow, who at that time was a postdoctoral fellow in Luetscher's laboratory. We were both interested in injecting aldosterone in the renal artery to develop a better bioassay for this steroid, and we decided to pool our resources. The results (Ganong and Mulrow, 1958) were disappointing from the point of view of a workable bioassay, since even intrarenal aldosterone took 20–60 min to affect sodium excretion. However, the occurrence of the delay, as reported by us and by Barger *et al.* (1958), led Edelman and others to the discovery that aldosterone acts by way of an effect on nuclear DNA and the formation of mRNA.

Mulrow and I also studied the effect of hemorrhage on aldosterone secretion in normal and hypophysectomized dogs (Mulrow and Ganong, 1961) and found clear-cut evidence that hemorrhage stimulated the secretion of an aldosterone-stimulating factor other than ACTH. We therefore set out to determine the site of origin and nature of this factor.

We turned our attention to the kidneys at this point, in part because of the reports of Hartroft and Hartroft (1955) and others that there was a parallelism between the size of the zona glomerulosa and the granularity of the juxtaglomerular cells in the kidney. The reports of Laragh *et al.* (1960) and Genest *et al.* (1961) were not available when we started our experiments, but I remember I had a strong hunch that somehow nephrectomy was the next experiment to do. What we found, of course, was that the kidney was the source of the aldosterone-stimulating factor in hypophysectomized dogs.

We had our first data on the effect of nephrectomy at the time of the International Endocrine Congress in Copenhagen in 1960, but wanted to collect more to be sure of the observation before making it public. I remember very well that my wife and I were sitting in a cafe on the Champs Elysée in Paris before the Congress testing the hypothesis that if you sit there long enough, someone you know will walk by. The one who walked by was James Davis, and, in the course of our conversation, he told me he had found that nephrectomy abolished the aldosterone response to hemorrhage in hypohysectomized dogs. I was crushed to learn that our discovery had independently been duplicated by someone else, but I consoled myself with various aphorisms about "ideas whose time has come."

Mulrow and I subsequently published our data in *Nature* (Ganong and Mulrow, 1961), while Davis and associates published their data in the *Journal of Clinical Investigation* (Davis *et al.*, 1961). Both laboratories then proceeded to demonstrate that the aldosterone-stimulating factor in kidney extract was renin, acting via generation of angiotensin II. By this time, Mulrow had moved to Yale, and we decided to continue our collaboration on a transcontinental basis. Animal experiments were regularly carried out in San Francisco, and the samples were analyzed in New Haven. For several years, we did a great deal to subsidize the dry ice, air freight, and long-distance telephone industries, but the collaboration worked well, in large part due to the good nature and perseverence of Dr. Mulrow. Incidentally, it was at this time that I really learned the difference between renin and rennin. To explore the possibility that renin was the renal factor regulating aldosterone, we looked through a biological supply catalogue and found rennin. We actually purchased some of this material and injected it into several dogs before realizing that we were injecting the milk-coagulating enzyme from calf stomach rather than the renal hormone. Ever since that experience, I have been careful to pronounce renin with a long *e* and spell it with one *n*.

It was subsequently shown that the increase in aldosterone secretion in sodium depletion and a number of other conditions was associated with an increase in plasma renin activity. However, correlation does not prove cause and effect. To provide more direct proof that the rise in renin caused the increase in aldosterone secretion, my graduate student Thomas Lee and I prepared antibodies to renin and injected these antibodies into dogs fed a low-sodium diet. Edward Biglieri, a long-time friend and associate, helped us measure aldosterone secretion by isotope dilution in the injected dogs. We found that the antirenin antibodies decreased aldosterone secretion (Lee *et al.*, 1965). However, we were using an impure antigen to raise our antibodies. It was therefore reassuring to see the same drop in aldosterone secretion produced some years later by injection of saralasin, the competitive antagonist to angiotensin II (Johnson and Davis, 1973).

We next turned our attention to the control of renin secretion and were able to show that the sympathetic nervous system played an important role in the process. We also demonstrated that the excitatory effect of the sympathetic nervous system was mediated via β-adrenergic receptors, and was blocked by propranolol and potentiated by α-adrenergic blocking drugs. Much of the research was carried out by three energetic and productive postdoctoral fellows, Tania Assaykeen, Kensaku Otsuka, and Stanley Passo. Dr. Assaykeen first presented the data on β-adrenergic blockade at the International Congress of Nephrology in 1969 (Assaykeen *et al.*, 1969). The same year, Winer and his associates in Kansas City also reported that β-adrenergic blocking drugs decreased renin secretion (Winer *et al.*, 1969). However, they reported that, in addition, renin secretion was inhibited by α-adrenergic blocking drugs, an observation we have never been able to confirm. The work demonstrating neural regulation of renin secretion was important not only because of its theoretical and clinical implication but also because it proved that those who thought in 1955 that the regulation of aldosterone secretion was a neuroendocrine problem were right, even if for the wrong reasons.

More recent research in my laboratory conducted by my graduate students, postdoctoral fellows, and other associates makes it increasingly clear that the juxtaglomerular cells are indeed endocrine cells that receive a direct excitatory β-adrenergically mediated input via the sympathetic nervous system. Work by others on the sympathetic and cholinergic innervation of the pancreatic islets has paralleled ours, and outstanding work has been done by Axelrod and associates on the mechanism by which norepinephrine secreted from sympathetic neurons acts via β-adrenergic receptors to increase the synthesis of melatonin in the pineal gland. It seems to me that this type of work is important not only because it adds fundamental information about molecular mechanisms that operate in secretory cells but also because it emphasizes the ubiquity of neural control in the endocrine system. Some think of neuroendocrinology as that discipline concerned with the regulation of pituitary secretion by the hypothalamus, and others limit it even further to the regulation of anterior pituitary secretion. Actually, there are few, if any, endocrine glands in the body that are not affected by the nervous system. I have made this point in somewhat more detail, along with documentation, in a minireview in *Life Sciences* (Ganong, 1974). In addition, hormones act on the nervous system in multiple, complex, and important ways. Thus the nervous and endocrine systems act on each other, and neuroendocrine interactions are of fundamental importance throughout life to the integrative function of the organism.

Another current aspect of our research on renin secretion is the role of brain amines in the process. The source of my interest in the relation of brain amines to endocrine function was actually the report by Tullner and Hertz (1963) that the monoamine oxidase inhibitor α-ethyltryptamine inhibits stress-induced ACTH secretion. Barbara Zipf, a medical student research fellow, confirmed this observation in my laboratory in the summer of 1963, and we then embarked on a search for the mechanism by which the inhibition was produced. On the basis of work carried out by Leola Lorenzen, another medical student research fellow, and several other associates, it appeared that the property of α-ethyltryptamine that best correlated with inhibition of ACTH was its ability to elevate blood pressure. I therefore advanced the hypothesis that increased blood pressure stimulated the arterial baroreceptors, which in turn activated neural pathways that inhibited ACTH secretion. This was reasonable, since a drop in blood pressure in the carotid sinus stimulates ACTH secretion (Biglieri and Ganong, 1961), and increased baroreceptor discharge inhibits vasopressin secretion (Share, 1969). However, my graduate student Glen Van Loon doubted this hypothesis and he proceeded to do two experiments that made it untenable. I suppose this proves that one should listen to graduate students even when they attack one's pet theories. It turned out that the blood pressure rise was a red herring and that, instead, as Julius Axelrod suggested when he heard our data, catecholamine-releasing drugs act in the brain to inhibit ACTH secretion. My associates and I have now collected a large amount of evidence indicating that, at least in the dog, adrenergic neurons in the hypothalamus inhibit ACTH secretion. The site of this action is above the blood–brain barrier, probably in the more dorsal portions of the hypothalamus, and the receptors involved have the properties of α-adrenergic receptors (Ganong *et al.*, 1976).

I have also had a continuing interest in reproductive endocrinology, and in this area I had the good fortune to work with M. T. Clegg. It was Clegg who, among other things, goaded me into seeing whether light penetrated into the brain of living mammals. I had been telling students that, although light acted directly on the hypothalamus in birds, it certainly could not penetrate the hair, skin, and skull of mammals. To prove this was true, we implanted an electronic light sensor in the hypothalamus of sheep and smaller mammals, and, to my surprise, light penetrated to the infundibulum with ease. I do not consider the resulting paper (Ganong *et al.*, 1963) one of my major contributions to science, but it has certainly attracted more attention, at least among pseudoscientists, than anything else I have written.

Finally, I remain fascinated by the problem of the mechanism responsible for the onset of puberty, and with several graduate students and postdoctoral fellows have explored various aspects of this problem (Gellert and Ganong, 1960; Gellert *et al.*, 1964; Kragt and Ganong, 1968; Bloch and Ganong, 1971; Rabii and Ganong, 1976). Despite excellent research in a number of laboratories and the development of several ingenious theories, I do not think the problem has been solved; indeed, I think the really exciting results are yet to come. However, the time has regrettably come when neuroendocrinology has grown to the point that specialization is essential, and I will have to leave the answers to the questions about puberty to those who specialize in the secretion of gonadotropins rather than ACTH and renin.

REFERENCES

Assaykeen, T. A., Goldfien, A., Otsuka, K., and Ganong, W. F. (1969). Effect of α and β blocking agents on the increase in plasma renin activity provoked in dogs by insulin-induced hypoglycemia. In *Abstracts of the Fourth International Congress of Nephrology*, p. 37.

Barger, A. C., Berlin, R. D., and Tulenko, J. F. (1958). Infusion of aldosterone, 9-α-fluorohydrocortisone and anti-diuretic hormone into the renal artery of normal and adrenalectomized dogs: Effect on electrolyte and water excretion. *Endocrinology* **62**:804.

Biglieri, E. G., and Ganong, W. F. (1961). Effect of hypophysectomy on adrenocortical response to bilateral carotid constriction. *Proc. Soc. Exp. Biol. Med.* **106**:806.

Bloch, G. J., and Ganong, W. F. (1971). Lesions of the brain and the onset of puberty in the female rat. *Endocrinology* **89**:898.

Cowie, A. T., Ganong, W. F., and Hume, D. M. (1954). The eosinopenic response to graded doses of hydrocortisone in the adrenalectomized dog with and without surgical trauma. *Endocrinology* **55**:745.

Dallman, M. F., Engeland, W. C., and Shinsako, J. (1976). Compensatory adrenal growth: A neurally mediated reflex. *Am. J. Physiol.* **231**:408.

Davis, J. O., Carpenter, C. C. J., Ayers, C. R., Holman, J., and Bohn, R. C. (1961). Evidence for secretion of aldosterone-stimulating hormone by the kidney. *J. Clin. Invest.* **40**:684.

Farrell, G. (1958). Regulation of aldosterone secretion. *Physiol. Rev.* **38**:709.

Ganong, W. F. (1954). The effect of ACTH on adrenal size in hypophysectomized rats after removal of one adrenal. *Endocrinology* **55**:117.

Ganong, W. F. (1974). The role of catecholamines and acetylcholine in the regulation of endocrine function. *Life Sci.* **15**:1401.

Ganong, W. F., and Hume, D. M. (1954). Absence of stress-induced and "compensatory" adrenal hypertrophy in dogs with hypothalamic lesions. *Endocrinology* **55**:474.

Ganong, W. F., and Hume, D. M. (1955). Effect of hypothalamic lesions on steroid-induced atrophy of adrenal cortex in dog. *Proc. Soc. Exp. Biol. Med.* **88**:528.

Ganong, W. F., and Hume, D. M. (1956*a*). Concentration of ACTH in cavernous sinus and peripheral arterial blood in the dog. *Proc. Soc. Exp. Biol. Med.* **92**:621.

Ganong, W. F., and Hume, D. M. (1956*b*). The effect of graded hypophysectomy on thyroid, gonadal, and adrenocortical function in the dog. *Endocrinology* **59**:292.

Ganong, W. F., and Mulrow, P. J. (1958). Rate of change in sodium and potassium excretion after injection of aldosterone into the aorta and renal artery of the dog. *Am. J. Physiol.* **195**:337.

Ganong, W. F., and Mulrow, P. J. (1961). Evidence of secretion of an aldosterone-stimulating substance by the kidney. *Nature* (*London*) **190**:1115.

Ganong, W. F., Zucker, E., Clawson, C. K., Voss, E. C., Klotzbach, M. L., and Platt, K. A. (1953). The early field diagnosis of epidemic hemorrhagic fever. *Ann. Int. Med.* **38**:61.

Ganong, W. F., Fredrickson, D. S., and Hume, D. M. (1955). The effect of hypothalamic lesions on thyroid function in the dog. *Endocrinology* **57**:355.

Ganong, W. F., Lieberman, A. H., Daily, W. J. R., Yuen, V. S., Mulrow, P. J., Luetscher, J. A., Jr., and Bailey, R. E. (1959). Aldosterone secretion in dogs with hypothalamic lesions. *Endocrinology* **65**:18.

Ganong, W. F., Shepherd, M. D., Wall, J. R., Van Brunt, E. E., and Clegg, M. T. (1963). Penetration of light into the brain of mammals. *Endocrinology* **72**:962.

Ganong, W. F., Kramer, N., Salmon, J., Reid, I. A., Lovinger, R., Scapagnini, U., Boryczka, A. T., and Shackelford, R. (1976). Pharmacological evidence for inhibition of ACTH secretion by a central adrenergic system in the dog. *Neuroscience* **1**:167.

Gellert, R. J., and Ganong, W. F. (1960). Precocious puberty in rats with hypothalamic lesions. *Acta Endocrinol.* **33**:569.

Gellert, R. J., Bass, E., Jacobs, C., Smith, R., and Ganong, W. F. (1964). Precocious vaginal opening and cornification in rats following injections of extracts of steer median eminence and pars tuberalis. *Endocrinology* **75**:861.

Genest, J., Bairon, T., Kiow, E., Nowaczynski, W., Boucher, R., and Chretien, M. (1961). Studies of pathogenesis of human hypertension: The adrenal cortex and renal pressor mechanism. *Ann. Int. Med.* **55**:12.

de Groot, J., and Harris, G. W. (1950). Hypothalamic control of the secretion of adrenocorticotropic hormone. *J. Physiol.* **111**:335.

Hartroft, P. M., and Hartroft, W. S. (1955). Studies on renal juxtaglomerular cells. II. Correlation of the degree of granulation of juxtaglomerular cells with width of the zona glomerulosa of the adrenal cortex. *J. Exp. Med.* **102**:205.

Hume, D. M. (1949). The role of the hypothalamus in the pituitary-adrenal cortical response to stress. *J. Clin. Invest.* **28**:790.

Hume, D. M., and Ganong, W. F. (1956). A method for accurate placement of electrodes in the hypothalamus of the dog. *Electroencephalogr. Clin. Neurophysiol.* **8**:136.

Hume, D. M., and Wittenstein, D. J. (1950). The relationship of the hypothalamus to pituitary-adrenocortical function. In Mote, J. R. (ed.), *Proceedings of the First Clinical ACTH Conference*, Blakiston, Philadelphia, p. 134.

Johnson, J. A., and Davis, J. O. (1973). Angiotensin II: Important role in the maintenance of arterial blood pressure. *Science* **179**:906.

Kragt, C. L., and Ganong, W. F. (1968). Pituitary FSH content in female rats at various ages. *Endocrinology* **82**:1241.

Laragh, J. H., Angers, M., Kelly, W. G., and Lieberman, S. (1960). Hypotensive agents and pressor substances. *J. Am. Med. Assoc.* **174**:234.

Lee, T. C., Biglieri, E. G., Van Brunt, E. E., and Ganong, W. F. (1965). Inhibition of aldosterone secretion by passive transfer of antirenin antibodies to dogs on a low sodium diet. *Proc. Soc. Exp. Biol. Med.* **119**:315.

Lown, B., Ganong, W. F., and Levine, S. A. (1952). The syndrome of short P-R interval, normal QRS complex and paroxysmal rapid heart action. *Circulation* **5**:693.

Lown, B., Arons, W. L., Ganong, W. F., Vazifdar, J. P., and Levine, S. A. (1955). Adrenal steroids and auriculoventricular conduction. *Am. Heart J.* **50**:760.

Mulrow, P. J., and Ganong, W. F. (1961). The effect of hemorrhage upon aldosterone secretion in normal and hypophysectomized dogs. *J. Clin. Invest.* **40**:579.

Rabii, J., and Ganong, W. F. (1976). Responses of plasma "estradiol" and plasma LH to ovariectomy, ovariectomy plus adrenalectomy, and estrogen injection at various ages. *Neuroendocrinology* **20**:270.

Saffran, M., Schally, A. V., and Benfey, B. G. (1955). Stimulation of the release of corticotropin from the adenohypophysis by a neurohypophysial factor. *Endocrinology* **57**:439.

Share, L. (1969). Extracellular fluid volume and vasopressin secretion. In Ganong, W. F., and Martini, L. (eds.), *Frontiers in Neuroendocrinology, 1969*, Oxford University Press, New York, p. 183.

Steenburg, R. W., and Ganong, W. F. (1955). Observations on the influence of extra-adrenal factors on circulating 17-hydroxycorticoids in the surgically stressed, adrenalectomized animal. *Surgery* **38**:92.

Stillman, J. S., and Ganong, W. F. (1952). The Guillain-Barre syndrome: Report of a case treated with ACTH and cortisone. *N. Engl. J. Med.* **246**:293.

Tullner, W. W., and Hertz, R. (1963). Suppression of endogenous ACTH by 3-(2-aminobutyl)-indole (Monase) in the dog. In *Endocrine Society Abstracts*, p. 19.

Winer, N., Chokshi, D. S., Yoon, M. S., and Freeman, A. D. (1969). Adrenergic receptor mediation of renin secretion. *J. Clin. Endocrinol.* **29**:1168.

13

Monte Arnold Greer

Monte Arnold Greer was born in Portland, Oregon, in 1922. He attended Oregon State College and received the bachelor's degree from Stanford University in 1944 and the M.D. from Stanford in 1947. After internship in San Francisco he served as a research fellow with Dr. E. B. Astwood in Boston until 1951, when he became associated with the National Institutes of Health for 4 years. There he worked in Dr. Roy Hertz's unit at the National Cancer Institute. He became director of the Radioisotope Unit at the VA Hospital in Long Beach, California, in 1955, but after 1 year moved to Portland, Oregon, where he became head of the division of Endocrinology of the Department of Medicine at the University of Oregon Medical School. He rose to the rank of professor of medicine by 1962. He has remained there ever since and has done pioneering work in neuroendocrinology, particularly as related to control of TSH and ACTH secretion. He received the Ciba Award of the Endocrine Society in recognition of his research accomplishments in 1958. He served on the editorial boards of *Endocrinology* from 1960 to 1973 and *Neuroendocrinology* from 1965 to 1975. He was co-editor of the volume on the thyroid for the *Handbook of Physiology* and is a member of the Endocrinology Study Section of the National Institutes of Health. He is currently vice-president of the Endocrine Society. He has published widely in the endocrine literature and also has many chapters in books. He has been an invited speaker at numerous symposia and visiting professor at schools throughout the world.

13

Why I Am Still Waiting for a Free Trip to Stockholm

MONTE ARNOLD GREER

The longer I live, the more convinced I am that one's course in life is determined 99% by chance. This is especially true of my own. Although there was no family tradition of medicine or science, the earliest childhood ambitions that I can remember were both to be a physician and to do research. With an unforseeable combination of fortuitous events, I have been able to reach both goals.

I had the usual child's chemistry set in the basement and, being an only child, often spent rainy afternoons (and it rains a lot in Oregon) trying to amuse myself there. I had a fascination with making, or attempting to make, gunpowder and employing it in various pyrotechnic displays. I don't know what instigated this obsession, but perhaps I prematurely realized that the world was against me and I was mapping techniques for revenge. Looking back, I am amazed that I survived this period. I was able to find the necessary instructions for making small bombs, roman candles, etc., from the public library and purchased the ingredients from a local pharmacy. However, I could not find an adequate protocol for mixing the components. I decided that the most satisfactory technique would be to grind the materials vigorously with a mortar and pestle until they formed a homogeneous powder. The first important element of chance must have come during these early scientific experiments, because I neither blew myself or the house up nor even lost a hand or my eyesight.

This phase of my career was forcibly terminated when I was 14. My parents were sitting around the kitchen table with guests on Christmas day. To help celebrate the occasion, I excitedly ignited in an ashtray a small flare which I had just made; it was my first attempt with this technique and was a spectacular triumph. A flash of brilliant red light flew into the air and

MONTE ARNOLD GREER • Division of Endocrinology, Department of Medicine, University of Oregon Medical School, Portland, Oregon 97201.

settled on the front of my father's new white Christmas sweater, increasing its sartorial splendor by allowing his chest to be seen through a unique, brown-edged circular opening.

After graduating from high school, I entered Oregon State University for premedical training. Two young members of the zoology faculty, Dr. Ernst Dornfeld and Dr. Clifford Grobstein, had a great influence in molding the direction my career would take. Dornfeld was later to become head of the Zoology Department at Oregon State, and Grobstein head of the Biology Department at Stanford and dean during its formative years of the School of Medicine at the University of California at San Diego. Both were superb, enthusiastic teachers, and we became good friends. Cliff Grobstein had just obtained his Ph.D. with Bennett Allen at UCLA and was interested in hormonally induced morphological changes in swordtail fish. He therefore decided to offer the first elective course in endocrinology at Oregon State. I knew nothing about the subject and wasn't even sure what the word meant, but it sounded like an esoteric new area so I enrolled in the course. It was an excellent introduction to the discipline and raised many unanswerable questions.

The next term, I did my first research project in biology with Cliff. Having attended a seminar in which I learned that limb regeneration will occur in tadpoles but not in adult frogs, and having learned from the endocrinology course that thyroxine causes amphibian metamorphosis, I collected two dozen bullfrog tadpoles from the local ponds, cut off one leg or a wedge of tail, and put half the animals in a tank containing thyroxine while the other half had only normal pond water. Unfortunately, the concentration of thyroxine chosen was too strong. Metamorphosis accelerated so rapidly that the tail disappeared, instead of regenerating, before the legs became large enough to allow the animals to swim. The hapless creatures lay like thalidomide-treated babies at the bottom of their tanks, four tiny stumps protruding from their swollen bodies, until they drowned. Their gills resorbed before they were able to reach the top of the tank to breathe air.

By this time, we were well into World War II. I decided to enter Stanford Medical School, primarily because Peggy Johnson, whom I had met the previous summer and would marry later that year, was living in Berkeley. During medical school, I was uncertain what specialty I wished eventually to enter. No active endocrinologist was on the Stanford faculty to keep alive the spark ignited at Oregon State. C. H. (Tom) Sawyer was one of the instructors in my anatomy course. However, although he is now recognized as one of the world's foremost neuroendocrinologists, at that time he had not yet begun to work in endocrinology and was interested primarily in myoneural junctions. Since I supposedly had had fragilitis osseum and one extremity or another was in a cast much of my early life, I was leaning toward orthopedic surgery. During my senior year, I applied for, and was granted, a surgical internship at Stanford. But about a month before the internship was to start, I began to wonder if I could survive standing in an operating room several hours a day. I obtained a reluctant release from the Department of Surgery and instead began a rotating internship at the San Francisco General Hospital in 1946. During my internship, I became convinced that I would prefer internal medicine to surgery.

Because of the logistic needs of the defense forces, medical schools were on an accelerated program during the war. To prevent personal financial disaster, I and almost all my classmates were put in uniform and our school expenses were paid by the government. I therefore expected to spend the required 2 years repaying my military obligations at the end of my internship. However, the climate in San Francisco (where Stanford Medical School was located at that time) was too foggy for my constitution. I developed severe asthmatic attacks which would sometimes necessitate overnight hospitalization. At the mustering-out center, after my case had been mulled over for several days, it was eventually decided not to reward me with a commission but instead to discharge me completely. My ecstasy was tempered by the realization that I now needed to find a position for the next year. I applied for, but was unable to obtain, a first-year residency in internal medicine at the San Francisco General, primarily because the war had just ended and many physicians were returning from the service to training positions previously promised to them.

I had become interested during my internship in studying the effect of ACTH on blood platelets. This was stimulated by an article I had read by Dougherty and White (1944) showing that ACTH would cause profound changes in blood lymphocyte concentration. I had had an unfortunate young patient with thrombocytopenic purpura who died in spite of all the maneuvers attempted to save her. In reading about the disease, I noticed that some authors described an increase in the platelet count following acute stress. Since ACTH supposedly increased during stress, I thought that ACTH itself might be responsible. I was able to con two dozen bottles of "ACTH" solution from one of the pharmaceutical companies and spent all my spare time doing platelet counts on volunteer patients and student nurses (preferably the latter) before and after injecting them intragluteally with ACTH. Only later did I learn that the "ACTH" provided me had the equivalent potency of tap water.

Because of the interest I had developed in the effect of ACTH on blood platelets and my previous interest in hormones stimulated by the course with Cliff Grobstein, and because I was not anxious to take a second-class residency, I decided to see if I could spend a year doing research in endocrinology. Not knowing which laboratories might be suitable, I visited one of my former physiology professors at Stanford to ask him for suggestions. He kindly gave me the names of several individuals to whom he felt it would be appropriate to write. At the end of the interview, I thanked him profoundly for his help. He said he was glad to give me advice, but crushingly added, "Why do you want to spend a year in endocrinology research? Do you think it will help you get into medical school?"

I wrote to the various people who had been suggested. Several replied with offers of positions in their laboratories. However, the opportunity for a funded fellowship was even bleaker then than it is at present. Of those who responded positively, only one offered to pay me a stipend. This was Edwin B. (Ted) Astwood, who had been recommended as an up-and-coming young man in Boston. Although his name had not been mentioned in my endocrinology course at Oregon State, I was vaguely aware that thiouracil was being explored as a substitute for surgery in treating hyperthyroidism and that Astwood was the primary instigator of this therapy. Thus, because of the excitement of going to work with Astwood, but primarily because it was the only opportunity that offered a salary, we drove off to Boston in the summer of 1947, immediately following the completion of my internship.

This was another critical spin of the wheel. Astwood was one of the most stimulating, friendly, and helpful individuals I have ever known. Sally Astwood ran the household in a charming and witty fashion and tried to ensure that Peggy and I were introduced to the strange environs of New England with a minimum of trauma. Since Ted was originally from Bermuda and Sally from South Carolina, they had some appreciation of the pangs of culture shock.

There was a relatively small group then working with Astwood at the New England Medical Center. Bill VanderLaan had just finished his fellowship and had joined the clinical staff. Malcolm Stanley was in his second year of fellowship. Robert Reiss and I came simultaneously as the "new boys." The laboratory at that time was primarily oriented toward thyroid research. Radioiodine had just become available for investigative use through the courtesy of the Atomic Energy Commission, and Astwood was busily learning about radioactivity and pioneering in techniques to utilize the isotope in studying thyroid physiology and pharmacology.

During my initial interview with Ted, he asked me what I would like to work on. I had had no interest in doing anything connected with the thyroid following my failure in solving tadpole regeneration; however, I was very eager to continue my studies on the effect of ACTH on the blood platelets. Astwood obviously did not share my enthusiasm, but he did not directly discourage me from studying the problem. His policy was to let each fellow pursue his own interests with a minimum amount of interference. Guidance was freely given, but usually only if specifically requested. One might even spend a year in the laboratory doing essentially nothing without hearing angry screams from the chief. On the other hand, such a performance was not rewarded with an offer to stay on for another year.

In the case of my interest in the relation of ACTH to platelets, Ted informed me that I could go ahead with this, but I had to realize that the ACTH which had been provided me was inactive. If I was really intrigued with the problem, I would have to make my own ACTH. This was a horrifying prospect since my experience with organic chemistry and biochemistry was close to nil. However, I looked up what was available on ACTH preparation and ran across a newly described method which used a Sohxlet extraction apparatus. Astwood had a large quantity of dried bovine pituitary powder he had obtained from one of the pharmaceutical companies which I was able to use as starting material. Unfortunately, there was no simple assay procedure for ACTH at that time. I happened to read of some studies just reported in abstract form by a young man named George Sayers, working in C. N. H. Long's laboratory at Yale, who was utilizing as an ACTH assay the fall in adrenal ascorbic acid induced by the hormone in hypophysectomized rats. This sounded like a vastly improved technique, but there weren't any hypophysectomized rats commercially available at that time. In order to use the assay, I would have to learn how to hypophysectomize rats myself.

This proved to be a rather formidable task. Bill VanderLaan had mastered hypophysectomies the year before; otherwise, no one was really familiar with the technique. Bill was too busy with his newly acquired staff duties to be of much help, so I had to learn the operation myself by trial and error after reading through some of P. E. Smith's old papers. Eventually, I was getting a 25–50% success rate with reasonable survival of the animals, and it looked like we might be in business. However, at that time, we learned that Armour had begun releasing an active ACTH preparation for both laboratory and clinical trial. Ted suggested that he might obtain some of the Armour ACTH and this could be used to get an answer to my question more rapidly. The gift from Armour ended my abortive attempt to purify ACTH. To my sorrow, even Armour ACTH had no effect on blood platelet concentration (Greer and Brown, 1948), but valuable side products of the experience were that I learned what a Sohxlet apparatus was and how to hypophysectomize rats.

An interesting additional offshoot of my negative experiment was that a young man, Henry Foster, who had recently graduated from veterinary school, was interested in establishing a commercial supply of laboratory rats for the Boston area and was planning to start a breeding colony. He correctly surmised that many investigators might desire animals that could be supplied with various endocrine organ ablations so that the scientists could be spared the necessity of arduously learning the surgery themselves. He had heard that I had "mastered" hypophysectomies, so came to the laboratory to learn it from me. He became a spectacularly successful entrepreneur. His company is now internationally respected as the Charles River Breeding Laboratories.

The following year, Maury Raben and Dick Payne joined the laboratory. By this time, Ted Astwood had persuaded me that a golden future wouldn't be found in curing thrombocytopenia with ACTH, so I had joined the group attack on the mysteries of the thyroid gland.

During my second fellowship year, Payne and I became interested in the effects of obesity on the reproductive cycle. We were impressed that many obese young women developed amenorrhea or oligomenorrhea and thought that it would be interesting to make rats obese to see what effect it had on the estrus cycle. The first step was to learn how to make and interpret vaginal smears; the next step was to learn how to make fat rats. We started out by buying a hand-operated "gun" used in service stations to grease cars, making up a high-calorie liquid formula, and force-feeding the animals slowly increasing amounts 3 times a day through a stomach tube attached to the gun. Although the patient mortality was very high, we finally became adept at this technique. However, although we were able to bring some of the animals up to approximately 1 kg in weight, the experiment took so long and was so arduous that we wanted to try a simpler fattening method. Ted told us that a better technique might be to make hypothalamic lesions to produce obesity. He didn't know anything about making such lesions himself, but he knew that John Brobeck at Yale was familiar with stereotaxic machines. He made arrangements for Dick and me to go to New Haven to see the machine in action. Brobeck courteously demonstrated the apparatus, and gave us references to a small Krieg machine especially designed for rats. But we were warned that no such instrument was commercially available. If we wanted one, we would have to build it ourselves. The demonstration of the wonderful world of stereotaxy excited us. However, neither of us felt possessed of sufficient mechanical ability to build a machine, so we let the project drop.

By this time, I had realized that rather than going back into a residency in internal medicine and heading for practice, endocrine research had proven sufficiently exciting in itself that I wanted to make this my career. Having had only a rotating internship before joining Astwood's laboratory, I realized that I needed further training in internal medicine if I were to feel comfortable in also caring for patients. At the end of my second year of fellowship, I therefore began a residency in internal medicine with Chester

Keefer at the Massachusetts Memorial Hospitals. Keefer's program for residents included an assignment to spend approximately half of one's time working on research projects with a staff member to whom each resident was assigned for the year. Since I had spent 2 years in endocrinology, I was assigned to work with Charles H. Burnett, who was really more interested in kidney physiology than in endocrinology. He had trained with Fuller Albright and was actively engaged in studying the effect of adrenal hormones on kidney function when I began my service.

I knew essentially nothing about kidney physiology and felt rather lost in the group laboratory discussions. Although I read through Homer Smith's book and tried diligently to study the excretory process, my basic purpose in obtaining the residency was to learn general internal medicine. I didn't feel that it was satisfying my needs to spend most of my time learning minutiae in another subspecialty. Therefore, when Joe Ross, who was the chief of Hematology (and the only staff member with previous experience with radioactivity), suggested that I help him establish a radioiodine laboratory for studying thyroid function, this seemed like a more reasonable direction to go. I could draw on my previous experience and look like a hero with a minimum of effort. This plan was only partially successful, because a major share of the responsibility for setting up the thyroid radioisotope laboratory was given to me without first clearing the switch in my assignment with Burnett. Instead of my subspecialty obligations being decreased, they were actually doubled. This made it even more difficult to get in the time on the general medicine wards that I desired.

About half way through the year, I decided that being a resident wasn't giving me much more general medicine than being a fellow, but it was more confusing and less productive. Therefore, I decided to go back to spend more time in Astwood's laboratory. The availability of fellowship funds had not improved appreciably, but there was some opportunity of obtaining direct postdoctoral support from a few organizations. Astwood suggested that I apply for a fellowship from the National Cancer Institute. At that time, money for endocrinological research was supported primarily by cancer groups because hormones were considered to have an important role in regulating both normal and abnormal growth. The question was what project I should write up to try to obtain the funds.

I happened to read at that time an article in *Physiological Reviews* by some person previously unknown to me named Geoffrey Harris. He had marshalled evidence in favor of an important role of the central nervous system and the hypothalamus in regulating anterior pituitary function neurohumorally through the hypophyseal portal system. This seemed highly speculative, but it struck a spark of inspiration for an idea I might use as a winning ploy to secure my prospective fellowship funds. Through my previous experience, I had learned a fair amount about thyroid physiology and how to do vaginal smears, and had at least had a first-hand demonstration of a stereotaxic machine. I therefore submitted a proposal to study the effect of hypothalamic lesions on TSH and gonadotropin secretion in the rat.

The lag between application for a fellowship and notification of the result must have been shorter in those days than it is now. As I recall, I learned within 3 months that my fellowship application had been successful. Now the problem was how to obtain a stereotaxic apparatus. Ted Astwood had a well-equipped workshop in his basement. I suggested that he might allow me to use it to make a machine, following the Krieg article with which Brobeck had previously acquainted me. Ted readily acquiesced. However, since I was still fully occupied with my triple-threat residency, and since he was a much better craftsman than I anyway, he constructed the machine largely by himself.

During this time, H. D. Purves had come from New Zealand to spend several sabbatical months working with Ted. He was quite intrigued with the stereotaxic machine, especially since he was planning to study the influence of the hypothalamus on endocrine function himself. When Ted saw his keen interest in our creation, he graciously suggested, "Why don't you take it back to Dunedin with you?" Not wishing to offend Astwood's feelings, Dick Purves politely replied, "Well, if you insist." Therefore, when I finished my residency and went back to start my third fellowship year, there was an unavoidable delay until I could build another stereotaxic instrument.

Finally, I was able to finish the construction of the apparatus and begin getting first-hand experience using the machine. There were a number of problems to overcome because I was unable to locate anyone in Boston who had ever used one. There were difficulties in locating the proper wires to use for electrodes and learning how to insulate them with enamel. Not the least of the obstacles was the fact that there was neither equipment nor technician for histology in the laboratory. In order to ascertain the location of the lesions, I had to pay a technician in the general pathology laboratory to section the brains, if and when she had the time. Since she had never made serial or brain sections previously, the results were somewhat less than ideal.

In spite of all the stumbling blocks, I was finally able to get everything together. The initial experiments were designed to get back to the question that Dick Payne and I had tried to attack with the force-feeding experiments earlier: what is the effect of obesity on the reproductive cycle? Dick was no longer interested in the problem. He was fully engaged in a project with Astwood and Raben in purifying ACTH on a large scale. The hormone had by this time been shown to be extraordinarily effective in treating rheumatoid arthritis and other diseases (but *not* thrombocytopenia).

Because we had originally learned the principles of stereotaxy in order to obtain fat rats more easily than through force-feeding and because it was known that lesions in the ventromedial nucleus produced obesity, I aimed at this nucleus first. I was delighted to find that not only could I obtain a high proportion of obese rats but also they showed significant changes in their reproductive cycles. A large number developed persistent estrus or diestrus. However, when I tried to correlate the obesity with the changes in the reproductive cycle, there seemed to be no relation. Some obese rats had normal cycles and some had persistent estrus; the same was true for rats which did not gain weight.

In reading about various forms of persistent estrus, I found the reports of John Everett relating his experience with a spontaneous persistent estrus that occurred in a strain of rats in his laboratory. Ovulation could be induced in these animals by injection of a small dose of progesterone which was insufficient to have any direct effect on either the uterus or vagina. If repeated daily injections of these small doses of progesterone were made, the animals would resume normal estrous cycles. I considered that a spontaneous hypothalamic dysfunction might have occurred in his animals. It therefore seemed of interest to try progesterone injections in my rats with persistent estrus following hypothalamic lesions to see if normal cycles could be induced. They could (Greer, 1953). Large numbers of corpora lutea regularly appeared in the rats injected daily with progesterone, while only follicles were seen if no progesterone was given. Surprisingly, normal estrous cycles persisted in about half the animals once the injection of progesterone was stopped. Unfortunately, the lesions abolished mating behavior, even in the animals with normal cycles during progesterone injection, so I could not determine if pregnancies could be successfully maintained. However, the animals did become pseudopregnant and developed deciduomata with uterine trauma.

I then went on to study the effect of hypothalamic lesions on thyroid function. The VanderLaans had shown that there was a marked increase in the activity of the thyroid iodide pump in animals chronically treated with propylthiouracil. This was measured by determining the ratio of radioactive iodide in the thyroid compared to that in the serum [$T/S(I^-)$]. The $T/S(I^-)$ was increased tenfold with propylthiouracil treatment for 10 days. Bill VanderLaan and I subsequently showed that this increase in the activity of the iodide pump was dependent on an augmented secretion of TSH. Since no sensitive or reliable assay for plasma TSH was readily available in those days, my plan was to make relatively large bilateral lesions in various areas of the hypothalamus and to then administer propylthiouracil for a period of 10 days after the animals had recovered from the operation. Thyroid weight, thyroid histology, and the $T/S(I^-)$ would be measured to assess the degree of TSH secretion.

This experiment was a shot in the dark because at that time there was no compelling evidence for any neural control of TSH secretion. When the experiment was completed in the spring of 1951, I was overjoyed and amazed to discover that many of the rats with hypothalamic lesions did not develop thryoid hypertrophy, although they did develop as high a $T/S(I^-)$ as the goitrous control rats fed propylthiouracil.

The Korean War had broken out a few months earlier, and all young physicians who had not previously spent time in the service were being called up for active duty. Upon inquiring at the induction board, I was told that asthma would definitely exclude one from serving in the armed forces. However, unbeknownst to me (and to the induction board medical examiner), the regulations had been changed since I had been refused a commission at the end of World War II. Asthma excluded one from military service only if he were *not* to serve as a physician.

This startling development started me on a new line of investigation oriented toward finding the assignment in which I might best serve the needs of my nation. I was influenced by the sage counsel of Bill Baker, who had been one of my fellow residents at the Massachusetts Memorial. He had repaid his military obligation in the U.S. Public Health Service at the National Institutes of Health in Bethesda. In 1951, the NIH was not the internationally renowned elite research center that it is today. The Clinical Center had not yet been built and many of the current institutes had not yet been established. In fact, I was only dimly aware that there was an institution of that name and I wasn't sure where it was located. However, Bill assured me that it was an exciting place to be and that I would probably be able to continue with some sort of research activities if I would obtain a position there.

Ted Astwood was a friend of Roy Hertz, who was the chief of the Endocrinology Section of the National Cancer Institute. Since my fellowship stipend was from the National Cancer Institute, this made a natural contact. After preliminary discussions between Astwood and Hertz, I was invited for an interview and was successful and extremely fortunate in obtaining a position with him. In July 1951 I left for Bethesda to start a new phase of my career.

Before leaving Boston, I had been anxious to repeat the study on the effect of hypothalamic lesions on inhibiting goitrogenesis in propylthiouracil treated rats. However, because of the unavoidable delays in obtaining histology, there was not enough time left to repeat the experiments there. Astwood and I discussed this and decided that, even though it was only a single experiment, the results looked so clear-cut and so unusual that it would be worth publishing as a preliminary note (Greer, 1951). Perhaps I would be able to continue the study when I moved to Bethesda.

Roy Hertz was very good to me. Although he took no active part in my neuroendocrine investigations, he allowed me to continue my research and provided space, material, and financial support. Bill Tullner and Morris Graff were working in the laboratory at that time, as was George Fisher, who arrived as a fellow recruit with me.

One of the exciting surprises was that Cliff Grobstein, who had initially taught me endocrinology, had in the meantime joined the NIH and had a laboratory adjacent to Hertz's. Although Cliff had drifted away from any primary interest in endocrinology, he was doing some elegant work in experimental mammalian morphogenesis and had learned the technique of transplanting various tissues to the anterior eye chamber.

A small group of young endocrinologists from various institutes at the NIH met informally once a week to talk over their personal experiments. The group included Si Wollman, Bob Scow, and Samuel McDonald McCann. For some reason, McCann preferred to be called Don rather than Sam or "Big Mac." Through interactions with this group, I learned that Scow and I had interned at the San Francisco General at the same time (although he was on the University of California service and we never had met previously), and that he, like me, had originally started in surgery. He dropped out of this discipline because he didn't like smelly patients, not because he was concerned about his ability to keep on his feet all day in the operating room. However, he had maintained and developed great technical prowess and had just perfected a method for hypophysectomizing mice. He and Si Wollman had been doing a series of collaborative experiments studying the importance of TSH in controlling thyroid function in this species.

I was interested in hypothalamic control of the pituitary, Cliff Grobstein could transplant tissue into mouse eyes, and Bob Scow could hypophysectomize mice. It therefore seemed natural to arrange a collaborative study where Cliff would transplant pituitaries into the eye, Bob would hypophysectomize the animals, and I would study their thyroid function under various conditions. The data obtained were similar in the hypophysectomized mice with intraocular pituitary transplants to those in rats with hypothalamic lesions. Thyroid function, as measured by radioiodine metabolism, was maintained at a much higher level than was the ability to develop thyroid hypertrophy in response to antithyroid treatment. However, TSH secretion by the pituitary transplants (measured indirectly) could be increased or decreased by altering plasma thyroid hormone concentration. This suggested that negative feedback control was maintained over the heterotopic pituitary. Eleanor Siperstein joined us for the last phase of these studies and made some elegant observations on morphological and cytological changes in the pituitary transplants as she attempted to teach me pituitary cytology. She failed in the latter project.

These results suggested that there might be two pituitary thyrotropic factors, one which promoted thyroid growth and was dependent on contiguity with the hypothalamus and one which promoted thyroid iodine metabolism and was independent of the hypothalamus. Subsequent work has shown that the stimulation of the thyroid iodide pump is due primarily to depletion of intrathyroidal organic iodide stores and is in part due to intrathyroidal autoregulatory mechanisms. The observations can thus be adequately explained by postulating only a single thyrotropin. However, at the time, the existence of two thyrotropins seemed a reasonable possibility and I was willing to suggest it as a plausible hypothesis (Greer, 1952).

However, during my first months at NIH, the problem which disturbed me the most was that I was unable to confirm my initial observations that hypothalamic lesions prevented thyroid hypertrophy induced by antithyroid drug treatment. I had been able to get another Krieg machine built in Bethesda and had tried to map the "thyrotropic" area in the hypothalamus by making much smaller bilateral lesions in various locations. I was unsuccessful in producing any inhibition of thyroid hypertrophy in the vast majority of several hundred rats thus prepared. I was about ready to disembowel myself for having published a false report when an abstract appeared by Bogdanove and Halmi confirming and amplifying the data I had obtained in Boston. This renewed my personal faith. After correlating the histology of the hypothalamic lesions with the degree of "goiter block" in my accumulated series, I decided that a relatively large bilateral lesion between the region of the paraventricular nuclei and the median eminence was required to inhibit TSH secretion effectively.

A corollary of the decrease in thyroid function produced by hypothalamic lesions would be to produce an increase in thyroid function by hypothalamic stimulation. I investigated several possibilities of stimulating the hypothalamus through direct electrode implantation, but none of these seemed suitable for chronic stimulation. The NIH had a superb instrument shop. Since miniaturization of electronic equipment with the use of transistors was just beginning, we decided to construct small, subcutaneous radiofrequency receivers with electrodes which could be implanted directly into the hypothalamus, thus permitting free movement of the rats at all times. David Hume and W. F. (Fran) Ganong were using implanted radio receivers in dogs for a similar purpose. After seeing their technique and apparatus in Boston, we designed a similar system for rats. We could vary the frequency and amplitude of the current entering the hypothalamus, using either bilateral or concentric bipolar electrodes and timed pulses. The technique worked beautifully. Unfortunately, we were unable to produce any definite endocrine changes, but we did produce certain exciting behavioral changes. The most memorable was the production of immediate drinking behavior in occasional rats as a response to the onset of stimulation. In one animal, the stimulus was so strong that it drank a quantity of water exceeding its body weight over a 24 hr period (Greer, 1955). The high cost of the receivers and the minimal return in data eventually forced us to give up this technique.

Since I had continued to be active in a number of thyroid studies employing radioiodine, although these were not directly related to neuroendocrinology, in 1955 I moved back to California to accept a position as chief of the radioisotope unit at the Long Beach Veterans Administration Hospital. This migration was arranged primarily by my friend and mentor, Joe Ross, who had previously moved from Boston to become associate dean at UCLA and was in charge of the regional radioisotope program there. One particularly attractive prospect relating to the move was that the physical facilities at the UCLA School of Medicine had not been completed at that time. The recently constituted Department of Anatomy had an extremely strong group in neuroendocrinology and they were doing all their work at the Long Beach VA, directly adjacent to the radioisotope laboratory where I would be based. This seemed like a potentially fruitful opportunity for collaboration, especially since my old friend from Stanford, Tom Sawyer, was one of the most active members of the anatomy group. Warner Florsheim was in the radioisotope unit when I joined it. One of his major goals while we were cohabiting was to prove my "two-thyrotropin" hypothesis was wrong. He helped considerably to bury the concept. Curt von Euler and Marthe Vogt were spending sabbatical years in Long Beach. H. W. Magoun, who was then chairman of Anatomy, had plans for von Euler and me to work together to study the control of TSH secretion, because Curt had previously done some beautiful work in this area. However, Magoun's plans never materialized. Von Euler had shifted his interest to temperature regulation and was not particularly excited about going back to TSH control. Vaughn Critchlow was a graduate student with Sawyer. Many years later, Vaughn and I renewed our friendship and developed an active collaboration when he came to Portland as chairman of the Anatomy Department.

Shortly after arriving in Long Beach, I was offered a position as head of the Division of Endocrinology at the University of Oregon Medical School. Although there was a great deal of development to do in the north, it seemed a challenging opportunity and one that would permit me to return to the Pacific Northwest and my native city.

During my tenure in Oregon I have been blessed with an outstanding group of young co-workers. My first research fellow was Takashi Yamada, who came from Gunma University. Upon his arrival in 1957, he announced that during his fellowship he would clear up all the mysteries about the neural control of TSH secretion and identify the nature of the elusive humoral "thyrotropin-releasing factor." Unfortunately, he did not achieve his goal entirely, but he was able to complete a number of elegant studies. Perhaps the most important of these was the demonstration that injection of minute quantities of thyroxine into either the hypothalamus or pituitary would inhibit TSH secretion and for which he was awarded the Van Meter prize of the American Thyroid Association (Yamada, 1959). Unfortunately, there were certain ambiguities to these experiments which have still not been resolved. It is presently uncertain whether there is any direct depression of TSH secretion induced by an increase in local hypothalamic concentration of thyroid hormone or whether thyroid hormone introduced into the hypothalamus must be transported through the hypophysial portal system to inhibit TSH secretion by directly interacting with adenohypophysial cells.

In 1960 a major effort of our laboratory shifted to neural control of ACTH secretion. This was primarily brought about by the fortuitous simultaneous arrival of John Kendall and Kunio Matsuda as research fellows. John had previously trained with Grant Liddle at Vanderbilt and was fully conversant with techniques for measuring adrenal hormones. Kunio was from Shizume's group at the University of Tokyo and had developed superb surgical skills. We jointly decided that it would be worthwhile to investigate the effects of graded brain removal in the rat on adrenal corticosterone secretions. This interest was stimulated by the work of Egdahl in dogs with extensive brain removal that suggested there was considerable autonomous ACTH secretion after the entire forebrain, or even the entire central nervous system, was removed.

Our studies both confirmed and were somewhat different from Egdahl's. Gradual accumulation of our data indicated that certain substances transported through the bloodstream, such as ether, appeared to stimulate ACTH secretion through a direct action on the median eminence, while other stimuli, such as trauma to an extremity, activated ACTH secretion by ascending through the spinal cord to enter the hypothalamus (Matsuda *et al.*, 1964). We also found that dexamethasone would suppress ACTH secretion (measured indirectly with corticosterone) even in rats with all forebrain removed and only an isolated pituitary island remaining (Kendall *et al.*, 1964).

Finley Gibbs, as a graduate student in physiology in our laboratory, extended the studies to show that the neural signal generated by traumatic stress which stimulated ACTH secretion ascended contralaterally in the spinal cord to the level of the pons (Gibbs 1969). Further experiments indicated that the pathway entered the contralateral anterior hypothalamus from an anterior direction and that the main path probably was through the medial forebrain bundle (Allen *et al.*, 1973).

The major difficulty in interpretation of our studies was that we were not able to measure plasma ACTH directly. In 1969, Lesley Rees joined us from Landon's group in London to spend 1 year while her husband took a fellowship in cardiovascular surgery in Portland. She set up a radioimmunoassay for ACTH which worked equally well in the rat and in man. This enabled us to investigate problems of rhythmic secretion of ACTH and the importance of neural structures in negative feedback control in adrenalectomized animals, which we had previously been incapable of solving. Most recently, Kazuo Takebe spent a sabbatical year from Hokkaido University and developed an *in vitro* assay for CRF using cultured rat adenohypophysial cells and measuring ACTH secreted from these cells with the radioimmunoassay (Takebe *et al.*, 1975). This work has been continued and expanded in our laboratory by Naoki Yasuda.

I don't feel it appropriate to make any concluding remarks, because work is still progressing in our laboratory. Although this book is to be written by aging, decaying "pioneers," I have been extremely fortunate that our laboratory has continued to be inhabited by fresh blood and young, vigorous intellects who can either argue down my objections to a projected experiment or ignore them entirely to carry out their plans in secret. Progressively rotting into oblivion as I am, I hope that I can continue to participate in the excitement of young people unraveling the mysteries of neuroendocrinology and developing ten new questions crying for a solution for every one they answer. It is this enthusiasm with which the torch is carried on by my younger associates that enables me to endure the unbearable pain of accelerating dissolution as the maggots slither and gnaw in my brain. At least I still do the best hypophysectomies in town!

REFERENCES

Allen, J. P., Allen, C. F., Greer, M. A., and Jacobs, J. J. (1973). Stress-induced secretion of ACTH. In Brodish, A., and Redgate, E. S. (eds.), *Brain-Pituitary-Adrenal Interrelationships*, Karger, Basel, pp. 99–127;

Dougherty, R. F., and White, A. (1944). Influence of hormones on lymphoid tissue structure and function: The role of the pituitary adrenotropic hormone in the regulation of the lymphocytes and other cellular elements of the blood. *Endocrinology* **35**:1.

Gibbs, F. P. (1969). Central nervous system lesions that block release of ACTH caused by traumatic stress. *Am. J. Physiol.* **217**:78.

Greer, M. A. (1951). Evidence of hypothalamic control of the pituitary release of thyrotrophin. *Proc. Soc. Exp. Biol. Med.* **77**:603.

Greer, M. A. (1952). The role of the hypothalamus in the control of thyroid function. *J. Clin. Endocrinol.* **12**:1259.

Greer, M. A. (1953). The effect of progesterone on persistent vaginal estrus produced by hypothalamic lesions in the rat. *Endocrinology* **53**:380.

Greer, M. A. (1955). Suggestive evidence of a primary "drinking center" in the hypothalamus of the rat. *Proc. Soc. Exp. Biol. Med.* **89**:59.

Greer, M. A., and Brown, B. R. (1948). Concerning the relation between pituitary adrenocorticotrophin and the circulating blood platelets. *Proc. Soc. Exp. Biol. Med.* **69**:361.

Kendall, J. W., Matsuda, K., Duyck, C., and Greer, M. A. (1964). Studies of the location of the receptor site for negative feedback control of ACTH release. *Endocrinology* **74**:279.

Matsuda, K., Duyck, C., Kendall, J. W. and Greer, M. A. (1964). Pathways by which traumatic stress and ether induce increased ACTH release in the rat. *Endocrinology* **74**:981.

Takebe, K., Yasuda, N., and Greer, M. A. (1975). A sensitive and simple *in vitro* assay for corticotropin-releasing substances utilizing ACTH release from cultured anterior pituitary cells. *Endocrinology* **97**:1248.

Yamada, T. (1959). Studies on the mechanism of hypothalamic control of thyrotropin secretion: Comparison of the sensitivity of the hypothalamus and of the pituitary to local changes on thyroid hormone concentration. *Endocrinology* **65**:920.

14

Roger and Lucienne Guillemin

Roger Guillemin was born in Dijon, France, on January 11, 1924. He received the M.D. degree from the Faculty of Medicine at Lyons in 1949 and the Ph.D. degree in physiology from the University of Montreal in 1952. He then joined the Baylor College of Medicine in Houston, Texas, where he was a professor of physiology. In 1970 he went to The Salk Institute for Biological Studies in La Jolla, California, to establish the Laboratories for Neuroendocrinology. He served on several advisory groups of the NIH from 1959 to 1969. He is a member of the National Academy of Sciences and of the American Academy of Arts and Sciences. He has received several honorary degrees as well as national and international prizes in recognition of his contributions to physiology and medicine. In 1977, Roger Guillemin was awarded the National Medal of Science, by the President of the USA, and the Nobel Prize in Physiology or Medicine, shared with Andrew Schally and Rosalyn Yalow.

He writes: "The following pages like all the essays of this series really deal with only a part of all of us involved; our professional lives. There are other aspects in our lives which are just as important, if not more so. I, for one, have been blessed with the presence, understanding, and fortitude of a gentle companion, the mother of our six children. My wife is a distinguished musician and harpsichordist; all our children, taking after her, are also in the arts. In our successive homes over the years, from the Chateau de Prunay in Louveciennes, France, to 17W Shady Lane, Houston, Texas, to 7316 Encelia Drive in La Jolla, California, as well as in the humble log cabin in the mountains of Truchas, New Mexico, we have entertained and received many of our friends who are writing in these two volumes. Also as part of our lives, our home has been gradually filled (if not overfilled) with contemporary French and American paintings, Pre-Columbian artifacts of the Mayas, the Huastecas, the West Coast peoples of Jalisco, Nayarit, and Colima, the fierce pieces of New Ireland and New Guinea, the moving Santos of Mora and Taos, and pottery of the Rio Grande pueblos. One day I hope to publish a small book on this simple family collection."

Pioneering in Neuroendocrinology 1952–1969

ROGER GUILLEMIN

Let me say at the outset of this essay that I am somewhat dubious about any historical or scientific significance of the type of microautobiographies that the previous volume has shown us in its tone, scope, and style. I said so to my old friend Joseph Meites when he asked me to contribute to this second volume. To have declined his invitation, however, would have probably been in bad taste in view of the list of contributors: all were friends and contemporaries; we had grown up together pioneering in the new field of neuroendocrinology. Thus I accepted the task of writing a short essay.

On a scholarship provided in part by the Fulbright exchange program and by some of my own funds from the Salk Institute, our present staff in the Laboratories for Neuroendocrinology acquired in the fall of 1975 a young associate, Bruno Latour, with a solid background in general philosophy and epistemology. His full-time interest and effort is to look at the epistemology of neuroendocrinology. At the time of writing this essay, it is already obvious that several extensive monographs will result from this unusual collaborative effort, in fact encompassing much more than had originally been expected. The essay I will contribute here is of no comparable sophistication; a simple recounting.

While I had always been interested in endocrinology as a medical student, my interest in neuroendocrinology undoubtedly started later in my colleaguial and friendly contacts with Claude Fortier when I joined in 1948 the group of young people attracted to the just-created Institute of Experimental Medicine and Surgery at the Université de Montréal. Its young director, Hans Selye, then 43, was at the peak of his attractive powers over young(er) minds intrigued by his extraordinary experimental abilities, the novelty of his observations, and the far-reaching implications he derived from

ROGER GUILLEMIN • Laboratories for Neuroendocrinology, The Salk Institute, La Jolla, California 92037.

them. Right after the end of World War II, medical literature from the United States started to trickle to Europe. One day, in Dijon, I heard that Selye would be giving several lectures on "stress" and the "diseases of adaptation" in Paris. I decided to go to hear him. Selye lectured, in French, at la Pitié (a charity and teaching hospital built in the seventeenth century where the cardiologist Paul Lian had organized for a number of years an annual teaching event with distinguished invited lecturers). The magnetism of the man was extraordinary. For me, just out of medical school, 5 years of which 4 had been in Nazi-occupied France, with teaching entirely directed toward medical care and no laboratory opportunities whatsoever, the lectures of Selye were from a different world. I went to talk to him after one of his lectures. After some sort of an interview in the office of Robert Courrier in the old building of Collége de France, and much in the style of what I would do later as a member of the Admissions Committee of the Medical School at Baylor, Selye assured me of a (modest) fellowship ($120 per month) on his own research funds to come to Montreal for at least a year.

A couple of months at that institute in the midst of the other young people from Canada, England, Holland, the United States, and Brazil convinced me that I had grossly miscalculated my abilities. There was no way I could ever reach their ease at handling the knowledge, techniques, and concepts which I had never heard of in these dark years of my medical "studies." I was thus inclined to regard the year as a rather unique escapade and then go back to the practice of medicine in Burgundy, probably in that medieval little town where I had already established a modest reputation as a young and alert practitioner in part-time assignments during my last year of medical schooling.

Selye had asked me to set up some technique to keep rats alive long enough after bilateral nephrectomy to see whether large doses of desoxycorticosterone acetate would still produce the vascular lesions he had shown earlier to be regularly produced by the mineralocorticoid in unilaterally nephrectomized rats.

While he was a remarkably lucid and elevating lecturer, Selye was not a teacher of graduate students or postgraduate fellows in the sense that he would make specific efforts to spend any of his own time to teach one anything. The tools, the environment were there and available, but it was left to everyone to make the best of it. Personal contacts, discussions, collaboration among the younger people were of major importance. It was through these that I learned of Fortier's interest in elucidating the mechanisms involved in the physiological control of ACTH secretion, one of the primordial events, as had been shown by Selye, in the response to stress. I read the available literature, and observed what Fortier was doing at that time transplanting the anterior lobe of the pituitary in the anterior chamber of the eye. Geoffrey Harris came to Selye's institute as a Claude Bernard Lecturer in 1949 and spent a week with us. Very different from Selye as a scientist, he convinced those of us interested in that field that there were major problems of classical physiology to be answered in the elucidation of the physiological mechanisms involved in the hypothalamic control of adenohypophysial secretions. Somewhat after that visit, the idea occurred to me to use what Selye had called "adaptation to nonspecific stress" in an effort to dissociate the nonspecific release of ACTH by drugs exerting other, very specific pharmacological actions, drugs such as antihistaminics. Why antihistaminics? Probably because we had had another set of Claude Bernard lecturers, Bernard Halpern and Jean Hamburger, who had lectured to us about the new phenothiazine derivatives that the French pharmaceutical industry had recently made available to them and that they had found to be powerful antihistaminics. Moreover, I had read reports from Kahlson's laboratory in Lund indicating large amounts of histamine in hypothalamic extracts.

My idea was to see whether one could reach a "stage of adaptation," in Selye's terminology, by repeated injections of Phenergan after it had lost its ability to induce (nonspecific) release of ACTH while still retaining its specific pharmacological activity as an antihistaminic. Thus we could ascertain whether the ultimate mechanism triggering the release of ACTH was histamine, or at least some endogenous substance specifically affected by Phenergan. I discussed this with Fortier and we started a few experiments along these lines. There was soon no doubt that repeated injections of Phenergan would each stimulate less and less ACTH release as judged by the adrenal ascorbic acid test of Sayers, and that a time could be reached when injections of Phenergan were no longer followed by acute release of ACTH. In these animals the drug retained its normal, potent antihistaminic property and would completely prevent the acute release of ACTH normally induced by injection of enormous doses of histamine. I then observed that in such a preparation any other type of stress agent such as surgery, injection of formalin, or forced immobilization would still stimulate the normal release of ACTH. Thus the "first mediator" of Selye's stress syndrome likely was not histamine. This was reported in a short note with Claude Fortier (*Trans. N.Y. Acad. Sci.*, Vol. 15, 138–140, 1953) and was my first contribution to the field of neuroendocrinology. Similar studies were conducted somewhat later with anticholinergic and adrenolytic drugs, with similar results. I also ascertained that the drugs were indeed acting within the central nervous system demonstrating their ability to inhibit focal cortical seizures recorded by electrocorticography and induced by depositing, on the parietal cortex of the rat brain, small paper pledgets soaked with histamine, carbachol, or serotonin. The conclusion was thus reached that none of the classical neurotransmitters studied here (acetylcholine, epinephrine, serotonin, histamine) was the exclusive ultimate mediator of the (stress-induced) release of ACTH. Some other (unknown?) substance of hypothalamic origin had to be postulated as the ultimate mediator.

Somehow, it was now 1952. I was still in Montreal, having enrolled in 1949 in an extraordinary series of courses in endocrinology (in which I met Murray Saffran as one of the undergraduate students) jointly offered by McGill and the Université de Montréal, and was now completing a dissertation for the Ph.D. degree in physiology (actually called experimental medicine and surgery). The experimental work for that degree dealt with the mechanisms involved in the production of hypertension and kidney lesions by desoxycorticosterone acetate—the topic Selye had assigned to me when I arrived in Montreal. Nothing earthshaking is to be found in that dissertation, but I had learned the fundamentals of experimental endocrinology, how to design an experimental protocol, and how to be critical of oneself (and of others). These had been 4 extraordinary years and I have always been grateful to Hans Selye for having given me the opportunity. While I knew that I enjoyed the life of an investigator, I was not too sure that I had what was necessary to be a meaningful one. I knew, though, that if there were to be a way of mine, it would be different from the ways of Selye. I had come to recognize that Selye's style was absolutely unique and probably not to be emulated. It would always be dealing with a purely descriptive phenomenology, with more than a touch of the dramatic and a quasiparanoid need to be read and/or presented as generating unified theories of medicine. Moreover, with the exception of a few early and elegant studies on the neuroendocrinology of the milk-letdown reflex, very classical in their approach, Selye's descriptive phenomenology, as I called it above, was the result of experimental decisions of such extremes as to make one wonder about their relevance not only to physiology but also to the causes of diseases of man. I would probably best fit as a "traditional" physiologist.

After I visited C. N. H. Long at Yale (when I first met Al Brodish, still a graduate student, as I remember it), he offered me a job in his Department of Physiology, which I accepted. Two days later, through a peculiar set of circumstances connected with my having received one of the early scholarships of the Markle Foundation the year before, I was asked to join the Department of Physiology at Baylor University College of Medicine in Houston, Texas, as an assistant professor to teach endocrinology. Hebbel Hoff had just moved to Baylor from McGill as chairman of the department. I flew to Houston and met him and Michael deBakey. There were at Baylor space, money, an incredibly open future, and also azaleas and live oaks. Somehow, I sensed that all that meant more than the Ivy League. I never regretted that decision. I sent a cable of apologies to C. N. H. Long, who, always the gentleman, never held it against me, as I was to know later on many occasions. In September of 1953 I joined the faculty of Baylor and started to teach endocrinology. With my money from the Markle Foundation, I also immediately started a modest research project to complete the "adaptation" studies started in Montreal on the mechanisms of ACTH release.

I stayed at Baylor and worked and taught physiology in Hoff's department for almost 20 years. Hoff was probably the most considerate and the most generous chairman I could ever have hoped for. He created an environment in that department where I could work with a minimum of encumbrances. His teaching load was immeasurably greater than mine for all these years. He accepted generously that I would prefer to spend so much of my time and effort in my laboratory with a few graduate students and postdoctoral fellows rather than with the medical students. He shielded me from too much involvement in the internal problems of the school (thanks to him, I barely knew when there were problems at the school). Hoff also easily enrolled me in his scholarly interest in the history of medicine. All the work on Claude Bernard's manuscripts, on the early time-recordings of natural phenomena, on the history of blood transfusion in man, in which I was rather deeply involved at some time—all that was seeded, started, and nurtured by the prodigious energy and the encyclopedic mind of this great scholar of modern American physiology. Hoff revolutionized the teaching of modern physiology to medical students by his early introduction, with Leslie Geddes, of simple and rugged electronic instrumentation in the early 1950s, to replace Ludwig's kymograph and the accompanying smoked drum that were still in use at that time. All of the current instrumentation is so obvious now that we forget it was not always like that. Hoff could have chosen to spend more of his time in his own laboratory to pursue his earlier work with John Eccles in Sherrington's department on the electrophysiology of the pacemaker of the heart or later with John Fulton on the physiology of the respiratory center. He chose instead to let me, along with other younger people in the department, spend my time in my laboratory. My debt of gratitude to Hebbel Hoff is equaled only by my feelings of affection for him and my respect for his scholarly mind.

The man who had been in charge of the department before Hoff became chairman was A. D. Keller. Keller and Breckenridge had reported a series of experiments in which they had observed persistence of hypophysial functions after "extirpation" of the pituitary stalk and also after partial hypophysectomy. Breckenridge was still in the department. I got to know him very well. A gentle man who decided to get an M.D. degree (which he obtained when he was in his late 40s), Breckenridge was still interested in the physiology of the control of the pituitary secretions. He showed me how to hypophysectomize dogs by an elegant transpalatinal approach that I showed later to one of my postgraduate fellows, Harry Lipscomb.

One day, I went to Galveston to visit the tissue culture laboratory of Charles Pomerat and also to see some of his famous watercolors of churches in Mexico. Pomerat showed me around, showed me his time-lapse movie photography of various cells, particularly of neurons, speaking either in English or in his slow, perfect French, with always the most exquisite and exacting choice of the *mot juste*. He introduced me to a young undergraduate, Barry Rosenberg, who had been culturing adenopituitary cells as his assignment for an M.Sc. degree. Then Pomerat said to me, "You know these pituitary cells which grow so well, for some reasons which we don't understand, do not seem to secrete hormones." I asked what hormones and he said, "Gonadotropins. We are testing the fluids by injecting them into mice prepared as for pregnancy tests." I immediately told Pomerat, "I think I know why this is so: Your pituitary cells *in vitro* are lacking some substance of hypothalamic origin. Could one culture jointly pituitary and hypothalamus?"

Pomerat was not particularly impressed. In a short discussion later on that day, I described to him the rationale for my statement, but I did not convince him. I left, intrigued, and kept thinking about this strange observation and my proposal. A week or so later, I went back to see Pomerat and asked him whether I could set up some simple experiments in his laboratories to test the idea I had mentioned to him the previous week. I do not remember exactly what happened; Pomerat[1] was no more enthusiastic than on the first occasion, but he suggested that perhaps Rosenberg could join me at Baylor in Houston (since Rosenberg had applied to Baylor College of Medicine, as I seem to recall) to study that very problem.

Barry Rosenberg came to Baylor a few weeks later and showed me how to do tissue cultures with the clot and coverslip method, and shortly thereafter, as I recall, he left to go to medical school in New York. A few months later, I knew that the pituitary cultures released ACTH in the culture medium but only for the first few days following transplantation, and that they would release ACTH again if and when co-cultured with fragments of median eminence or ventral hypothalamus. I had set up the adrenal ascorbic acid bioassay for ACTH, Sayers's assay, as I thought that it would be more sensitive than the assay for gonadotropins used earlier by Rosenberg, and also probably because studies on the mechanisms of response to stress (the release of ACTH) were much in my mind after 4 years in Selye's

[1] A couple of years after that episode, Pomerat left Galveston for the Scripps Institute of Oceanography in La Jolla. He died shortly thereafter of a malignancy.

laboratory. I still remember going home that evening of the day on which I had seen for the first time the effect of adding a culture of hypothalamus tissue to the pituitary cells and had observed the depletion of adrenal ascorbic acid in every animal of the bioassay, the sign of the presence of ACTH. I remember well telling my young wife, "I have made an observation today of such importance that you will never have to worry about our future in academic medicine." I also remember, during the next few weeks, the extraordinary excitement of learning the methods of pharmacological assays for vasopressin, oxytocin, histamine, epinephrine, and norepinephrine, and getting them to work in the one-room laboratory; the elating sensation of learning so many new things, getting them to work for new goals—things which had never been done by others; the thrill of discovering, of realizing the pregnant future, known only by those who have really experienced it. I am happy to see that still happening to the younger people in the laboratory around me when they come up with a good new idea or make an unexpected observation, confirm it, and expand it. It is certainly a different feeling from what one feels when putting the final mark on the solution of a problem which has taken years to solve. The former is all action, movement, and expectation. The latter is a feeling of achievement, of having reached the goal, of finally breathing, and of "what next."

That summer (1955) I spent a month in the laboratories of Geoffrey W. Harris at the Institute of Psychiatry in London. I showed the early results of combined tissue cultures to Harris. He was rather skeptical. During that short stay at the Institute, I met Seymour Reichlin, Bernard Donovan, Keith Brown-Grant, and H. J. Campbell; Claude Fortier was there also.

Later that fall, David Hume came to Houston for the meeting of the University Surgeons and he talked on kidney transplants, his new interest. I showed Hume the results of the combined tissue cultures with hypothalamus and pituitary. Hume was impressed and, I remember, encouraged me to go on with the *in vitro* method.

After a seminar I gave at Baylor, I was approached by Walter Hearn, another young fellow like me, from the Department of Biochemistry. He proposed that we should work together to isolate the hypothalamic hypophysiotropic substances. I was delighted by the proposal. I had by that time pretty much decided that the most important contribution to understand the mechanisms whereby the hypothalamus controlled the secretion of the pituitary was to establish the molecular structure of the hypothalamic factors involved. Anything short of that would be beating around the bush. Once the hypothalamic hypophysiotropic neurohormones were isolated and characterized, all the real physiological as well as clinical studies could proceed with synthetic replicates of the neurohumors in unlimited amounts.

In one of his lectures at McGill, David Thompson had once generalized that all hormones secreted by cells of ectodermic origin were proteins or polypeptides, those secreted by cells of endodermic origin were proteins or derivatives of amino acids (thyroxine), and those secreted by cells of mesodermic origin were steroids. Perhaps naively, I had formed the hypothesis that the hypophysiotropic hypothalamic substances would be peptides, probably small as were oxytocin and vasopressin. Since du Vigneaud had characterized and synthesized these two, the others would be synthesized also.

Hearn and I worked for 2 years (when he left Baylor for another job at Ames, Iowa—where one of his technicians, later graduate student, would be Roger Burgus). Joined by several graduate students and postdoctoral fellows, William Cheek, Buford Nichols, Dwight (Gene) Householder, Sidney Levine[2], and later Harry Lipscomb, we purified extracts of hypothalamic tissues (a few fragments, maybe 10–50, collected locally from a kosher operation) and later of posterior pituitary tissues; both had ACTH-releasing activity. We convinced ourselves that the ACTH-releasing activity both in the hypothalamic tissues and in the pituitary extracts was due to some substance that certainly was different from oxytocin and vasopressin, known since John Abel in 1924 (and as seen again in my bioassays) to be also in these hypothalamic extracts.

Why the work on posterior pituitary? Early in 1955 there had appeared a note by Samuel McCann and John Brobeck reporting that injections of relatively large doses of Pitressin (a commercial clinical preparation of vasopressin) would release ACTH in rats having a large lesion of the median eminence produced by electrocoagulation through electrodes located with a stereotaxic instrument. Such a lesion had been reported by McCann to inhibit or prevent the acute release of ACTH that takes place upon exposure to any sort of stressful situation. McCann in these early studies showed how such lesions could be produced routinely in the rat.

Sam McCann was born and reared in Houston; I had met him after my early studies with Fortier in Brobeck's department during a short visit in Philadelphia and that same evening at Brobeck's house, I knew that McCann was to be taken seriously. I had just received some highly purified vasopressin from du Vigneaud and had observed that it would not stimulate

[2] There were already, at that time, good reasons to suspect that the hypothalamus should have neurohumoral control of all adenohypophysial secretions, not just of ACTH. Sid Levine and I tried for more than a year to observe whether the co-cultures with rat hypothalamus would lead to increased secretion of gonadotropins and/or thyrotropin as assessed by simple bioassays. While there was always a trend of numbers in that direction, no experiment ever yielded a statistically significant set of results. I have come back many times to these protocols and the experimental results. Nothing statistically defensible could have been interpreted in these results; nothing was ever published.

release of ACTH from the pituitary tissue cultures. I confirmed McCann's and Brobeck's report when I observed that Pitressin would release ACTH from the pituitary tissue cultures whereas the high purified synthetic vasopressin (LVP1) from du Vigneaud would not. My conclusion was that some substance other than vasopressin in the relatively crude Pitressin was the responsible hypophysiotropic agent.

That was the beginning of an extraordinary series of experiments and spirited exchanges between McCann and myself that would last for almost 5 years. There would be the "vasopressin school," with McCann as its leader, followed by an ever-increasing number of people, pharmacologists, physiologists of sorts, and clinicians, all satisfied that vasopressin could be and was the physiological mediator of the stress-induced release of ACTH. The evidence appeared overwhelming, but was circumstantial. Against all these would be the other school (the "CRF school" as it would be named later), proposing that vasopressin could not be the physiological mediator of ACTH release (too-high doses of vasopressin were necessary) and instead was controlled by a corticotropin-releasing factor, possibly related chemically to vasopressin but different: Hearn and I had obtained "fraction D" from hypothalamic extracts which had no or little pressor or antidiuretic activity, which had a chromatographic behavior different from that of vasopressin, and which released ACTH. Finally, both McCann and I agreed in 1959 that vasopressin was not the *exclusive* mediator of stress-induced release of ACTH after a series of experiments I reported with Buford Nichols showed in non-anesthetized trained dogs that one could totally dissociate release of ACTH, measured by plasma 17-hydroxycorticosteroid levels, from the release of vasopressin as shown by concomitant antidiuresis; also, in several hundred animals (rats) stereotaxically placed hypothalamic lesions would or would not inhibit stress-induced release of ACTH, as assessed by the new fluorometric method I had devised with George Clayton to measure plasma corticosterone, with no correlation whatever with presence or absence of diabetes insipidus. Writing this in 1976, I am wondering again whether the whole thing may not have to be reconsidered, particularly in view of the new results of John Porter showing elevated concentrations of vasopressin in the portal vessels of the pituitary stalk. Vasopressin may have more opportunities to act as a CRF than I thought in 1959–1960. Its possible role in the stress-induced release of ACTH cannot be exclusive, however.

What was the origin of that name "corticotropin-releasing factor" or "CRF"? In 1955 I had organized with Charles Carlton and William Fields the third Annual Meeting of the Houston Neurological Society to be devoted to the hypothalamus. The only previous meeting on that subject had been the imposing meeting of the Association for Research in Nervous and

Mental Diseases (ARNMD) that had taken place in New York in December, 1939. The lean volume (as compared to the 1000-or-so-page *The Hypothalamus* published by the ARNMD in 1940) entitled *Hypothalamic Hypophysial Interrelationships* that was edited by the three of us was the first volume ever devoted to the neuroendocrinology of the hypothalamus.

To my surprise, I learned at that meeting from Geoffrey Harris, who came with John Green whom he had been visiting in California, that somebody else had been doing *in vitro* work with pituitary and hypothalamus. Harris had visited Saffran at McGill a few weeks earlier, who had shown him results of short-term incubation (a few hours vs. the days and weeks of my tissue cultures) of rat pituitary. Saffran had been incubating rat adenohypophyses with fragments of hypothalamus, brain cortex, or posterior pituitary. It was the first I had heard of Saffran's work on the control of ACTH secretion. Harris also said that Saffran, too, had concluded that vasopressin was not the releaser of ACTH—rather, it was another peptide that he called "corticotropin-releasing factor," in short "CRF," after a nomenclature proposed by Robert Cleghorn in whose department Saffran was working. Harris added that Saffran had almost isolated the corticotropin-releasing factor, which was present (probably stored) more in the posterior pituitary than in the hypothalamus. All of that was in remarkable agreement with what I had independently concluded.

When I got in touch with Saffran and exchanged information and results with him, it became obvious to me that Saffran's methodology was much more quantitative than mine, with his use of hemipituitaries (rather than my coverslip tissue cultures) and with a well-characterized incubation medium (rather than the tissue culture fluid with chick embryo extract and calf serum). There were also some peculiar discrepancies in results. The most potent extracts Hearn and I had made were of hypothalamic origin. Saffran's co-incubations with hypothalamus were inactive in releasing ACTH or were replaceable by brain cortex, and required the presence of a catecholamine; his best results (the most potent additive to stimulate release of ACTH) were obtained with rat posterior pituitary. Saffran's results (reported with Bruno Benfey and Andrew Schally) on the pituitary incubation and extracts, together with the whole thrust of the *in vivo* experimental results from McCann, and my own *in vitro* observations of ACTH release with Pitressin but not with synthetic vasopressin, led to intense work on extracts of the posterior pituitary.

On the appeal of the quantitative aspect, I shifted rapidly to the short-term hemipituitary incubation. Following discussions with Claude Fortier, who had just joined the staff of the Department of Physiology at Baylor, we so modified its design and calculation to make it even more amenable to sound statistical calculation—simple but powerful modifications that I was happy to learn from Saffran shortly thereafter were immediately incorporated by himself and Schally in their subsequent studies.

Hearn, Householder, and I observed corticotropin-releasing activity in fractions from posterior pituitary extracts with mobilities identical to those of the hypothalamic materials in several systems; all were different from the vasopressins or oxytocin. In studies of tissues other than hypothalamus and posterior pituitaries, I was surprised to observe occasional release of ACTH by fractions of brain cortex extracts as well as by fractions of relatively crude substance P of brain or gut origin. In all cases, the active fractions (releasing ACTH) behaved on two or three chromatographic systems identically to the ACTH-releasing fraction of hypothalamus or posterior pituitary origin.

The most purified materials (fraction D) active *in vitro* at 1 μg/ml were obtained in minute amounts (less than 100 μg); they were not homogeneous. When we tried to obtain them in greater purity by paper electrophoresis, we would regularly lose activity and peptide.

Hearn left Baylor for Iowa State at Ames. I immediately contacted some of the younger biochemists in the Medical Center in Houston to pursue the isolation of CRF. I was not (and still am not) a biochemist. Obviously, the isolation of CRF required chemical knowledge that I lacked, and the isolation of CRF was the most important thing to pursue and complete.

One day, I received a letter from Andrew Schally, writing from Saffran's laboratory and inquiring about the possibility of joining me at Baylor. My first meeting with Schally was at the following Federation Meeting in Atlantic City. Schally had written that he would be getting his Ph.D. degree the next summer or fall and that he would like to come and work in my laboratory without delay after that to complete the isolation and characterization of CRF and, he hoped, move on later to other suspected hypothalamic factors. The Atlantic City meeting confirmed these goals and Schally's interest.

In that first conversation, I found Schally to be an intense younger man who, to me, the physiologist, appeared to be a qualified biochemist already with knowledge and technical know-how on peptides and more particularly on CRF from his training with Saffran. Working together appeared to be a sure bet to finish the isolation of CRF which I had started with Hearn. The same conclusion had obviously been reached by Schally on the basis of his work with Saffran, and of what he had read of my papers. Somewhat surprised that he would not do all he could to keep him, I wrote Saffran to tell him of the letter and conversation with Schally. Saffran was affable as always, gave Schally a good though guarded recommendation, and told me that he (Saffran) was leaving on a sabbatical to work on projects unrelated to CRF.

Schally joined me in Houston in 1957. We worked together very well, very hard, with never an unpleasant word, on the isolation of CRF, which we saw within reach in another few months—repeatedly every 6 months for the next 4 years.

There is no doubt in my mind that Selye's studies on stress and his remarkable observations of the involvement of the pituitary-adrenal axis in response to stress were powerful and persuasive incentives for the early studies by Geoffrey Harris in London; David Hume, Don Nelson, and Fran Ganong at Harvard; Claude Fortier in Montreal; Evelyn Anderson, Gordon Farrell, and Sam McCann then at NIH; and later myself. To characterize the "first mediator" of the whole endocrine response to stress was quite a challenge in which many were interested. Selye's concepts made it of physiological interest, with possible clinical significance, when J. S. L. Browne and Eleanor Venning started to show in the early 1950s that the endocrine response to stress in man also involved the pituitary-adrenal cortex system. Selye, through his stress concept, had thus a major stimulating role in orienting the early efforts in neuroendocrinology toward the study of the hypothalamus–pituitary ACTH–adrenal cortex functional relationships.

Strangely enough, and unwittingly on Selye's part, this is probably about the worst thing that happened to nascent neuroendocrinology. The search for CRF was to prove so complex and baffling that it is still not satisfied at the writing of this essay, more than 20 years after it started.

The lack of an answer (isolation of CRF) after the first 3–4 years of early work—in fact, the lack of clear-cut progress toward isolation of CRF, multiple statements to the contrary—raised in the minds of many biologists, aware of the success of others in isolating biologically active substances, grave doubts about the validity of these early concepts of neuroendocrinology; the same doubts were also directed, with concern, at the few people involved in these unsuccessful attempts at characterizing CRF. One can reasonably and musingly post-pose that, had we started to work on the hypothalamic control of thyrotropin secretion, or even of gonadotropin secretion, this most likely would not have happened. While the isolation of TRF and the characterization of its molecular structure took from 1962 to 1969, the sequence of events involved was always logical and constructive and the reasons for slow progression were always totally understood. Not so for CRF. I have indeed wondered for some time whether the paradigm of an immediate neurohumoral peptidergic hypothalamic control of the secretion of each and all adenohypophysial hormones may well not be operative or not be sufficient in the case of adrenocorticotropin. The commonality of biosynthetic origin of ACTH with [β-lipotropin–β-endorphin] only recently recognized (*Proc. Natl. Acad. Sci. USA*, Vol. 74, 3014–3018, 1977), along with the demonstration of the presence of ACTH and α- and β-endorphin in the brain, may signal the opening of totally new concepts in understanding the mechanisms of control of ACTH secretion.

In June 1960 I assumed the post of associate director of the Laboratory for Experimental Endocrinology, of which Robert Courrier was the chairman, at the Collége de France, in Paris. On the insistence of Hebbel Hoff, I maintained my laboratory at Baylor operative, funded, and active with Harry Lipscomb and Andrew Schally. I literally commuted between Paris and Houston. Three years later, I decided to come back to Houston; local circumstances in Paris had been such that I could not reconcile them with my goals or my ethics in science.

It was in Paris, during those 3 extraordinary years, that Edouard Sakiz and I obtained our first solid evidence for the presence of a luteinizing hormone releasing factor (LRF) in hypothalamic extracts and reported its early purification by gel filtration and ion-exchange chromatography (*C. R. Acad. Sci. Paris*, Vol. 256, 504, 1973). The very same methodology was to be used 10 years later in the final isolation of LRF by my laboratory as well as that of Schally. Sam McCann and Geoffrey Harris also had obtained similar evidence for the existence of LRF at about the same time. I have already given in some detail a historical account of the search for LRF, its purification, and the involvement of both my laboratory and Schally's in its isolation and synthesis (see *Am. J. Obstet. Gynecol.*, Vol. 129, 214–218, 1977).

The most important achievement of these 3 years in Paris was the report with Sakiz, Yamazaki, and Jutisz of the first uncontrovertible evidence of a thyrotropin-releasing factor (TRF) in hypothalamic extract (*C. R. Acad. Sci. Paris*, Vol. 255, 1018–1020, 1962), its first purification, and, with Don Gard, the early evidence of the mode of action of TRF in competition with thyroid hormones at the pituitary level (*Endocrinology*, Vol. 73, 564–572, 1963). Large-scale collection of sheep hypothalamic fragments also was started in Paris. When I returned to Houston in 1963, I carried with me half a million fragments of sheep hypothalamus, dissected, trimmed, and lyophilized, ready for work. By 1962, I had definitely concluded that enormous quantities of hypothalamic tissues would be necessary to complete the work involved in the chemistry of isolating and characterizing the hypothalamic hypophysiotropic factors. Of their existence, there was no doubt in my mind. Over 3 years, I collected 5 million fragments of sheep hypothalamus.

I went to about every one of the largest slaughterhouses in the Midwest and Southwest, spending 1 or 2 days working on the floor with the local people to make clear what I wanted. There were some colorful episodes; my French accent was of little help in Paris (Texas). The summer after he became a graduate student at Baylor, Wylie Vale's assignment was to work full time on the killing floor of one of the slaughterhouses in San Antonio which had agreed to donate the tissues if we could furnish the labor. That was hard labor for him.

Eventually, every one of the fragments was redissected in the laboratory to ascertain anatomical correctness of the section involved and to trim away peripheral tissues (du Vigneaud had once told me that, in the isolation of oxytocin, the most efficient purification step had been that of separating the pituitary from the cow); more than 50 tons of fresh frozen tissues was handled, processed, lyophilized, and extracted in the laboratory from 1964 to 1967.

When in November 1963 I returned from Paris to Houston, Schally was no longer in my laboratories at Baylor. Deeply disturbed at our inability to solve the problem of the nature of CRF in the 4 years of collaborative efforts, I had told him during one of my many commuting visits from Paris to Houston that perhaps some reappraisal of our collaborative arrangement would be necessary in the future. Late in 1962 or early 1963 Schally went to the the Veterans Administration Hospital in New Orleans to set up a unit of research on polypeptides. In his own essay in this series Schally probably will recount how this proposal came to him.

In our working together, Schally and I had learned a lot about the strategy of an isolation program that would be of use to both of us in future endeavors. As I said above, it is perhaps unfortunate for neuroendocrinology that we did not first look for a TRF or a LRF in hypothalamic tissues. The hard work that Schally and I had devoted unremittingly to the characterization of CRF had little immediate reward. But this is the usual case with hindsight.

Edouard Sakiz and Eichi Yamazaki had both decided to come with me to the United States from Paris. At the last minute, Yamazaki told me that he wished to abandon science for the spreading of true Buddhism; he became (and still is to my knowledge) a high-ranking member of Soka-gakai. Sakiz went to Houston ahead of me by a few months to get things going at our usual pace. From Paris to Houston, Sakiz and I worked together for almost 7 years, with a bond of friendship and of intellectual commensality that I never encountered with anybody else. That warm friendship is still very much alive. Sakiz has been for some time, and with great success, in charge of all research and development in one of the largest pharmaceutical industries in France.

Upon return to Houston, we worked hard at purifying both LRF and TRF, in consultation with Darrell Ward across the street at M.D. Anderson Hospital. I organized the handling and cutting of the hypothalamic fragments and their lyophilization in industrial-size desiccators; I also performed the first solvent extraction and stockpiled the products of the first steps of purification of the extract, chromatography on gigantic gel filtration columns of 15 cm × 2 m heights, in batches of 100,000 fragments, followed by ion-exchange chromatography on columns also of respectable size. Meanwhile, Sakiz was running the bioassays; he was also writing more and more sophisticated computer software for our statistical analyses in experimental endocrinology. In spite of all that hard work, efficiency, and enthusiasm, there is no doubt in my mind that the move from Paris back to Houston, with the accompanying physical and emotional strain, was a terrible drawback for our work in the laboratory.

In my historical account (mentioned above) of the isolation of LRF, I explained why I decided early in 1965 to shelve the isolation of LRF until a better bioassay than the ovarian ascorbic depletion assay became available (that was not to arrive until 1969 from our work with Max Amoss). Because the bioassay for TRF I had designed in Paris in 1962 was so reliable, it meant that we should devote our full-time effort toward characterizing TRF.

With Sakiz, Pierre, and Simone Ducommun, who had joined us after 3 years in Fortier's department in Quebec, and young Wylie Vale, we also conducted physiological studies on the mechanisms of simultaneous secretion of ACTH and TSH, and on the mechanisms of secretion of TSH as modified by thyroid hormones and crude preparations of TRF. With Wylie Vale we showed the effects of elevated K^+ on TSH secretion, the role of Ca^{2+}, the antagonism by thyroid hormones, and the rapid degradation in plasma of (still-crude) TRF. From these times came the elegant study that was part of Wylie's dissertation showing the dissociated effects as a function of time of cycloheximide vs. actinomycin D on thyroxine inhibition of TRF-induced release of TSH.

Soon after my return from Paris I had been inquiring for a chemist to join our group, as Ward could not devote much of his time and efforts to this project and I knew I did not have the competence to bring to completion the isolation of TRF and its structural characterization, particularly in view of the submilligram quantities that were expected to constitute the final yield. Anand Dhariwal, who had left Houston with Schally, had returned to my laboratories at Baylor a few months later, but not for long. He then went to McCann's laboratory in Dallas. In September of 1965 Roger Burgus joined us at Baylor. While he had been working on the chemistry of cobalamines in the preceding few years, he had trained with Hearn at Ames, Iowa, and had been involved in Hearn's own efforts, after he had left Baylor, at preparing CRF from posterior pituitary powders.

The arrival of Burgus was taking place at a propitious and also critical time: propitious because I had accumulated large amounts of the hypothalamic extract and Sephadex fraction containing TRF for a meaningful attack on its final isolation; critical because Sakiz, Ward, and I had recently started to wonder whether TRF was a polypeptide or at least a homomeric peptide. A short note stating that question (*C. R. Acad. Sci. Paris*, Vol. 262, 2278–2281, 1966) was prepared some months after Burgus's arrival. Burgus had carefully insisted that the text clearly read that, while the results recently obtained were compatible with such a proposal, they did not exclude the possibility that TRF could be a peptide, although a somewhat unusual one (by our thinking, of the time).

Then followed the series of experiments that led in 1968 to the isolation of TRF in the laboratory at Baylor. The decisions as to the chemical steps, the handling of the minute amounts of pure material generated, and the final approach with mass spectrometry of the synthetic pGlu-His-Pro-NH_2 and finally of the native ovine TRF were those of Roger Burgus, while the performance and appraisal of the bioassays and the surrounding biology were my decisions, together with the participation of Wylie Vale, who was to receive his Ph.D. in physiology in the summer of 1968.

I have already given, with Burgus and Vale, a careful historical account of these 4 years; of how we purified, isolated, and characterized the structure of TRF, in an extensive and critical review (*Vit. Horm.*, Vol. 29, 1–39, 1971). To this day, we have nothing to add to, remove from, or qualify in that review. It goes with great technical detail into the unerring approach that we set and followed in what should be a classic of the strategy used in the isolation and characterization of a natural product of complicated biological assay that is present in minute amounts (for the technology of that time) in an unusual starting material. References to the various pertinent technical papers are extensive and exhaustive, not only for the publications of my own laboratory but also to the few other groups working on the isolation of TRF, particularly Schally's group.

Let me add here only a few details in a lighter vein. With Roger Burgus, Thomas Dunn, and Wylie Vale, I submitted to *Science*, in February 1969, a manuscript describing the TRF biological activity of the protected (treatment with acetic anhydride) synthetic tripeptide (R)Glu-His-Pro-OH, and showed, in the same note, the absence of TRF biological activity of all the other protected isomer tripeptides. This was without any possible argument the first evidence of a known peptide (H-Glu-His-Pro-OH) with no TRF activity in itself and showing the generation of TRF activity solely upon protection of its NH_2-terminus as we knew native TRF to be protected. Moreover, synthesis of the series of all tripeptides composed of His, Pro, Glu, produced in record time at my request by Rolf Studer and his collaborators at Hoffman-LaRoche in Basel, had been triggered by our recently acquired knowledge that the whole of the molecule of TRF could be accounted for by the three amino acids His, Pro, Glu as we had just publicly reported—Schally and Folkers in the audience—in January 1969 at the Tucson meeting (see J. Meites, ed., *Hypothalamic Hypophysiotropic Hormones*,[3] Williams and Wilkins, Baltimore, 1970, pp. 21–35). That manuscript was rejected by *Science* on the comments of one referee who said (or words to this effect) that "these results were of much less importance than recent reports of LaBella showing stimulation of release of TSH *in vitro* by nanograms of oxytocin or vasopressin and that anyway

[3] The published proceedings of that Tucson meeting make interesting reading for the historian of neuroendocrinology. One finds in it the extensive description by Schally and Arimura of their isolation of "GHRH" and its biological activity (pp. 208–226); the claims of FRF free of LRF activity (p. 248), both being proposed as polyamines (pp. 248–252); the claims of the isolation of MIF and MRF (rather, MSH-RIH and MSHRH) (pp. 171–183)—all from the laboratories of Schally. There also is the paper with Burgus showing the isolation of ovine TRF (pp. 227–241) and the original figures showing its composition: 81.6% of the weight was accounted for in terms of the three amino acids His, Glu, Pro; theoretical ponderal contribution of the amino acids for a tripeptide isolated as a monoacetate is 86%. Three months after our first note on the TRF activity of protected Glu-His-Pro-OH appeared, Schally was still *concluding* in 1969 that TRF was not a homomeric peptide, and that the nonpeptidic moiety of the molecule of TRF (66% of its weight) was responsible for the biological activity of TRF: the ultimate proof of that conclusion, so said Schally *et al.* in *J. Biol. Chem.*, Vol. 244, 4077, 1969, was that, indeed, all tripeptides composed of Glu, His, Pro had no biological activity. Schally had had such peptides prepared by the group of Merck as early as 1966. In fact, when Burgus and I became interested in the very same compounds, I wrote to my friends at Merck asking for these, only to be told that none was left as the whole lot had been given to Schally. I then wrote Schally asking for some small aliquots of these peptides, pointing out that this request and a cooperative response on his part would be not only in good scientific spirit but also for the best use of our taxpayers' grant moneys. Schally did not see fit to go along with my request, with the excuse that "the FDA did not allow such transfers across state lines" (*sic*). The data on the homogeneity of our latest batch of ovine TRF reported by Burgus at the Tucson meeting, showing it to be a peptide composed exclusively of the three amino acids, had been obtained and definitely ascertained only a few days before the meeting. In view of the results published by Schally's group, our new observation was obviously a turning point, of which Burgus and I fully realized the implications. We never considered not to reveal it at the Tucson meeting. But Burgus and I knew that time would be short for us, as soon as we had given it away. It was from Tucson that I made by phone my first request to the group at Hoffmann-LaRoche for the synthetic tripeptides. The conclusion is inescapable that Schally and his collaborators never isolated TRF as a single entity, either in 1966 or later. I have previously said so in *C. R. Acad. Sci. Paris*, Vol. 269, 1870–1873, 1969; also in *Vit. Horm.*, Vol. 29, 1–39, 1977. I do not doubt that their porcine TRF was probably obtained as a peptide practically free of other peptides; it was, however, so contaminated with other nonpeptidic side products that Schally was led to believe that the peptide in their TRF was not the principal component. These nonpeptidic components turned out to be classical contaminants, cellulose, dextran, myrystoleic acid, leached from the equipment used in the last stages of purification, as finally recognized by Schally on p. 1103 of Vol. 9, *Biochemistry*, 1970, in a footnote acknowledging that their preparation of TRF was never better than 65% pure.

these hypothalamic releasing factors were not much else than a lasting fancy of Guillemin's vivid imagination." Having anticipated that this might well be the fate of this revolutionary manuscript, I had sent in early March a short note to the French Academy of Sciences (that was to appear as *C. R. Acad. Sci. Paris*, Vol. 268, 2116–2118, 1969). I returned the manuscript to *Science* with what I thought was a clear and careful rebuff of such ridiculous comments, and also with new data, only to have it returned a month later with a new comment by the same referee that he had just read (?) the note in French describing the same results; *Science* did not deal in repetitions. That fateful referee's knowledge of French had to be about as profound as his knowledge of and, I would guess, his contribution to the field as shown by his earlier comments: the new version to *Science* had the evidence of the pyro-Glu NH_2-terminus and other additional new data leading to conclusions proposed but not proven in the earlier note in French. There is no doubt in my mind that had *Science* published that note when we sent it the unpleasant exchanges that were to follow later on in the year with Folkers and Schally about the priority of our characterizing the primary structure of TRF would never have occurred (or who knows!).

May I also reflect here for a few sentences about the too-prevalent inexactitude of "authoritative" reviews, my equivalent of Oscar Wilde's complaints (in "The Decay of Lying") about historians who had fallen into "careless habits of accuracy." Volume IV, Part 2, *Endocrinology*, in the remarkable series of *Handbooks* published by the American Physiological Society, deals in three different chapters with various aspects of the isolation of TRF. Each gives a different version, none totally correct if correctness is to be judged by at least accurate quoting of the published literature—all the published literature—obviously a time-consuming endeavor not for the "casual reader." One of these chapters (pp. 563–586) is specifically devoted to the chemistry of the hypothalamic hypophysiotropic peptides. It is so inaccurate both in what it says (for instance, and in several places, that treating Glu-His-Pro with *glacial acetic acid* generates pGlu-His-Pro—one of many more such inaccuracies) and in what it does not say (the exhaustive review on the isolation and characterization of TRF in *Vit. Horm.*, Vol. 29, 1–39, 1971, that I mentioned above *is not even quoted*, publications from our laboratory on isolation and structure of ovine LRF are not even mentioned, etc.) that it should have been corrected in print by its author. His letter of apologies to me upon my calling his attention to some of these gross inaccuracies was a friendly but personal gesture. History of science in the making is obviously fraught with the same problems as those encountered in history of science as made in a far ago past. Excuses for laxity in reporting the former are far less acceptable than difficulties in dealing with the latter. The protagonists of the various episodes involved, their scripts, and their personal approach were all available. I wish the author of that review had contacted me or Roger Burgus before the text in the *Handbook of Physiology* went to press. I trust that he would have thoughtfully rewritten the inaccurate parts of it; thus would have been avoided at least one instance of one of the not so minor plagues of the current prolific scientific literature: that of the inaccurate magisterial review, forever imparting inexact knowledge to at least one generation of students.

I have said and written on several occasions that I consider the isolation and characterization of TRF as the major event in modern neuroendocrinology, the inflection point that separated confusion and a great deal of doubt from real knowledge. Modern neuroendocrinology was born of that event. Isolations of LRF, somatostatin, and the recent endorphins were all extensions (as there will be still more, I am sure) of that major event—the isolation of TRF, a novel molecule in hypothalamic extracts, with hypophysiotropic activity, the first so characterized. I am happy that Geoffrey Harris was still alive when that happened. I have a letter from him in which he expressed in very friendly terms his great satisfaction with that happening. The event was the vindication of 14 years of hard work[4] within the paradigm of a hypothalamic neurohumoral control of adenohypophysial secretions. From observation of what has happened in neuroendocrinology since 1969, the isolation of TRF was also the vindication of my early decision, as a physiologist, that the most heuristic event in neuroendocrinology would be the isolation and characterization of the first one (any one) of the then-hypothetical hypothalamic hypophysiotropic factors.

After TRF, pioneering in neuroendocrinology ceased and became the harvesting of a new and expanding science.

What about the cost? Over the years (from 1953 to 1969) the funding of my laboratory was almost exclusively by the National Institutes of Health (NIH), more specifically the National Institute for Arthritis, Metabolic, and Digestive Diseases (NIAMDD), with a few tactically important contributions from the Ford Foundation, the Population Council, and the Markle Foundation. Making rather simple assumptions based on the budget of my own laboratory and what I was reading in newspapers (assuming that all work and expenses from 1953 to 1969 were necessary for, involved in, and responsible for the isolation of the first 1 mg of ovine TRF), I once calculated that that 1 mg of native, pure, ovine TRF made a kilogram of pure, native TRF, 2–5 times *more* expensive than a kilogram of moon rock brought back from the Apollo XI mission. Today, the cost of synthetic TRF is a few cents per milligram.

15

Béla Halász

Béla Halász was born on July 4, 1927, in Kalaznó, Hungary. He received his M.D. from the University Medical School, Pécs, Hungary, in 1954, after which he joined the Department of Anatomy of the Medical University. After working in this department as assistant professor and later as associate professor, in 1971 he became professor and chairman of the Second Department of Anatomy, Histology, and Embryology, Semmelweis Medical University Budapest, Hungary. As a Ford Foundation Fellow he spent 1 year (1964–1965) in the Department of Anatomy and Brain Research Institute, University of California, Los Angeles. He received a Doctor of Science degree in 1972. In 1976 Dr. Halász was given the Ulf von Euler award of the International Society of Endocrinology.

The Hypophysiotropic Area

BÉLA HALÁSZ

As a first-year student of the Medical University of Pécs, Hungary, I was greatly impressed by the brilliant and most stimulating lectures on human anatomy, histology, and embryology delivered by J. Szentágothai, then professor and chairman of the Department of Anatomy. Instead of pure descriptive morphology, he always discussed structure in relation to function on a broad biological basis and in a very colorful manner. He never failed to distinguish between what was already known and what awaited future clarification.

I had the good fortune to join the staff of this department in 1950, as a second-year medical student. At that time, research there was partly concerned with neurohistology and neuroembryology and partly with neuroendocrinology. I took more interest in the latter field and became increasingly attached to it. B. Flerkó and B. Mess had already worked there doing neuroendocrine studies. At that time, the department had only one technician. Although she did her best, she was not able to keep pace with all the histological work. As a young student, my first "neuroendocrine" task was to cut hypothalami into serial sections and to stain them with hematoxylin and eosin for histological control of hypothalamic lesions.

Since I had no idea what kind of research to do, very soon Szentágothai suggested that I study pituitary ACTH secretion. This was for a very simple reason. B. Flerkó was already engaged on pituitary gonadotropic function and B. Mess on the control of thyrotropic secretion, while the person who was dealing with the control of ACTH had just left the department. So I took over this last field and was soon fascinated by its possibilities.

The first experiment I did in collaboration with a very bright and skillful friend, L. Szöllössy, was proposed by Szentágothai. Previous experiments in the department (Fülöp, 1952) had shown that electrolytic lesions of

BÉLA HALÁSZ • Second Department of Anatomy, Histology, and Embryology, Semmelweis University of Medicine, Budapest, Hungary.

the hypothalamic ventromedial area resulted in enlargement of the nuclei of the zona fasciculata cells of the adrenal cortex. Our task was to investigate the role of the peripheral nervous system in this effect. We interrupted the splanchnic nerves on one side and placed bilateral electrolytic lesions in the area of the ventromedial nucleus (Halász and Szöllössy, 1953). Szentágothai taught me how to make hypothalamic lesions and to carry out dissections of the animals. Among much else, I learned from him that one should get as much information as possible from any one animal. Therefore, in every case, the pituitary and all endocrine target glands had to be checked. This approach proved very advantageous later when I became engaged in intrahypothalamic pituitary implantation studies and experiments with hypothalamic deafferentation. Neuroendocrinologists might obtain more useful information from their studies if they did not strictly limit their interests to one trophic hormone. After all, in trying to evaluate all the data available on the neural structures involved in the control of the various pituitary trophic hormones, it becomes evident that approximately the same regions control the secretion of ACTH, gonadotropic hormones, growth hormone, etc. But this fact is not well reflected in most publications, because most authors deal only with one hormone. We are still unable to answer the simple question of whether there are distinct regions within the hypothalamus and the limbic system which control the synthesis and release of the different trophic hormones or whether a more diffuse control system has separate outputs for the control of various trophic mechanisms. It also is possible, indeed likely, that several of the controlling elements are involved in other functions of the central nervous system as well. Szentágothai imbued me with a fondness for research, and taught me exact scientific thinking, the manner in which a scientific question should be put and examined. His ingenious, often bold, ideas, outstanding personality, and human qualities had a very profound effect on me.

Another influential university teacher was Sz. Donhoffer, professor of pathophysiology. I learned from him the need for a critical attitude in the interpretation of experimental findings. In his lectures he usually preceded step by step in mentioning the data which brought more and more light to a question. When the students thought that the subject matter was clarified, he began to refer to findings that did not fit in with the concept formulated on the basis of the previously cited results. This mode of presentation left me with a feeling for the complexity of biological mechanisms. Sz. Donhoffer's clarity and logical way of thinking deeply affected my approach to a problem.

In the 1950s the equipment for neuroendocrine research in the Department of Anatomy at Pécs was rather limited (Szentágothai, 1975). Hypothalamic lesions, hormone treatments, and implantations of endocrine tissues were the main experimental procedures. Evaluation of the hypothalamic-pituitary-target gland system was based primarily on weight and histology of the endocrine glands, and on some quantitative morphological methods such as karyometry, which is a fairly time-consuming technique. Having become interested in the location of the hypothalamic structures involved in the control of pituitary ACTH secretion, we adopted an indirect approach and studied the effect of adrenalectomy, ACTH and cortisone treatment, formalin stress, etc., on the size of the cell nuclei of the various hypothalamic cell groups. There was experimental evidence that the size of the nuclei of the neurons might be a good measure of the functional activity of the nerve cells (Falk, 1954; Hertl, 1955), and we obtained quite unexpected results. The cell nuclei of the ventromedial nucleus were shrunken following ACTH treatment or formalin stress, and, conversely, were swollen after adrenalectomy or cortisone treatment. These findings could not be explained by changes in ACTH secretion because ACTH treatment and formalin stress had the same effect; this also holds true for adrenalectomy and cortisone treatment. On the basis of our observations, we speculated in the following way: The common feature after ACTH treatment and after administration of stressful stimuli was that in both cases the adrenals were enlarged, while following cortisone administration the adrenals were smaller, and after adrenalectomy they were not present. We suspected that there might exist nervous connections between the adrenal cortex and the hypothalamus over which such information might be conveyed centripetally. Further studies, which demonstrated that unilateral adrenalectomy caused different changes in the hypothalamic ventromedial nucleus of the two sides (Halász and Szentágothai, 1959) supported this view. This series of experiments shows well how an original idea in a research project can give rise to quite unexpected findings.

I did a number of experiments to gain more direct information on the functional significance of the nervous connections between the adrenal cortex and the hypothalamus, but without success. It should be mentioned that the most recent studies of Engeland and Dallman (1975) and Engeland *et al.* (1975) are promising in this respect. According to them, the pathway between the adrenal and the hypothalamus is of great importance. They demonstrated (1) that corticosteroids and ACTH are not required for an increase in adrenal weight (Engeland *et al.*, 1975), and (2) that unilateral electrolytic lesioning of the ventromedial nucleus of the hypothalamus (Engeland and Dallman, 1975) or (3) hemitransection of the spinal cord (personal communication) prevents compensatory hypertrophy of the adrenal gland following unilateral adrenalectomy.

I also studied the site of corticoid feedback action (Halász, 1955) and the time course of nuclear size changes in the cells of the adrenal cortex following ACTH, desoxycorticosterone, or cortisone treatment (Halász, 1958). The results, although providing us with some new information, were not satisfactory. Therefore, Szentágothai became concerned about the prospects of further research along this line. He even suggested that I should consider switching from neuroendocrinology to another field. Although taking his comments seriously, I was already so deeply interested in neuroendocrinology that I could not easily change.

In the 1950s, B. Török, working then in our department, studied the vascular supply and blood flow of the dog hypophysis. He observed (Török, 1954) that there was some upward blood flow, toward the hypothalamus, in the vessels of the subependymal layer of the median eminence that have connections with the portal vascular system. This raised the possibility that pituitary hormones might act directly on the hypothalamus. To test this hypothesis, we implanted pituitary tissue into the hypothalamus of non-hypophysectomized animals. The intrahypothalamic implantation experiments with ovarian tissue by Flerkó and Szentágothai (1957) were already under way, and their findings appeared to be very exciting. We assumed that the grafted pituitary might release some ACTH, and this, if it had a direct action on the hypothalamus, would lead to inhibition of ACTH secretion by the animal's own hypophysis. We could use only a very crude and indirect method for evaluation; we determined the mean size of the nuclei of the zona fasciculata cells and regarded that as an indirect measure of adrenocortical activity. The brains bearing the implants were serially sectioned and initially stained only with hematoxylin-eosin, because we were not interested in the histology of the grafted pituitary tissue. We only wanted to know the exact location of the implants and whether or not the graft had taken. We found that if the implant was in the medial basal hypothalamus the size of the nuclei of the zona fasciculata cells decreased (Halász and Szentágothai, 1958, 1960). This was the first experimental evidence for direct feedback on the hypothalamus of ACTH (internal or short-loop feedback).

When looking at the serial sections of the brains bearing pituitary tissue, I found, to my great surprise, that the histological structure of the grafts was strongly dependent on their exact location. Implants situated in the medial basal hypothalamus contained, in addition to chromophobe and eosinophilic cells, numerous large basophilic cells, whereas this last cell type was not present in the grafts situated outside that area.

This finding deflected my interest from the not-too-well-documented ACTH internal feedback to the question of the site of production of the hypothalamic releasing and inhibiting factors. The elegant studies of Nikitovitch-Winer and Everett (1958, 1959), demonstrating the structural and functional restitution of pituitary grafts retransplanted from the renal capsule under the median eminence, were already published. I started, together with L. Pupp, S. Uhlarik, and L. Tima, all of whom proved to be excellent collaborators, an intensive study to find out which region of the central nervous system was capable of maintaining the structure and function of pars distalis tissue grafted into the brain of hypophysectomized animals. These experiments confirmed our previous observations and indicated that the histological integrity and trophic functions of the pituitary implants were preserved only when the tissue was situated in the medial basal hypothalamus. Grafts located in this region, which we called the "hypophysiotropic area" (Halász *et al.*, 1962), contained a large number of periodic acid-Schiff (PAS) positive basophilic cells, whereas this cell type did not appear in implants in any other location.

It should be mentioned that recent immunocytochemical studies of intracerebral pituitary grafts made by Dubois and by Sétáló (personal communications) have shown that each cell type in the anterior lobe, including LH- and TSH-positive cells, occurred in every intracerebral graft irrespective of its location. The only difference was in the size of the cells. They were much larger in pituitary implants located in the medial basal hypothalamus than in grafts situated outside this region. Thus, interestingly enough, if we had used a more refined and specific technique in our intracerebral pituitary implantation studies, the existence of a hypophysiotropic area in the hypothalamus would not have appeared so pronounced as it did when working with less specific methods. The occurrence of all immunoreactive cell types in the intracerebral grafts may be due to the fact that heterotopic pituitary transplants (either beneath the renal capsule or in the anterior ocular chamber) are able to synthesize small amounts of all pituitary trophic hormones (for references, see Szentágothai *et al.*, 1968), and there may be enough for the immunocytochemical reaction.

The data obtained in animals bearing pituitary grafts in the medial basal hypothalamus suggested to us that this hypothalamic area might be the major site of the production of the releasing and inhibiting factors (hormones), and thus might play a key role in the control of the pars distalis. However, there was no indication whether this area simply transmitted the information from other hypothalamic and extrahypothalamic structures to the pituitary or whether it could exert a regulatory influence on the pars distalis by itself.

It was evident that investigation of pituitary function after the isolation of this region from the rest of the brain could be informative. To make an island of the medial basal hypothalamus in the rat, the removal of the forebrain, already done by Matsuda *et al.* (1963), did not seem to be a good approach because of the very short survival time (a few days only) of such animals. Therefore, a technique was needed to interrupt the connections between the medial basal hypothalamus and the rest of the brain without removal of the forebrain and severance of the pituitary stalk.

My first idea was that this could be done by a special, funnel-shaped, thin plate system slightly compressed from both sides to match the elongated shape of the hypophysiotropic area. However, the construction of such a device appeared to be extremely complicated, so I dropped this idea and tried to think of a simpler approach. Having pondered about this problem for several days, I woke up one night and told my wife that I knew how the hypophysiotropic area could be isolated without severing the pituitary stalk. Then I described the knife assembly which we (L. Pupp and I) constructed the next day (Halász and Pupp, 1965).

After several trials and changes in coordinates, we were able to cut around the medial basal hypothalamus completely. When looking at such brains under the dissecting microscope, we were pleased to see that the knife cut was visible and that the isolated area could be detached easily from the brain by a gentle push with a fine needle. Of course, our big concern was still not solved, because we did not know whether the rats would survive after such an operation. Much to our surprise, they did well, and the mortality rate was rather low. Our surprise was even greater 3 weeks after surgery when we autopsied the animals. There was very little or no atrophy of the endocrine glands.

At the beginning of our hypothalamic deafferentation studies, we were so excited and interested in the functional capacity of the isolated rat medial basal hypothalamus that we completely forgot to take the blood supply of this region into account. We learned only later that the vessels reach the hypophysiotropic area from the base of the brain and that this region can be cut around without causing serious damage to the area.

Shortly after we had worked out the technique of hypothalamic deafferentation, there was a change in the chairmanship of the department. In 1963 J. Szentágothai accepted the chair of the First Department of Anatomy, Histology, and Embryology at the Semmelweis University Medical School in Budapest. B. Flerkó was appointed to replace him at the Department of Anatomy in Pécs. However, this personal change did not affect the direction of the department. Flerkó most valiantly continued the work of his predecessor. The standards of experimental work, as well as the friendly atmosphere in the department, did not change at all. Flerkó provided me with his greatest support and ensured very good possibilities for my further research.

When we started our complete deafferentations of the medial basal hypothalamus, I did not think that this operation also could be used for studying the functional significance of afferents to the medial basal hypothalamus. This became evident only when we observed that basal secretion of pituitary trophic hormones was maintained following the neural isolation of the medial basal hypothalamus, but that cyclic gonadotropic function as well as diurnal fluctuations in pituitary ACTH secretion failed (Halász and Gorski, 1967; Halász *et al.*, 1967*a,b,c*). The excellent and very thorough studies of Everett (1969) suggested that the afferents to the medial basal region of the hypothalamus critical for ovulation probably reach this area from an anterior direction. Therefore, it was to our surprise that neither what we called "incomplete deafferentation" (interruption of the bilateral, superior, and posterior connections of the medial basal hypothalamus) nor a frontal cut behind the optic chiasma (severing only anterior pathways of the hypophysiotropic area) blocked ovulation. At first I thought something was wrong with our deafferentation technique. But later, when looking at the histology of the brains with partial deafferentation, I found that the lack of blockade of ovulation following partial deafferentation was probably due to the fact that fibers from an anterolateral direction were not severed either by incomplete deafferentation or by a frontal cut. This assumption could then be tested by performing more extended frontal transections. None of the animals ovulated after this intervention (Halász and Gorski, 1967).

I performed a considerable part of the experiments with partial or total deafferentation of the medial basal hypothalamus as a Ford Foundation Fellow of Professor C. H. Sawyer, and worked together with R. A. Gorski at the Department of Anatomy and Brain Research Institute, University of California at Los Angeles. The year 1964–1965 I spent there was a most profitable one. I was very fortunate to have been able to work there shortly after introducing the hypothalamic deafferentation technique. At Los Angeles I had the chance to collaborate with distinguished neuroendocrinologists. Besides R. A. Gorski and C. H. Sawyer, I did some joint work with W. H. Florsheim, D. S. Schalch, M. Slusher, J. Vernikos-Danellis, and A. Yokoyama. During my stay, I attended several meetings held at various United States cities and visited a number of laboratories. In this way I was able to meet numerous neuroendocrinologists and to establish personal contact with them. I consider this to be essential for an investigator. This is one of the best ways to exchange ideas and to discuss various matters. I learned a great deal from the seminars given by visitors or staff members and was very impressed by the high standard of scientific work at UCLA. All these activities, and the outstanding hospitality of Professors Sawyer and Gorski, made my stay at Los Angeles unforgettable and had a great influence on my further research.

After I returned from Los Angeles, I continued to work along the same lines, using primarily the deafferentation technique. With K. Köves we began to search for the location of the neural structures triggering ovulation in the rat. On the basis of the available data (Everett, 1969), the medial preoptic area appeared to be one of the most probable sites. However, it was extremely hard work to prove this (Köves and Halász, 1970), and at the beginning we had no idea of the difficulties. We had cut the connections of the medial preoptic area from the front, from both sides, and from above in 20 adult female rats in order to see whether these animals ovulated after the intervention. But much to our regret, none of these rats was alive 5 days after surgery. We repeated the experiments, but in spite of all kinds of treatments the result was the same. We did not give up, but decided to perform the surgery in two stages; we first cut the connections on one side and after recovery we operated on the other side. This approach slightly improved the rate of survival, for the mortality rate of the animals could be reduced to 90%! In spite of this, several months of hard work was required to complete these experiments because hundreds of rats had to be prepared in order to have about 20 animals which survived and bore histologically proven knife cuts in the right places.

After working with the deafferentation technique for several years, I realized that it can also be used in another field, namely, for investigating the structural organization of the hypothalamus and other brain regions. In recent years we have made several studies of this type.

In 1971 I was appointed to the chair of the Second Department of Anatomy, Histology, and Embryology, Semmelweis University of Medicine, Budapest, where I had to organize a new neuroendocrine research group. Presently, we are primarily interested in the structural organization and functional significance of the various nervous elements involved in the control of the anterior pituitary. Further, we are studying the neurotransmitters participating in mediation of the impulses between the controlling elements, and are happily provided with substantial support by the Hungarian Academy of Sciences. I was greatly attached to my original University at Pécs and its Anatomy Department, but the loss of this contact has since been compensated by working in the same building with Szentágothai, sharing some facilities and developing research groups for neurobiology and neuroendocrinology.

REFERENCES

Engeland, W. C., and Dallman, M. F. (1975). Compensatory adrenal growth is neurally mediated. *Neuroendocrinology* **19**:352.

Engeland, W. C., Shinsako, J., and Dallman, M. F. (1975). Corticosteroids and ACTH are not required for compensatory adrenal growth. *Am. J. Physiol.* **229**:1461.

Everett, J. W. (1969). Neuroendocrine aspects of mammalian reproduction. *Physiol. Rev.* **31**:383.

Falk, G. (1954). Zur Frage eines nervösen Sexualzentrums bei der Ratte. Med. Diss. Marburg a.d. Lahn.

Flerkó, B., and Szentágothai, J. (1957). Oestrogen sensitive nervous structures in the hypothalamus. *Acta Endocrinol.* (*Kbh.*) **26**:121.

Fülöp, T. (1952). Veränderungen der Kerngrösse in der Nebennierenrinde nach Hypothalamus-laesionen. *Acta Morphol. Acad. Sci. Hung.* **1**:41.

Halász, B. (1955). Die Rückwirkung von Cortisonzufuhr auf den Bau der Nebennierenrinde nach Hypothalamus-laesion. *Acta Morphol. Acad. Sci. Hung.* **6**:119.

Halász, B. (1958). Der zeitliche Ablauf von Veränderungen des Kernvolumens in der Nebennierenrinde. *Acta Morphol. Acad. Sci. Hung.* **8**:193.

Halász, B., and Gorski, R. A. (1967). Gonadotrophic hormone secretion in female rats after partial or total interruption of neural afferents to the medial basal hypothalamus. *Endocrinology* **80**:608.

Halász, B., and Pupp, L. (1965). Hormone secretion of the anterior pituitary gland after physical interruption of all nervous pathways to the hypophysiotrophic area. *Endocrinology* **77**:553.

Halász, B., and Szentágothai, J. (1958). Über die unmittelbare Rückwirkung einer von Hypophysenvorderlappen erzeugten Substanz auf den Hypothalamus. *Acta Physiol. Acad., Sci., Hung. Suppl.* **14**:6.

Halász, B., and Szentágothai, J. (1959). Histologischer Beweis einer nervösen Signalübermittlung von der Nebennierenrinde zum Hypothalamus. *Z. Zellforsch. Mikrosk. Anat.* **50**:297.

Halász, B., and Szentágothai, J. (1960). Control of adenocorticotrophic function by direct influence of pituitary substance on the hypothalamus. *Acta Morphol. Acad. Sci. Hung.* **9**:251.

Halász, B., and Szöllössy, L. (1953). Einfluss peripherischer Denervation auf den hypothalamischen Kernvergrösserungseffekt der Zona Fasciculata der Nebennierenrinde. *Acta Morphol. Acad. Sci. Hung.* **3**:1.

Halász, B., Pupp, L., and Uhlarik, S. (1962). Hypophysiotrophic area in the hypothalamus. *J. Endocrinol.* **25**:147.

Halász, B., Florsheim, W. H., Corcorran, N. L., and Gorski, R. A. (1967*a*). Thyrotrophic hormone secretion in rats after partial or total interruption of neural afferents to the medial basal hypothalamus. *Endocrinology* **80**:1075.

Halász, B., Slusher, M., and Gorski, R. A. (1967*b*). Adrenocorticotrophic hormone secretion in rats after partial or total deafferentation of the medial basal hypothalamus. *Neuroendocrinology* **2**:43.

Halász, B., Vernikos-Danellis, J., and Gorski, R. A. (1967*c*). Pituitary ACTH content in rats after partial or total interruption of neural afferents to the medial basal hypothalamus. *Endocrinology* **81**:921.

Hertl, M. (1955). Das Verhalten einiger Hypothalamuskerne der weissen Maus während verschiedener Entwicklungs- und Funktionsphasen des weiblichen Genitalapparates. *Z. Zellforsch. Mikrosk. Anat.* **42**:481.

Köves, K., and Halász, B. (1970). Location of the neural structures triggering ovulation in the rat. *Neuroendocrinology* **6**:180.

Matsuda, K., Kendall, J. W., Jr., Duyck, C., and Greer, M. A. (1963). Neural control of ACTH secretion: Effect of acute decerebration in the rat. *Endocrinology* **72**:845.

Nikitovitch-Winer, M., and Everett, J. W. (1958). Functional restitution of pituitary grafts re-transplanted from kidney to median eminence. *Endocrinology* **63**:916.

Nikitovitch-Winer, M., and Everett, J. W. (1959). Histologic changes in grafts of rat pituitary on the kidney and upon retransplantation under the diencephalon. *Endocrinology* **65**:357.

Szentágothai, J. (1975). Under the spell of hypothalamic feedback. In Meites, J., Donovan, B. T., and McCann, S. M. (eds.), *Pioneers in Neuroendocrinology*, Plenum, New York, p. 297.

Szentágothai, J., Flerkó, B., Mess, B., and Halász, B. (1968). *Hypothalamic Control of the Anterior Pituitary: An Experimental-Morphological Study*, Akadémiai Kiadó, Budapest, p. 110.

Török, B. (1954). Lebendbeobachtung des Hypophysenkreislaufes an Hunden. *Acta Morphol. Acad. Sci. Hung.* **4**:83.

16

Luciano Martini

Luciano Martini was born on May 14, 1927, in Milan, where he received his education in classics and medicine, and was graduated in medicine in 1950. In 1952 he was appointed assistant professor in the Department of Pharmacology of the University of Milan, and later became full professor of pharmacology in 1968, when he was appointed chairman of the Department of Pharmacology of the Medical School of the University of Perugia. He returned to the University of Milan in 1972 as full professor of endocrinology and chairman of the newly created Department of Endocrinology.

Dr. Martini is a member of many scientific societies, as well as being vice-president of the International Society for Psychoneuroendocrinology, secretary-treasurer of the International Society of Neuroendocrinology, and treasurer of the International Brain Research Organization. He has served on many national and international committees (WHO, Scientific Group of Human Neuroendocrinology and Reproduction, Karolinska Symposia, etc.), and is currently chairman of the Organizing Committee of the Vth International Congress on Hormonal Steroids and chairman of the Scientific Committee of the Italian Endocrine Society.

Dr. Martini has served the Italian Public Service as a member of the Evaluation Committee for New Drugs, and is the editor of several books on neuroendocrinology, the physiology of reproduction, and steroid pharmacology.

16

A Retrospect with Nostalgia

LUCIANO MARTINI

When I received an invitation to contribute a chapter to this book, I was very flattered to be considered a pioneer in neuroendocrinology. However, at the same time, I was very reluctant to accept the assignment. To write a chapter on my early years as a scientist, I would have had to fill the manuscript with lots of "I's." But I have a sort of allergy to using this pronoun in scientific publications and in all my papers I have made a point of never using expressions like "I have shown" or "I have reported" or "I have done." I have always adopted more general sentences, like "it has been reported from the Milan laboratory," "it has been found by this group," "it has been shown by this department." This is not because of modesty (a scientist cannot be modest), but because I have always felt (and still do) that, in today's science, organized teams are more productive than individuals. When an experimental project is terminated, it is indeed always very difficult (sometimes impossible) to establish who the real "author" is. The person who had the first idea? The man or woman who developed it? The group who discussed the data critically at several meetings? How much is due to serendipity? However, I eventually decided to write this piece because I felt it would be relaxing to forget, just for a few days, the many problems we are presently facing in this country, and return to the years of my youth, with a sort of romantic approach.

Let me then start from the very beginning. The first questions I have to answer are "How did I get into medical school?" and "Why did I take up scientific training?" After leaving high school at the age of 17, I had to make up my mind about my future. At that time, I was quite convinced I was going to make a career in music as a pianist. I had very good training from a very famous teacher at the Milan Conservatory. He felt I was gifted enough to devote myself to music, but we were right in the middle of World War II, and he underlined just how difficult it was to become a professional

LUCIANO MARTINI • Professor of Endocrinology and Chairman of the Department of Endocrinology of the University of Milan, Italy.

performer under those circumstances. On the other hand, because of the war, there was a great need for doctors. After a lot of thought and discussion with my family, I gave up the idea of being a pianist and decided to go to medical school. It was a "status symbol" to have a son at medical school, and my family were obviously in favor of a decision of this kind. I, of course, meant to get out of the medical profession and go back to music as soon as things quieted down again. Obviously, this never happened. Music is still very important to me, and has never been a mere hobby. I am still really happy only on those odd weekends when I can sit at my piano and play.

I said I was planning to get out of the medical profession as soon as I could. As a matter of fact, I never really got into it. In 1948 a young professor had been appointed chairman of the Department of Pharmacology at the University of Milan: Professor Emilio Trabucchi. Soon after moving to Milan, he had been requested to make a speech to country doctors. He knew that Giovanni Pascoli (one of the major Italian poets at the turn of the nineteenth century) had written about country doctors and he wanted to quote some of his verses. He could not find the poem in his books and asked a friend of mine, who was already working in his laboratory, how he could get hold of the Pascoli book. This friend of mine (who is now Professor Trabucchi's successor) knew that my father had a very large private library, and asked me if I could find the book Professor Trabucchi needed. The only reason, therefore, that I went to the department was to take the book. I was immediately fascinated by Professor Trabucchi's personality and captivated by his devotion to science, as well as by the enthusiasm which animated the young investigators in his laboratory. I was also attracted, I must confess, by the complexity of the equipment. At that time, my "technical knowledge" was limited to the differences between pianos made by Erard in Paris and by Bösendorfer in Vienna. The most sophisticated device I knew was the "double escapement" of my Steinway. In retrospect, this seems very naive, considering the complexity and style of the equipment in use in the 1970s. But my entire approach to life had been naive up to that stage. I immediately asked Professor Trabucchi whether I could prepare my graduation thesis in his department, and I was assigned a job on the pharmacology of a substance which had just been discovered by U. S. von Euler: norepinephrine.

My shift to endocrinology occurred a few years later. I had decided to stay in research even if I was faced every day with incredible financial problems. Research money was almost nonexistent in Italy at that time. Through the contacts Professor Trabucchi had with several drug companies, we were lucky enough to receive the animals free of charge and to obtain, from time to time, small donations. My enthusiasm certainly helped me in overcoming these problems. Obviously, one should not forget that the methods used in the 1950s were not as expensive as those utilized nowadays. In the department there was a group of investigators working on vasopressin. Under the influence of a paper by C. N. H. Long, of the theory of stress by H. Selye, and of some magnificent experiments made by C. Fortier which had just appeared, I was testing the effects of norepinephrine on the release of ACTH. At that time, I had a "good" method for measuring the release of ACTH: the drop in the number of peripheral eosinophilic cells. For some reasons, I was led to determine whether vasopressin injected into rats and dogs might decrease the number of eosinophils in the general circulation. This proved to be the case in both species after either systemic injection or injection into the cerebrospinal fluid of the cisterna magna (Fraja and Martini, 1952). Accordingly, I postulated that vasopressin might exert an ACTH-releasing effect. In the meantime, other methods became available to quantitate the release of ACTH. These included the evaluation of the fall in adrenal ascorbic acid and adrenal cholesterol induced by the hormone. I repeated the injections of vasopressin, using these two parameters as end points, and was able to confirm that vasopressin was a very potent ACTH-releasing agent (Bertelli and Martini, 1952; Martini and Morpurgo, 1955). At that time, Italy was very conservative, and it was considered rather odd to publish scientific papers in foreign journals. What is more, my English was almost nonexistent. Consequently, my observations were presented either to the Società Italiana di Biologia Sperimentale or to the Accademia Medica Lombarda, which published their journals in Italian. I soon became aware of the fact that other research workers were investigating the same problem and obtaining comparable results. I would like to mention them because they soon became very close friends: Drs. Smelik and de Wied, at that time in Gröningen, Professor Stutinsky, at that time in Paris, and Dr. McCann, at that time in Philadelphia. A little later, a very interesting paper was published from a Canadian laboratory. Drs. Saffran and Schally, at McGill University in Montreal, had started purifying posterior pituitary extracts with a complex paper chromatography system and were able to identify a fraction devoid of vasopressin activity which was still able to stimulate the release of ACTH. They named the fraction corticotropin-releasing factor (or CRF). Then a scientific fight started between people who believed that vasopressin was the corticotropin-releasing factor and people who believed that the ACTH-releasing effect of vasopressin was an artifact or a side effect. The dispute still continues 24 years later, and CRF remains the most elusive of the hypothalamic hypophysiotropic hormones (see Martini, 1966, for a review). In this context, I would like to recall a remark made at a meeting a few years ago by Dr. Guillemin. While speaking about CRF, he said this abbreviation may also derive from "Constant Research Frustration." This remark is still very appropriate. While mentioning Dr. Guillemin (another pioneer in this as well as in many other fields of neuroendocrinology), I would like to mention that he is usually given the credit for having invented the term "hypophysiotropic" for hypothalamic factors (Guillemin *et al.*, 1961). I believe I should share the credit, since in the same year I used the same term in a paper prepared in collaboration with Dr. Giuliani and others (Giuliani *et al.*, 1961). Probably our common background in European classical schools, where a lot of training in Latin and Greek is provided, explains the coincidence.

For the reasons mentioned, my data did not reach an international audience immediately, but I was given an excellent opportunity to present my work to a large audience in 1956. Italy was in the process of entering what was later known as the "Italian economic miracle." I was approached by the director of an Italian drug company who wanted to enter the field of what we now call "the hypothalamic hormones." He suggested that the organization of an international meeting on the hypothalamus might be very useful to his staff, and help them to get to know all the results available in this field. He also suggested that I act as scientific secretary of this symposium. I accepted the task without realizing that many problems would arise. My only experience of international meetings came from my participation in the Symposium on the Neurohypophysis, organized in Bristol by the late Professor H. Heller. I still recall the state of anxiety in which I lived during the last few weeks before the official opening of the Symposium Internazionale sul Diencefalo. However, the meeting was considered a success. In retrospect, the following were the three most important features of that conference for me. First of all, I met the late Professor G. W. Harris for the first time. He gave me a lot of useful suggestions on the program of the meeting when I showed him the draft I had prepared, and this was the beginning of a long and deep friendship. From that time, his advice was always crucial in the development of my scientific career. Second, the Symposium brought me into personal contact with many people interested in the hypothalamus at the time. I like to recall in particular the names of Drs. Assenmacher, Bargmann, Cross, Croxatto, Hess, Hume, Miahle-Voloss, and Saffran. One exception was Dr. McCann, who was forced to refuse my invitation because he was on his way back to the United States after spending a year in Sweden working with Bengt Andersson. But even more important, the meeting provided the first opportunity for three very distinguished Hungarian scientists to come to Western Europe after the war. I felt (and still feel) proud of having been able to open the "Iron Curtain" to Professors Kiss, Korpassy, and Szentágothai, even for only a few days. I was also fortunate enough to be able to make one of their dreams come true: to go to La Scala Opera House. I still remember Professor Korpassy, a sensitive musician, moved to tears while Giuseppe di Stefano and Maria Callas were singing the duet of Verdi's "Masked Ball." However, a few months later I learned that Professor Korpassy had committed suicide. I always wonder whether his sad decision had been brought on by his trip to Italy and by the new contacts he had had with freedom and prosperity.

As I said before, I had several contacts with Professor Harris in the preparation of the symposium. In this context, there is a short story I would like to tell. In 1954 I was granted a 6-month scholarship by the Italian Foreign Office for work in England. As soon as I got this news, I wrote to Professor Harris (at that time at the Institute of Psychiatry in London) asking whether I could join Drs. Donovan and van der Werff ten Bosch in his famous laboratory. I received a short polite letter back: admission was denied. I must confess I was really offended, the more so since at that time I could not understand the reason Harris gave for his refusal: he told me he could not accept anyone for less than 2 years because it would not be profitable, either to the new arrival or to his laboratory. Years have passed, and I now know just how right he was. Since 1967 I have been running a training program on reproductive endocrinology (supported by the Ford Foundation in New York), and I myself had to make a rule that no trainee could come to my laboratory for less than a 2-year period. I am indebted to Professor Harris in many ways. This is an example. Anyhow, Professor Harris's refusal put me in a rather difficult position. According to the rules of the Italian Foreign Office, I had to spend my scholarship in England, and my second choice was the Department of Pharmacology at Oxford, where I worked under Dr. John Walker. Fortunately, my stay was short, and consequently I published only an abstract. There would otherwise have been a publication by Martini and Walker (i.e., an Italian vermouth and a famous Scotch whisky: a sort of Manhattan cocktail). This would have been very embarrassing for me, especially after I had spent a year in Belgium (where I worked in classic pharmacology under Professor Corneille Heymans, the Nobel prize winner), and had published a couple of papers with Dr. van den Heuvel. Anyone who has been to Belgium knows this is the name of a very famous beer. At that time, I used to feel very frustrated to see my name associated with all kinds of drinks. Now, I must confess, every time I see the ad for the famous Italian vermouth in a new or old airport, I get the feeling it is giving me a "special welcome" to the country I am visiting.

Let's go back to the work in my laboratory. I was still interested in CRF and in the mechanisms of control of ACTH secretion, and was looking for a new method of evaluating the ACTH-releasing properties of different hypothalamic factors, all the more so since we had started a joint program with the Organon company and in particular with Drs. Tausk and

Overbeek. It was the time of the boom in the new synthetic corticoids. We decided to test whether some of these might block ACTH secretion better than corticosterone (the physiological corticoid in the rat), without interfering with the determination of this hormone. The end point of our assay was the peculiar fluorescence of corticosterone in sulfuric acid. While screening a large group of synthetic steroids, we found that dexamethasone possessed both attributes: it was a strong inhibitor of ACTH secretion, and it did not interfere in our fluorometric method. I think we were the first to make this observation even if we were again late in making it known abroad for the usual reasons (Giuliani *et al.*, 1961). The discovery was not only important to our work on CRF: dexamethasone also sparked off our interest in steroids, a field which we have never abandoned.

When I started working on steroids, I was attracted by the discovery of Dr. G. Pincus that human fertility could be controlled by appropriate combinations of estrogens and progestogens. The ethical and sociological implications of this finding were especially striking, and the interests of my laboratory were diverted more and more toward the study of processes involved in reproductive physiology, and of the pharmacological properties of possibly contraceptive sex steroids and their derivatives.

I thus discovered a totally new world. New and long-lasting friendships were also established, for instance, with Dr. Pincus himself and with Dr. E. Diczfalusy. What a great influence these two men have had on my scientific life. They have also facilitated the survival and the expansion of my laboratory by introducing me to the officers of the Ford Foundation and by making it possible for me to obtain financial support from this philanthropic organization. Without the generous support of the Ford Foundation, it would have been impossible to attract to the laboratory a large group of devoted Italian scientists from the new generation, and a series of foreign investigators whose presence in the laboratory has been extremely stimulating. With Goody (as his closest friends were allowed to call Dr. Pincus), I organized the First International Congress of Hormonal Steroids in 1962 in Milan. Everybody interested in this field knows that the International Congresses on Hormonal Steroids, organized every 4 years in different continents, are attended regularly by over 1200 scientists.

In 1962 Dr. McCann indicated that the hypothalamus contained a factor able to release LH from the anterior pituitary. He named this factor "luteinizing hormone releasing factor." I was interested in this new discovery, and immediately got my coworkers to start looking for other "hypophysiotropic" principles which might be present in the hypothalamus. As usual in science, the crucial point was the availability of a good method for evaluation of these possible hypothalamic factors. Radioimmunoassays were not available, and consequently pituitary hormones could not be detected in the blood with a satisfactory degree of certainty. Therefore, I decided to see whether hypothalamic extracts could deplete the pituitary gland of its hormones, because if a hypothalamic extract contained a releasing factor the stores of the appropriate pituitary hormone should be depleted after an injection of the extract. For a reason that I do not recall, we decided to inject the hypothalamic extracts into the carotid artery, a procedure which soon became standard in our and in several other laboratories. After some preliminary work, we standardized a procedure which permitted the simultaneous evaluation in a hypothalamic extract of LHRF, FSHRF, GHRF, and TSHRF. This procedure, which has passed into the literature as the "pituitary depletion method" (see Motta *et al.*, 1970, for review), was immediately adopted by several groups of investigators, including Schally's, who for several years used it as a screening method in the purification of hypothalamic principles. In this context, I would like to quote a surprising and indeed fascinating fact. For the evaluation of the LH content of the pituitary of the animals pretreated with different hypothalamic extracts, we needed a sensitive and specific method. Dr. Albert Parlow had elaborated for that purpose the "ovarian ascorbic acid depletion method" derived from Sayer's adrenal ascorbic acid depletion method, with which we were familiar. The only information we had on Parlow's method was the short abstract in the *Federation Proceedings*, and it always strikes me as remarkable that we were able to use this new method correctly after a few days. How Al could include all the necessary details in 200 words I do not know, but there was sufficient information. I keep telling my co-workers that any time they are faced with the difficult task of writing an abstract they should follow that example.

One of the major discoveries of the 1950s was 5-hydroxytryptamine. This substance was isolated from the gut of several species of vertebrates by Professor Erspamer (working in Parma, Italy), who named the new compound "enteramine." The same principle was found by others in the serum (and in the platelets) of mammals and consequently was called "serotonin." Finally, others found 5-hydroxytryptamine in the brain. The name "serotonin" prevailed, even if it is clearly not appropriate, because this compound exerts neurotransmitterlike effects in the central nervous system. Serotonin brought me to investigate a group of related compounds which were shown by Julius Axelrod and Richard Wurtman to be present mainly in the pineal gland. We could soon demonstrate that melatonin and other pineal indoles inhibit gonadotropin secretion whether given systemically or after direct placement into the hypothalamus and the midbrain. They are ineffective when applied to the anterior pituitary. We postulated an indirect (CNS) mode of operation of pineal principles, and this has been subsequently confirmed with more refined biochemical and electrophysiological approaches by other investigators (see Martini, 1969, and Martini *et al.*, 1968, for review). I believe that our studies, as well as those of Axelrod and Wurtman, influenced the revival of interest in the pineal gland.

Our pineal studies employed a traditional method in neuroendocrinology: the direct stereotaxic implantation of compounds into different regions of the brain and of the pituitary. We had used stereotaxic techniques extensively in our studies on feedback mechanisms which began when we discovered the potent ACTH-inhibiting activity of dexamethasone. Incidentally, we found that dexamethasone was able to block ACTH not only when implanted in the hypothalamus but also when implanted in the midbrain. This finding was very exciting, but could not be explained until it was shown (several years later) that the midbrain is a crucial area in the control of the hypothalamic-anterior pituitary complex because it is the site of origin of the now-famous ascending serotoninergic and noradrenergic pathways. An important extension of our feedback studies started when we came across a very interesting finding of Halász and Szentágothai (1960). They demonstrated that the implantation of fragments of the anterior pituitary gland into the infundibular recess of the third ventricle of the rat was followed by inhibition of the release of ACTH. The Hungarian authors argued that anterior pituitary hormones released by the implanted gland might inhibit the release of trophins from the intrasellar pituitary directly, i.e., without eliciting the secretion of hormones from the peripheral target gland. The concept of "short-loop feedback mechanisms," anticipated a few years before by the pioneer work of Sawyer and Kawakami, was definitely born. Through the courtesy of the National Institutes of Health, Bethesda, Maryland, we were able to obtain samples of purified anterior pituitary hormones, which we implanted in minute amounts directly in the median eminence of the hypothalamus and in the anterior pituitary of animals subjected to ablation of the peripheral target glands, in order to eliminate the possible interference of "long" feedback signals. Using this procedure, it was relatively easy to show that the intrahypothalamic implantation of ACTH resulted in the inhibition of ACTH release (Motta *et al.*, 1965) and that implantation of LH was followed by a significant decrease of LH secretion (David *et al.*, 1966). These data, and those obtained in subsequent experiments (see Motta *et al.*, 1969, for review), provided the first demonstration of the specificity of the "receptors" sensitive to "short feedback" signals. This is obviously an essential fact, if "short feedback mechanisms" are to be considered of physiological importance in the control of anterior pituitary function.

The idea of starting work in the area of the "short feedback mechanisms" developed after I read a paper published by two leading members of the Department of Anatomy of Pécs University. This reminds me of a story which Dr. Béla Flerkó (at that time associate professor and presently chairman of that famous department) and I have recalled very often: the way we first met. During my first trip to the United States (actually on my first day there), I was invited to visit Dr. Roy Greep (another person to whom I am deeply indebted) and his laboratory, and to attend a seminar by Dr. Béla Flerkó, who, as I learned later, was also on his first day in the United States. Béla did not know I was in the audience. He gave a wonderful presentation, and at the end a student of Roy's laboratory asked him a question. His answer was, "I wish Dr. Martini could be here to answer this question." I raised my hand, stated my name, and tried to provide the answer. Béla and the audience were astonished. It seemed a perfectly organized *coup de thèâtre*. It was not. Béla and I left the laboratory together, went out to dinner, and a new friendship began. From that day, the collaboration between our two laboratories has been close, and several of Flerkó's co-workers have joined my department for various periods of time. We particularly enjoyed having with us Dr. Béla Mess and Dr. Lajos Tima.

My story will end here. As a closing point I would like to cite two almost simultaneous events: the invitation to summarize the findings of my laboratory at the 1967 Laurentian Conference (Martini *et al.*, 1968) and the publication with Dr. W. Ganong of the two volumes of *Neuroendocrinology* which we edited for Academic Press, New York, in 1966 and 1967. At that time, Dr. Ganong and I felt that neuroendocrinology already was an established science which deserved a comprehensive review. We did not realize that our discipline was going to become even more prosperous and one of the fastest-moving fields in endocrinology. I don't want to imply that our two books were especially influential in this respect, but if they provided some stimulus for a few young investigators to enter our area of research, this is the best reward we might expect. Personally, I had an additional reward: being exposed to Fran Ganong's ideas and enthusiasm is an unforgettable experience.

At this point, I think I should reflect on the strongest stimulus which has kept me in the research field and has helped me to accept all the inevitable problems and frustrations. The answer lies in a single word which recurs throughout these pages: friendship. I hope this word will match the number of "I's" and "we's" used in this chapter. Because of the special nature of his work, a scientist meets colleagues from all over the world and all have the same devotion to their work and the same motivation. These are the common denominators on which really deep friendships will be founded. Wherever a scientist goes, he is never alone. He knows he can count on somebody who shares his interests, who knows what he likes, what he needs, in most cities in the world. Occasionally, it is necessary to argue with friends. However, I have always been astonished to see how soon a big fight between two scientists ends. At a meeting, you may hear two gentlemen at odds and may think the two will never speak to each other again. But, after a couple of hours, you go down to the bar of the hotel, and you find these very men chatting quietly together as if nothing had ever happened. I believe that scientific friendships have been and will continue to be essential for the promotion of real and peaceful international cooperation.

Let me go back for a moment to music, the point from which I began: I hope it has been a "short" feedback for the reader. There is one marked analogy between music and science: harmony. Harmony in science respects the same type of stringent rules which can be found in Bach's polyphony. In science, everything is predetermined, just as it is in Bach's music. In polyphony, one note automatically leads to another. In science, a discovery (major or minor, it does not matter) automatically leads to another advance, which those working in the area can easily predict. This is why several laboratories work on exactly the same subject at the same time. But it really does not matter, for the progress of science, whether you are personally exploiting one of your previous findings and working on the next discovery, or whether the next step is taken by others who have picked up your idea. Some people feel this is stealing from you. In my opinion, this is the best compliment one scientist can pay another: it provides the clearest indication that you were right, and that you have not "written on the water."

REFERENCES

Bertelli, A., and Martini, L. (1952). Caduta dell'acido ascorbico surrenalico in animali trattati con ormoni postipofisari. Nota 3. *Atti Soc. Lomb. Sci. Med. Biol.* **7**:430.

David, M. A., Fraschini, F., and Martini, L. (1966). Control of LH secretion: Role of a short feedback system. *Endocrinology* **78**:55.

Fraja, A., and Martini, L. (1952). Studi sui rapporti tra ipofisi anteriore ed ipofisi posteriore. *Boll. Soc. Ital. Biol. Sper.* **28**:407.

Giuliani, G., Martini, L., Müller, E., and Pecile, A. (1961). Utilizazione dell'animale pretrattato con desametazone per lo studio dei farmaci ad azione ipofisiotropica. *Folia Endocrinol.* **14**:248.

Guillemin, R., Jutisz, M., and Courrier, R. (eds.) (1961). *Etudes d'Endocrinologie*, Hermann, Paris.

Halász, B., and Szentágothai, J. (1960). Control of adrenocorticotrophic function by direct influence of pituitary substance on the hypothalamus. *Acta Morphol. Acad. Sci. Hung.* **9**:251.

Martini, L. (1966). Neurohypophysis and anterior pituitary activity. In Harris, G. W., and Donovan, B. T., (eds.), *The Pituitary Gland*, Vol. 3, Butterworths, London, pp. 535–577.

Martini, L. (1969). Action of hormones on the central nervous system. *Gen. Comp. Endocrinol. Suppl.* **2**:214.

Martini, L., and Morpurgo, C. (1955). Neurohumoral control of the release of adrenocorticotrophic hormone. *Nature* (*London*) **175**:1127.

Martini, L., and W. F. Ganong (eds.) (1966–1967). *Neuroendocrinology*, Academic Press, New York.

Martini, L., Fraschini, F., and Motta, M. (1968). Neural control of anterior pituitary function. *Recent Progr. Horm. Res.* **24**:439.

Motta, M., Mangili, G., and Martini, L. (1965). A "short" feedback loop in the control of ACTH secretion. *Endocrinology* **77**:392.

Motta, M., Fraschini, F., and Martini, L. (1969). "Short" feedback mechanisms in the control of anterior pituitary function. In Ganong, W. F., and Martini, L. (eds.), *Frontiers in Neuroendocrinology, 1969*, Pergamon Press, New York, pp. 211–253.

Motta, M., Piva, F., Fraschini, F., and Martini, L. (1970). "Pituitary depletion methods" for the bioassay of hypothalamic releasing factors. In Meites, J. (ed.), *Hypophysiotropic Hormones of the Hypothalamus: Assay and Chemistry*, Williams and Wilkins, Baltimore, pp. 44–59.

17

Samuel McDonald McCann

Samuel McDonald McCann was born in Houston, Texas, in 1925. After finishing premedical studies at Rice University in 1944 he entered medical school and received his M.D. in 1948 from the University of Pennsylvania School of Medicine. After an internship and residency in medicine at the Massachusetts General Hospital in Boston there followed a period of military service during which he was assigned to the Walter Reed Medical Center in Washington and carried out neuroendocrine research. Returning to academia at his alma mater, in 14 years he rose from instructor to professor and served 1 year as acting chairman of the Department of Physiology at Pennsylvania. This period included a year spent in the laboratories of Dr. Bengt Andersson in Sweden. These were the years when Don was a major force in weaning neuroendocrinology from the world of the anatomist to the laboratories of the physiologist. In 1965 he moved to his present position at Southwestern Medical School in the University of Texas system at Dallas to become chairman of physiology.

In 30 years of scientific activity Don has published over 150 papers, written and edited many chapters and books, served on a number of committees for the National Institutes of Health and the Endocrine Society, and helped form the International Society of Neuroendocrinology. He serves on the council of that society. He has served several terms on the editorial boards of *Endocrinology* and *Neuroendocrinology*. Several awards reflect his abilities and accomplishments. The first, the Graham Baker Prize for the student with the highest academic average for 3 years, was received at Rice, and others include the Lindback Award for distinguished teaching at the University of Pennsylvania and the Endocrine Society Ernst Oppenheimer Award for Research. He has traveled widely as an invited lecturer but is best known for his research activities, in conjunction with which he continues the tradition of stimulation, guidance, and training of a large number of young associates, many of whom have progressed to head their own neuroendocrine research units all over the world.

In Search of Hypothalamic Hormones

SAMUEL McDONALD McCANN

I was fortunate to be born into an intellectual family since my father was on the faculty at Rice Institute and my mother had taken graduate work in French. Furthermore, my grandfather, although not formally trained, took a lively interest in natural phenomena. For whatever reason, I early developed an interest in science and did a great deal of reading, first in astronomy and later in paleontology and ichthyology. I was particularly fascinated by the book called *Backyard Exploration* and used an old microscope to study organisms in pond water. I also chased butterflies and kept tropical fish. At various stages, I was interested in a career in mathematics, physics, and chemistry, but by the time I had reached college I had more or less settled on biochemistry as my major interest.

My interest in endocrinology actually began while I was a college student and I was fortunate enough to be able to write my principal paper in freshman English at Rice on the pituitary growth hormone. I began to do research while a freshman in medical school at the University of Pennsylvania working in the laboratory of Dr. William C. Stadie. I attempted to identify the bound form of acetylcholine reported to exist by Abdon and Hammarskjiold, Swedish editors of *Acta Physiologica Scandinavica*. Suffice it to say that this effort ended in failure since acetylcholine is sequestered in vesicles rather than being in the bound form, as was subsequently discovered. This early work was never published. While a sophomore, I began a project with Al Rothballer, one of my roommates (currently professor of neurosurgery at New York Medical College), and two staff people at Penn, Drs. Eleanor Yeakel and Henry Shenkin. We attempted to induce hypertension in Norway rats by subjecting them to the sound of an air jet. We were successful in this, and, influenced by the stress

SAMUEL McDONALD McCANN • Department of Physiology, Southwestern Medical School, University of Texas Health Science Center at Dallas, Dallas, Texas 75235.

concept of Selye, we adrenalectomized the animals to demonstrate that the hypertension disappeared (McCann *et al.*, 1948). Stress-induced hypertension apparently has become a useful tool in the intervening years, and I only realized this when I attended a meeting on stress in Czechoslovakia in 1973 and heard much more on it. The principal importance of this period was that it initiated my interest in the hypothalamic control of anterior pituitary hormone secretion. At this time (1948), I read with great interest the article in *Physiological Reviews* (Harris, 1948) on neurohumoral control of the anterior pituitary by the late Professor G. W. Harris. This paper was a major factor in stimulating my later work in this area.

After 2 years on the house staff at Massachusetts General Hospital in Boston, I returned to the Physiology Department at Penn only to find that I would be drafted in the doctor's draft. Consequently, I volunteered and was fortunate to be sent to Walter Reed in 1951 (through the efforts, of Dr. I. S. Ravdin), where I worked in the groups of Drs. Evelyn Anderson at NIH and D. M. Rioch at Walter Reed. With Gordon Farrell we demonstrated that intravenous epinephrine would elevate ACTH titers in blood (Farrell and McCann, 1952). At this time C. N. H. Long's group had proposed that epinephrine acted directly on the pituitary to release ACTH. In the meantime we had begun the study of the effects of hypothalamic lesions on the release of ACTH.

Because of difficulties with indices of ACTH secretion in the cat, which we first used to study the effects of hypothalamic lesions on ACTH secretion, I suggested to Dr. Anderson that it would be better to use the rat and to employ adrenal ascorbic acid depletion, which we were already using, as the index for ACTH secretion. Fortunately, we had an endocrine club which met weekly to discuss the literature, and I found out that Monte Greer had a stereotaxic instrument for the rat which he had developed himself. He was kind enough to show me how to use it in a few animals, so I made frequent journeys over to the National Cancer Institute to use his instrument after work and brought back the vicious, fighting animals to Building 3. In this connection, it is of interest that Building 3 contained quite an assortment of people at that time. Sharing our secretary was Dr. Arther Kornberg, an obscure young medically trained biochemist without a Ph.D., who subsequently won a Nobel Prize, and some of his co-workers, such as B. L. Horecker. In the basement was Chris Anfinsen, who later won a Nobel Prize. On the second floor was a technician named Julius Axelrod, who also later won a Nobel Prize. There were others of note, such as Dr. Robert W. Berliner, who is currently dean at Yale, and Dr. James O. Davis, who established himself as the world's authority on aldosterone control. Jim and I became close friends. We discussed endocrine problems at length, usually while I was tube-feeding hypothalamically lesioned aphagic cats.

At this point I had the good fortune to secure the help of Dr. Walle Nauta, who had developed the neurohistology lab for Dr. Rioch's group. He unstintingly examined at least 70 brains with me, and we concluded that the region involved in control of ACTH release was the median eminence (McCann, 1953). I asked Walle to join me in the paper, but with characteristic modesty he refused. When I submitted the paper to Dr. Rioch for his approval, he found considerable fault with it. At that point, he put the finger on each flaw in the paper such that when it was submitted to the *American Journal of Physiology* it was accepted without change. From that experience I developed a great admiration for Dr. Rioch's intellect.

As a result of finding that median eminence lesions blocked ACTH release in the rat, we quickly adjusted our coordinates in the cat and were able to find similar results using more indirect indices of ACTH secretion (Laqueur *et al.*, 1955). The lesions prevented the release of ACTH induced by epinephrine, which indicated that epinephrine did not stimulate the pituitary directly as postulated by Long's group. This was the first demonstration of the importance of the median eminence for ACTH release and fitted with the earlier work of Dey in Ranson's laboratory in which median eminence lesions had been found to inhibit gonadotropin secretion. We also noted the impairment in gonadotropin release.

When I returned to Pennsylvania I received a very hospitable welcome. I had my laboratory with Dr. Lukens in the Cox Institute (for study of diabetes) and my faculty appointment in Physiology. Even though Dr. Brobeck, my boss, did not believe in hypothalamic control of the pituitary except for a partial control exercised by epinephrine, he was willing for me to go on with our studies with hypothalamic lesions which had shown that these lesions blocked the pituitary's response to epinephrine. Similarly, Dr. Lukens supported my studies on hypothalamic control of ACTH even though he indicated that I was simply trying to add another pituitary on top of the pituitary gland. Twenty years later it is clear that we have in fact done just what Dr. Lukens had suggested.

Dr. Brobeck was the only person who was knowledgeable in the area of hypothalamic pituitary interrelationships at Penn, and he lent great encouragement during my years there. In fact, he is one of the finest, most self-effacing gentlemen that I have ever encountered.

The work on hypothalamic lesions begun while I was in the Army was the beginning of a major effort which has continued up to the present day in which we have studied effects of hypothalamic lesions on secretion of all pituitary hormones with the exception of TSH (I made an agreement with Greer to leave TSH to him) and MSH. Even in the latter case, we noted an increase in the size of the intermediate lobe of the pituitary in animals with median eminence lesions suggestive of increased MSH secretion (McCann,

1953). Initially, we made observations based on organ weights and histology and adrenal ascorbic acid depletion as an index of ACTH secretion. This was then extended to measurement of bioassayable adrenocortical hormones in the adrenal vein in the cat (Laqueur *et al.*, 1955), to measurement of bioassayable ACTH in blood of the rat (McCann and Sydnor, 1954), to bioassay of LH (Taleisnik and McCann, 1961) and FSH (Igarashi and McCann, 1964) in the blood of ovariectomized rats, and, with the advent of radioimmunoassay, to measurement of immunoassayable plasma LH, FSH, and prolactin (Bishop *et al.*, 1972*a*). It has become abundantly clear from these studies that median eminence lesions result in inhibition of the secretion of all pituitary hormones with the exception of prolactin and probably MSH.

We demonstrated the increase in prolactin release following such lesions (McCann and Friedman, 1960) and showed that the increase occurred in males as well as females, but this demonstration was antedated by the reports from Everett's lab (Everett, 1956) that pituitaries grafted under the kidney capsule in hypophysectomized animals probably secreted increased amounts of prolactin as judged by their ability to maintain corpus luteum function.

In other experiments we showed that suprachiasmatic lesions (Taleisnik and McCann, 1961) would induce constant vaginal estrus, in agreement with earlier work of Dey and Hillarp, and we further showed that these lesions permitted at least a partial maintenance of the release of LH following castration but a blockade of the LH release induced by progesterone in estrogen-primed animals. FSH release induced by progesterone was not blocked by the lesions (Bishop *et al.*, 1972*b*). These findings indicated that the stimulatory or positive feedback of progesterone on LH release was blocked by the lesions, whereas the stimulatory effect on FSH release was not blocked, presumably because the region concerned with FSH release was located more caudally than that concerned with LH release which was destroyed by the lesions. The negative feedback of steroids on FSH and LH release was still at least partially intact and might act on basal hypothalamic structures. These results supported the concept of Barraclough and Gorski that positive feedback of gonadal steroids was mediated rostrally on the preoptic region and that negative feedback was mediated in the basal tuberal region.

Interestingly, it was very apparent in our early studies that the lesions could produce selective effects on secretion of different pituitary hormones (McCann, 1953) as also observed by Bogdanove in the rat and Ganong in the dog. For example, ACTH release could be blocked whereas testicular and accessory sex organ weight remained normal and *vice versa*. Our studies with lesions indicated that hypothalamic control was mediated by a final common pathway through the median eminence.

The fact that the anterior pituitaries in these animals with lesions appeared to be well vascularized in spite of some damage to the primary plexus of the hypophysial portal system indicated that the effects were not mediated by vascular insult and were probably brought about by interruption of a neurohumoral pathway (McCann, 1953). Much later on, anterior pituitary blood flow was measured following these lesions in collaboration with John Porter and found to be, if anything, enhanced. The early finding of good pituitary vascularization in rats with median eminence lesions led, on my return to Pennsylvania in 1952, to a search for the neurohumors which might alter pituitary hormone secretion. Early attempts with rat and later beef hypothalamic extracts were unsuccessful. Initially, acetic acid extracts were tested in animals with median eminence lesions to block the ubiquitous stress-induced release of ACTH. I now know that this failure to obtain ACTH release was due to the relative insensitivity of the animals with chronic lesions which was probably caused by a combination of decreased responsiveness to corticotropin-releasing factor (CRF) and decreased responsiveness of the adrenals to the ACTH which might have been released. We later measured the decreased responsiveness to ACTH (McCann and Haberland, 1959).

In view of the reports that stress would induce ADH release, we began in 1953 to evaluate the effects of neurohypophysial peptides on ACTH release in the animals with lesions and discovered that large doses of Pitressin (Parke-Davis, vasopressin preparation) would release ACTH whereas Pitocin (Parke-Davis, oxytocin preparation) and a host of other putative agents were ineffective. At the same time we correlated the presence of diabetes insipidus, as measured by water intake, with the defect in ACTH secretion in the animals. We observed that there was an excellent correlation between severity of the diabetes insipidus and the impairment in ACTH secretion. In fact, animals which had water intakes of 200 ml or more per day failed to release ACTH in response to stress. We therefore concluded that vasopressin itself or a contaminant in the Pitressin accounted for the ACTH-releasing activity of Pitressin and postulated a possible role for the supraopticohypophyseal tract in the control of ACTH secretion (McCann and Brobeck, 1954).

At that time, I left for Sweden to work with Bengt Andersson for 1 year. While I was there, Saffran and Schally reported that neurohypophyseal extracts would release ACTH from pituitaries *in vitro* (but only in the presence of norepinephrine), in confirmation of our *in vivo* results, and they quickly reported that they could separate a factor from neurohypophysial extracts different from vasopressin which released ACTH (Saffran *et al.*, 1955). Also in this year (1955), Guillemin reported that pituitaries organ-cultured in a coculture system with hypothalami released additional ACTH (Guillemin and Rosemberg, 1955). Consequently, on my return to Philadelphia in September 1955 I took up the problem of the ACTH-releasing factor once more and attempted to confirm the work of Saffran and Schally. I was unable to obtain satisfactory mobility in the chromatographic system proposed by Saffran and Schally and turned then to one which had been proposed subsequently by Guillemin and Hearn in 1956 (Guillemin *et al.*, 1957). This chromatographic system worked well, but all of the activity *in vivo* was associated with the portion of the chromatogram which contained vasopressin and there was no activity in the fraction stated by Guillemin to contain CRF activity even when the dose of this fraction was increased by a factor of 10 above that required from the pressor portion of the chromatogram. On communication with Guillemin, a sample of his extract was tested and it had only borderline activity at doses as high as 1 mg. On the other hand, vasopressin was active at a dose of 0.2 μg (McCann, 1957).

In the meantime, we discovered that the animals with lesions were more sensitive to the ACTH-releasing activity when used 2 days after lesions rather than in the chronic situation used previously. We also tested synthetic lysine-vasopressin and found it to release ACTH in these animals. The vasopressin content of Pitressin was sufficient to account for its ACTH-releasing activity (McCann and Fruit, 1957). This was the basis for the vasopressin theory of ACTH release.

Because of a report by Royce and Sayers in 1958 that hypothalamic extract would release ACTH in our *in vivo* assay system, we returned to the problem and after much difficulty with ACTH contamination of beef hypothalami were able to show that beef and rat hypothalamic extracts, in contrast to neurohypophysial extracts, contained a CRF different from vasopressin since there was too little pressor activity in the extract to account for the ACTH-releasing activity. We therefore concluded that there was indeed a CRF distinct from vasopressin but that vasopressin also had this activity and might still play a physiological role since it could be released in high concentration into portal capillaries in the median eminence (McCann and Haberland, 1959).

In the meantime, Guillemin and Schally, and Saffran also reached the conclusion that neurohypophyseal extracts were not a good source to obtain CRF distinct from vasopressin and they also turned to fractionation of hypothalamic extracts. The early structures for CRFs, based on the work with the neurohypophysial extracts from Guillemin's laboratory, have never been verified.

Subsequently, we carried out a few additional studies on the possible role of vasopressin as a CRF. We found that there was a partial defect in ACTH secretion in animals with hereditary diabetes insipidus and lacking vasopressin and concluded, therefore, that CRF was the major factor which controlled ACTH release. In other experiments in collaboration with Yates, we obtained further evidence that vasopressin could release ACTH and that it might potentiate the action of CRF. We also found with Gale and Taleisnik that destruction of the median eminence in hypophysectomized animals led to development of diabetes insipidus, confirming early observations of Heinbecker in the dog which suggested that vasopressin probably was secreted into hypophysial portal vessels and that it might reach the pituitary in high enough concentration to release ACTH, as we originally suggested. It is of interest to note that Zimmerman and others have now demonstrated vasopressin-containing terminals in the region of the primary plexus of the portal vessels and he has detected a high concentration of vasopressin in portal blood in the monkey, which certainly indicates that vasopressin may in fact play a physiological role in the control of ACTH secretion (Zimmerman, 1976).

With the realization that there was indeed a corticotropin-releasing factor distinct from vasopressin, we quickly turned our attention to the search for other postulated releasing factors to alter secretion of other pituitary hormones and discovered in 1959 that median eminence extracts from the rat would deplete ovarian ascorbic acid in the immature PMS-HCG-primed rat, the most sensitive bioassay for LH. We continued this work with the arrival of Samuel Taleisnik in the laboratory and published our initial report on LH-releasing factor in 1960 (McCann *et al.*, 1960).

Since the initial work was based on ovarian ascorbic acid depletion in especially prepared test rats, an assay developed by Parlow, we quickly turned our attention to the evaluation of the effect of these extracts on plasma LH and were able to show that they would elevate plasma LH in ovariectomized, estrogen-primed rats. At that time, we made the first attempt to determine the localization of a hypothalamic factor and showed that LHRH was contained largely in the median eminence-arcuate region but that a small amount of activity was also found over the optic chiasm (McCann, 1962), findings that have now been confirmed after much disbelief by further bioassays and most recently by radioimmunoassay in our laboratory (Wheaton *et al.*, 1975) and later elsewhere. Most recently, the rostral localization has been confirmed by immunohistochemical techniques which indicate that there are nerve cell bodies in the preoptic-suprachiasmatic region which contain the hormone and that the hormone is also found in the organum vasculosum lamina terminalis, as we also observed from extraction of frozen sections cut through this region. During the early period, we developed with Ramirez what is still the most sensitive bioassay for the LH-releasing hormone: the ovariectomized, estrogen- and progesterone-treated rat.

There followed further experiments to look for other factors in hypothalamic extracts which would release anterior pituitary hormones. Matsuo Igarashi arrived in the laboratory in 1963 with a mouse uterine weight assay for FSH. After initial work to improve the assay, we were able to demonstrate the presence of an FSH-releasing factor in hypothalamic extracts of rat and beef origin (Igarashi and McCann, 1964).

In the meantime, we had begun attempts to purify the factors, and with the arrival of Lad Krulich from Czechoslovakia we evaluated the possible growth hormone releasing activity of the extracts. This was found to be present using an *in vivo* assay of pituitary growth hormone depletion. GH was measured by the tibial epiphysial cartilage assay.

While screening fractions from Sephadex columns for growth hormone releasing factor, Krulich discovered that certain fractions inhibited the release of growth hormone from pituitaries incubated *in vitro*, and a series of four papers was published on the growth hormone inhibiting factor (Krulich *et al.*, 1968, 1972; Dhariwal *et al.*, 1969; Krulich and McCann, 1969). In these papers we established that the factor would inhibit release of growth hormone *in vitro*, would antagonize the action of growth hormone releasing factor, had little action on the release of ACTH, FSH, and LH, and could be highly purified by further chromatography. In initial work we bioassayed the growth hormone released by pituitaries incubated *in vitro*. With the advent of radioimmunoassay for growth hormone, we confirmed both the growth hormone releasing and growth hormone inhibiting activities of fractions prepared by slightly different methods and also established the hypothalamic distribution of both the growth hormone releasing and growth hormone inhibiting factors.

In 1973 Guillemin's group confirmed the existence of the factor and took advantage of an extremely sensitive assay system which they had developed for another purpose to isolate and determine structure of the molecule, which turned out to be a tetradecapeptide (Brazeau *et al.*, 1973). This they renamed "somatostatin." The distribution of the growth hormone inhibiting factor has now been studied by immunohistochemical techniques. It has been found to be in the areas which we observed in the early experiments but also in ventromedial nucleus and many other sites. Our failure to pick it up in the ventromedial nucleus was probably caused by the presence of growth hormone releasing factor in this region which masked the activity. Somatostatin has now been shown to have inhibitory effects in a variety of sites outside the pituitary. For example, it inhibits insulin and glucagon release, gastrin release, acid secretion by the stomach, salivary secretion, and many other biological functions. It is found even in dorsal root ganglia.

in the spinal cord, as well as in the pancreatic islets and gastrointestinal tract and probably plays an important role as an inhibitory transmitter or local hormone in a whole host of tissues in the body.

During this time of examination for possible activities, we collaborated with Clark Grosvenor (Grosvenor *et al.*, 1965) to show that median eminence extracts could block the suckling-induced depletion of pituitary prolactin in the rat, which was the first *in vivo* demonstration of prolactin-inhibiting factor, a factor previously demonstrated on the basis of *in vitro* experiments by Pasteels and by Meites and co-workers (Talwalker *et al.*, 1963). Subsequently, we showed that the inhibitor would also suppress milk secretion and most recently have shown that it will also lower blood prolactin as measured by radioimmunoassay, and in particular that it can suppress the suckling-induced release of prolactin.

So from this phase of our work we were responsible for the discovery of the LH-releasing factor (although there is no doubt that Harris's group was working on this point at the same time and published their work in papers which appeared from 1961 on, e.g., Campbell *et al.*, 1964), the FSH-releasing activity of hypothalamic extracts, and the growth hormone inhibiting factor, and we provided the first *in vivo* evidence for the prolactin-inhibiting factor.

As I indicated earlier, we began attempts to purify these factors and initially Dr. Ramirez and I were able to show that it was possible to purify the LH- and FSH-releasing factors by chromatography on Sephadex G-25, but in these early experiments we could not separate the LH- and FSH-releasing activities. With the arrival of Anand Dhariwal in 1964, who had been trained in purification methods in several laboratories, including those of Guillemin and Schally, we increased our efforts along this line and were able to purify for the first time the growth hormone releasing factor, to separate the FSH-releasing factor from the LH-releasing factor, and to purify also the hypothalamic CRF, MSH-releasing factor, and PIF. All of these could be separated from vasopressin and from each other (McCann *et al.*, 1968).

During this time I had acquired more and more teaching and administrative responsibilities at Pennsylvania and so when an offer came to chair the Department of Physiology at the University of Texas Southwestern Medical School in Dallas in 1965, I accepted it with the hope that we would still be able to move ahead with the research in spite of the obvious increase in administrative responsibilities. The move did slow up our research somewhat but we were able to proceed without too much delay after moving all of our fractions down from Philadelphia. It has been a source of great satisfaction to see the enormous growth in the quality and size of the department and of this medical school in Dallas in the intervening 11 years.

When Peter Fawcett, who had previously worked with Harris on purification of LH-releasing factor, joined us in 1967, we increased our activities with the aim of isolating and eventually determining structure of the factor. There is no doubt that at this point we made an error in judgment in thinking that we could obtain sufficient amounts for determination of structure by collecting at the slaughterhouse only approximately 75,000 hypothalamic fragments per year. Our competitors, namely Schally and Guillemin, were utilizing approximately 10 times this rate of collection. Nonetheless, on the basis of various tests applied to highly purified fractions, Fawcett came to the conclusion that the molecule might contain histidine and tryptophan. We reasoned, on the basis of the previous finding that the TRF contained pyroglutamic acid as the *N*-terminal of the tripeptide, that LHRH might be a tripeptide composed of pyro-Glu-His-Trp-NH_2 or pyro-Glu-Trp-His-NH_2 and obtained these two tripeptides through the courtesy of Dr. Karl Folkers. Unfortunately, neither proved active. This was the year before the announcement of the structure of the molecule by the Schally group at the Endocrine Society Meeting in 1971 (Matsuo *et al.*, 1971). It was with considerable chagrin that we noted that the decapeptide contained at the amino-terminal end the tripeptide pyro-Glu-His-Trp-NH_2. (We were further chagrined to note a report by Guillemin claiming that this tripeptide had biological activity. This turned out later to be erroneous.)

Since that time we have continued to do purification studies aimed at isolation of the FSHRF, which has proved difficult to separate from the LHRH, and have conducted further purification of PIF in an effort to settle the question as to whether or not PIF is a peptide or whether the prolactin-inhibiting activity of hypothalamic extracts can be accounted for by dopamine as postulated by some. Our belief at the present time is that dopamine is not the sole PIF since we have not been able to block the actions of crude or partially purified hypothalamic extracts with dopamine receptor blockers.

In addition, we have recently found that incubates of ventromedial hypothalami can inhibit the release of insulin and stimulate the release of glucagon both *in vivo* and *in vitro* (Moltz *et al.*, 1977). This cannot be accounted for by the somatostatin content of the extract since somatostatin inhibits both insulin and glucagon release. Furthermore, although neurotensin and substance P also inhibit insulin and stimulate glucagon release, our extracts contain too little of these peptides to account for the activity. So we are actively working to determine if there is a peptide or peptides in hypothalamic extracts which can alter glucagon and insulin release.

On the subject of the releasing factors, we also have carried out considerable work on mechanism of action of the factors. We demonstrated with Zor and Field that hypothalamic extracts would increase cyclic AMP and adenylate cyclase in pituitaries incubated *in vitro* (Zor *et al.*, 1969) and also showed with Wakabayashi that elevated medium potassium would release not only LH but also FSH from pituitaries incubated *in vitro*. We confirmed the importance of calcium in the release process and showed that magnesium had an inhibitory effect. We have since carried out other studies on the role of cyclic nucleotides which suggest that the situation is quite complicated and cannot simply be explained by the cyclic AMP mechanism, but may also involve cyclic GMP (with Fawcett, Sundberg, and Nakano). We attempted to evaluate the role of microtubules by use of colchicine to disrupt them and found that instead of inhibiting the release process colchicine, rather surprisingly, stimulated it. This suggests that microtubules might be holding secretory granules in the interior of the cell rather than promoting their transport to the surface.

Additionally, we have carried out first by bioassay (1961–1966) and later by radioimmunoassay (1970 to present) a whole series of studies on feedback actions of gonadal steroids which point to both hypothalamic and pituitary sites for the feedback (see McCann, 1974, for review). The most recent work of this type concentrated on demonstrating that injection of estrogen in ovariectomized animals had a biphasic effect, first suppressing the response of the pituitary to LHRH and then augmenting it (Libertun *et al.*, 1974). At the time of the inhibitory pituitary response, there appeared to be also a suppression of LHRH release since intraventricular estrogen could suppress LH release even though pituitary responsiveness did not change. We further studied this by evaluating in detail changes in responsiveness to the releasing hormone both *in vivo* and *in vitro* during the various stages of the estrus cycle. These studies indicate that there is an increase in responsiveness to LHRH during proestrus in the rat, probably brought on by ovarian estrogens, that the responsiveness reaches a peak at the time of the preovulatory discharge, and that this further enhancement is probably brought on by a self-priming action of LHRH. Responsiveness then declines with respect to both FSH and LH release by the morning of estrus. Since FSH release is still augmented at this time in the face of diminished responsiveness to LHRH, these findings suggest that another factor, perhaps FSHRF, is driving FSH secretion at this time in the cycle. We worked out the details of these changes in the rat (Cooper and McCann, 1975), but Fink had clearly shown the self-priming action of LHRH, which we quickly confirmed and found to be prominent only on proestrus.

In the course of these studies, we did early bioassay experiments in which we demonstrated the preovulatory surge of LH by bioassay of plasma LH (with Ramirez) and the ability of progesterone to elevate plasma LH (with Nallar) as predicted by early studies of Everett. We also demonstrated the ability of estrogen to lower plasma LH within an hour in the ovariectomized animal and the ability of relatively small doses of progesterone to inhibit if the animal was also treated with low doses of estrogen. Furthermore, at that time Ramirez and I showed in both the male and female that castration resulted in elevation of plasma LH in immature as well as adult rats and that much smaller doses of steroids were required to hold LH secretion in check in the immature than in the adult animal. We postulated that puberty was brought about by a resetting of a hypothalamic gonadostat (Ramirez and McCann, 1963). A declining hypothalamic sensitivity to negative feedback of gonadal steroids would bring about the enhanced gonadotropin release responsible for puberty. This concept had been previously espoused by Hohlweg and by Donovan and van der Werff ten Bosch. With the advent of radioimmunoassay we confirmed the observations on plasma LH changes previously made by bioassay (Negro-Vilar *et al.*, 1973).

With the demonstration of the existence of LHRH, we began extensive studies to determine factors influencing its release and have concentrated on putative synaptic transmitters which might be involved. Early work in this area was carried out by Sawyer and Everett (Sawyer, *et al.*, 1947). We carried out pioneering experiments on the role of dopamine in stimulating LHRH release (Schneider and McCann, 1970). Since this is now controversial, we have further evaluated not only dopamine but also norepinephrine and epinephrine and have found that all three of these catecholamines can release LHRH in the animal under the influence of estrogen. We did many pharmacological studies (with Donoso, the Kalras, and Ojeda) which have pretty well established a role for nonepinephrine in the LH control system (McCann *et al.*, 1976*b*), but the role of dopamine remains controversial since Fuxe and some others are claiming that it has an inhibitory rather than a stimulatory effect. We are currently working on this and believe that the differences between the results may be accounted for by differences in the steroid background of the animal.

At the same time we carried out experiments which have clearly established that dopamine is an inhibitory transmitter to suppress prolactin release either by inducing the release of prolactin-inhibiting factor or by acting directly on the pituitary to suppress prolactin release after its uptake by portal vessels and delivery to the gland or by both mechanisms. A number of others have also been working in this area, including Drs. Meites and Porter and various clinical investigators. With the introduction of radioimmunoassay for prolactin in the human, various adrenergic and dopaminergic drugs have been administered to humans which reveal that the situation in man is apparently exactly the same as that in the rat.

We are currently attempting to delineate the role of norepinephrine in the release of LH, and Martinovic has recently shown that one can block the preovulatory discharge of LH by injecting 6-hydroxydopamine into the ventral noradrenergic tract, which carries axons of brain stem noradrenergic neurons to the hypothalamus, whereas control injections into the tract and elsewhere in the brain have no effect. This indicates that adrenergic control over LH may be mediated by noradrenergic neurons whose cell bodies lie in the brain stem.

Paul Harms, Sergio Ojeda, and Dave Sundberg have recently carried out a series of studies on the role of prostaglandins in controlling the release of pituitary hormones and have obtained evidence that prostaglandins of the E series are important in releasing gonadotropins and prolactin via a hypothalamic action (McCann *et al.*, 1976*a*). Ojeda has carried out a series of studies on the developmental changes in pituitary hormone secretion in the rat which we have recently reviewed (Ojeda, 1975).

Lastly, when in 1971 we were successful in recruiting Bob Moss, who was trained in the study of sex behavior and electrophysiology with Barry Cross, we began studies to determine if LHRH had any role in inducing mating behavior. I had developed the hypothesis that it might be involved since LHRH is localized to regions known to be involved in sex behavior and since mating behavior commences shortly after the presumed preovulatory discharge of LHRH. We were able to show that subcutaneous administration of LHRH would induce mating in ovariectomized, estrogen-primed rats which began fairly shortly after the injection and that LHRH mimicked the action of progesterone in such animals but with a shorter time course. The action was shown not to be due to pituitary or steroid hormones which might be released by LHRH and not to be shared by TRF (Moss and McCann, 1973). In subsequent experiments, Foreman and Moss showed that one can obtain the effect by microinjection of the neurohormone into hypothalamic sites. This was probably the most impressive demonstration of a behavioral effect of a releasing hormone. Others have shown that massive doses of TRF and GIF can have an alerting and depressive action, respectively. With the widespread distribution of the releasing hormones to sites in the brain outside the hypothalamus, it now appears that the releasing hormones may play important roles as either synaptic transmitters or modulators of synaptic function. This could open up completely new vistas in terms of our understanding of central synaptic transmission and might pave the way also to important new drugs active in mental illness.

Finally, I should mention a few other studies which I took part in. First, during my period in Sweden with Bengt Andersson in 1954–1955, I was involved in the microinjection of hypertonic saline into various goat hypothalamic loci, which induced drinking of very large amounts of water. We then went on to study the effects of electrical stimulation and outlined in detail the region in the hypothalamus which would provoke massive drinking (Andersson and McCann, 1955). We correlated points of stimulation which produced drinking with points which caused release of either ADH or oxytocin and were able to find that in some cases there was overlapping between the points which induced drinking and those which induced release of neurohypophysial peptides but that in other instances a selective effect was obtained either on drinking or on hormonal release. We proceeded to make hypothalamic lesions in the same region in dogs and found that we could abolish drinking in these animals. Bengt has continued to work in this area in subsequent years. I have not done very much along these lines except to show with Smith that lesions in the median eminence could induce polydipsia in rats which were nephrectomized to eliminate variable renal water loss. This suggested that these lesions might be destroying a satiety center for drinking analogous to the satiety center for feeding. I might say that we have just recently returned to this and are now evaluating the effects of these lesions in animals with hereditary diabetes insipidus, which would completely eliminate the possibility of any contribution of vasopressin to the effects.

At the time I was in Sweden, we noticed that there was a natriuretic effect of some hypothalamic stimulations. Much later when Raul Orias joined us, he studied natriuresis and was able to show that MSH in fairly large doses had a natriuretic effect. We further showed with Janice Dorn that carbachol injected into the ventricle produced natriuresis and have recently carried out with Mariana Morris, a graduate student, an extensive study of the effect of putative synaptic transmitters and adrenergic blocking drugs in altering sodium and potassium excretion by the kidney following their third ventricular injection. In this connection, we have found that median eminence lesions can block the natriuresis induced by intraventricular hypertonic saline (Morris *et al.*, 1976). Since the natriuretic effect of intraventricular injections persisted in animals with hereditary diabetes insipidus, we believe that the effects of median eminence lesions were not brought about by elimination of vasopressin or oxytocin secretion. In fact, the studies are consistent with the possible existence of a natriuretic hormone, but obviously much more work would need to be done to establish this.

It is important to point out that most of the actual experimental work which we have performed has been carried out by pre- and postdoctoral fellows in our laboratory. We have had the good fortune to have some 50 postdoctoral fellows who have come from all over the world. We have also been fortunate that almost all of these have continued in research after returning to their home countries and a number of them have made important contributions to neuroendocrinology.

Needless to say, the progress that we have made has been dependent on the generosity of various granting agencies. Because of the possible application of LHRH to fertility control, large amounts of money were poured into the search for it by the National Institutes of Health. Similarly, private foundations also were interested in the search for LHRH and we were the recipients of generous grants from both NIH and the Ford Foundation. We were even helped by a local foundation known as the Population Crisis Foundation of Texas, which was organized prior to the recession and flourished for several years thanks largely to the generosity of Mr. McAshan in Houston. The Welch Foundation also helped us for several years. There is little doubt that the channeling of money was extremely important in determining the rate of progress in the field and by whom the progress was made. Unfortunately, the climate for fundamental research has eroded considerably in the last few years. It is our hope that this situation will improve because further progress in the treatment of disease and the control of population is dependent on continued and improved support of basic research.

REFERENCES

Andersson, B., and McCann, S. M. (1955). Drinking, antidiuresis and milk ejection from electrical stimulation within the hypothalamus of the goat. *Acta Physiol. Scand.* **35**:191.

Bishop, W., Fawcett, C. P., Krulich, L., and McCann, S. M. (1972*a*). Acute and chronic effects of hypothalamic lesions on the release of FSH, LH and prolactin in intact and castrated rats. *Endocrinology* **91**:643.

Bishop, W., Kalra, P. S., Fawcett, C. P., Krulich, L., and McCann, S. M. (1972*b*). The effects of hypothalamic lesions on the release of gonadotropins and prolactin in response to estrogen and progesterone treatment in female rats. *Endocrinology* **91**:1404.

Brazeau, P., Vale, W., Burgus, R., Ling, M., Butcher, M., Rivier, J., and Guillemin, R. (1973). Hypothalamic polypeptide that inhibits the secretion of immunoreactive pituitary growth hormone. *Science* **179**:77.

Campbell, H. J., Feuer, G., and Harris, G. W. (1964). The effect of intrapituitary infusion of median eminence and other brain extracts on anterior pituitary gonadotrophic secretion. *J. Physiol.* (*London*) **170**:474.

Cooper, K. J., and McCann, S. M. (1975). Influence of ovarian steroids on pituitary responsiveness to LH-releasing hormone (LH-RH) in the rat. In Motta, M., Crosignani, P. G., and Martini, L. (eds.), *Hypothalamic Hormones*, Academic Press, London, pp. 161–168.

Dhariwal, A. P. S., Krulich, L., and McCann, S. M. (1969). Purification of a growth hormone inhibiting factor (GIF) from sheep hypothalamus. *Neuroendocrinology* **4**:282.

Everett, J. W. (1956). Functional corpora lutea maintained for months by autografts of rat hypophyses. *Endocrinology* **58**:786.

Farrell, G. L., and McCann, S. M. (1952). Detectable amounts of adrenocorticotrophic hormone in blood following epinephrine. *Endocrinology* **50**:274.

Grosvenor, C. E., McCann, S. M., and Nallar, R. (1965). Inhibition of nursing-induced fall in pituitary prolactin concentration in lactating rats following injection of acid extracts of bovine and rat hypothalamus. *Endocrinology* **76**:883.

Guillemin, R., and Rosemberg, E. (1955). Humoral hypothalamic control of anterior pituitary: A study with combined tissue cultures. *Endocrinology* **57**:599.

Guillemin, R., Hearn, W. R., Cheek, W. R., and Housholder, D. E. (1957). Control of corticotrophin release: Further studies with *in vitro* methods. *Endocrinology* **60**:488.

Harris, G. W. (1948). Neural control of the pituitary gland. *Physiol. Rev.* **28**:139.

Igarashi, M., and McCann, S. M. (1964). A hypothalamic follicle stimulating hormone releasing factor. *Endocrinology* **74**:446.

Krulich, L., and McCann, S. M. (1969). Effect of GRF and GIF on the release and concentration of GH in pituitaries incubated *in vitro*. *Endocrinology* **85**:319.

Krulich, L., Dhariwal, A. P. S., and McCann, S. M. (1968). Stimulatory and inhibitory effects of purified hypothalamic extracts on growth hormone release from rat pituitary *in vitro*. *Endocrinology* **83**:783.

Krulich, L., Illner, P., Fawcett, C. P., Quijada, M., and McCann, S. M. (1972). Dual hypothalamic regulation of growth hormone secretion. In Pecile, A., and Muller, E. E. (eds.), *Growth and Growth Hormone*, Excerpta Medica, Amsterdam, pp. 306–316.

Laqueur, G. L., McCann, S. M., Schreiner, L. H., Rosemberg, E., Rioch, D. M., and Anderson, E. (1955). Alterations of adrenal cortical and ovarian activity following hypothalamic lesions. *Endocrinology* **57**:44.

Libertun, C., Orias, R., and McCann, S. M. (1974). Biphasic effect of estrogen on the sensitivity of the pituitary to luteinizing hormone-releasing factor (LRF). *Endocrinology* **94**:1094.

Matsuo, H., Baba, Y., Nair, R. M. G., Arimura, A., and Schally, A.V. (1971). Structure of the porcine LH- and FSH-releasing hormone. I. The proposed amino acid sequence. *Biochem. Biophys. Res. Commun.* **43**:1334.

McCann, S. M. (1953). Effect of hypothalamic lesions on the adrenal cortical response to stress in the rat. *Am. J. Physiol.* **175**:13.

McCann, S. M. (1957). The ACTH-releasing activity of extracts of the posterior lobe of the pituitary *in vivo*. *Endocrinology* **60**:664.

McCann, S. M. (1962). A hypothalamic lutenizing hormone releasing factor. *Am. J. Physiol.* **202**:395.

McCann, S. M. (1974). Regulation of secretion of follicle-stimulating hormone and luteinizing hormone. In Greep, R. O., and Astwood, E. B. (eds.), *Handbook of Physiology, Sect. 7, Vol. 4, Pt. 2*, Americal Physiological Society, Washington, D.C., pp. 489–517.

McCann, S. M., and Brobeck J. R. (1954). Evidence for a role of the supraopticohypophyseal system in regulation of adrenocorticotrophin secretion. *Proc. Soc. Exp. Biol. Med.* **87**:318.

McCann, S. M., and Friedman, H. M. (1960). The effect of hypothalamic lesions on the secretion of luteotrophin. *Endocrinology* **67**:597.

McCann, S. M., and Fruit, A. (1957). Effect of synthetic vasopressin on release of adrenocorticotrophin in rats with hypothalamic lesions. *Proc. Soc. Exp. Biol. Med.* **96**:556.

McCann, S. M., and Haberland, P., (1959). Relative abundance of vasopressin and corticotrophin-releasing factor in neurohypophyseal extracts. *Proc. Soc. Exp. Biol. Med.* **102**:319.

McCann, S. M., and Sydnor, K. L. (1954). Blood and pituitary adrenocortocotrophin in adrenalectomized rats with hypothalamic lesions. *Proc. Soc. Exp. Biol. Med.* **87**:369.

McCann, S. M., Rothballer, A. B., Yeakel, E. H., and Shenkin, H. A. (1948). Adrenalectomy and blood pressure of rats subjected to auditory stimulation. *Am. J. Physiol.* **155**:128.

McCann, S. M., Taleisnik, S., and Friedman, H. M. (1960). LH-releasing activity in hypothalamic extracts. *Proc. Soc. Exp. Biol. Med.* **104**:432.

McCann, S. M., Dhariwal, A. P. S., and Porter, J. C. (1968). Regulation of the adenohypophysis. *Annu. Rev. Physiol.* **30**:589.

McCann, S. M., Ojeda, S. R., Harms, P. G., Wheaton, J. E., Sundberg, D. K., and Fawcett, C. P. (1976). Control of adenohypophyseal hormone secretion by prostaglandins. In Naftolin, F., Ryan, K. J., and Davies, I. J. (eds.), *Subcellular Mechanisms in Reproductive Neuroendocrinology*, Elsevier, Amsterdam.

McCann, S. M., Ojeda, S. R., Martinovic, J., and Vijayan, E. (1977). Role of catecholamines, in particular dopamine, in the control of gonadotropin secretion. *Adv. Biochem. Psychopharm.* **16**:109.

Moltz, J. H., Dobbs, R. E., McCann, S. M., and Fawcett, C. P. (1977). Effects of hypothalamic factors on insulin and glucagon release from the islets of Langerhans. *Endocrinology* **101**:196.

Morris, M., McCann, S. M., and Orias, R. (1976). Evidence for hormonal participation in the natriuretic and kaliuretic responses to intraventricular hypertonic saline and norepinephrine. *Proc. Soc. Exp. Biol. Med.* **152**:95.

Moss, R. L., and McCann, S. M. (1973). Induction of mating behavior in rats by luteinizing hormone releasing factor. *Science* **181**:77.

Negro-Vilar, A., Ojeda, S. R., and McCann, S. M. (1973). Evidence for changes in sensitivity to testosterone negative feedback on gonadotropin release during sexual development in the male rat. *Endocrinology* **93**:729.

Ojeda, S. R. (1975). Maturation of the control of gonadotropin and prolactin release in the rat. *Ann. Biol. Anim. Biochem. Biophys.* **16**:329.

Ramirez, V. D., and McCann, S. M. (1963). A comparison of the regulation of luteinizing hormone (LH) secretion in immature and adult rats. *Endocrinology* **72**:452.

Saffran, M., Schally, A. V., and Benfey, B. G. (1955). Stimulation of the release of corticotrophin from the adenohypophysis by neurohypophyseal factor. *Endocrinology* **57**:439.

Sawyer, C. H., Markee, J. E., and Hollinshead, W. H. (1947). The inhibition of ovulation in the rabbit by the adrenergic blocking agent Dibenamine. *Endocrinology* **41**:395.

Schneider, H. P. G., and McCann, S. M. (1970). Dopaminergic pathways and gonadotropin releasing factors. In Bargmann, W., and Scharrer, B. (eds.), *Aspects of Neuroendocrinology*. Springer-Verlag, Berlin, pp. 177–191.

Taleisnik, S., and McCann, S. M. (1961). Effects of hypothalamic lesions on the secretion and storage of hypophysial luteinizing hormone. *Endocrinology* **68**:263.

Talwalker, P. K., Ratner, A., and Meites, J. (1963). *In vitro* inhibition of pituitary prolactin synthesis and release by hypothalamic extract. *Am. J. Physiol.* **205**:213.

Wheaton, J. E., Krulich, L., and McCann, S. M. (1975). Localization of luteinizing hormone-releasing hormone in the preoptic area and hypothalamus of the rat using radioimmunoassay. *Endocrinology* **97**:30.

Zimmerman, E. A. (1976). Localization of hypothalamic hormones by immunocytochemical techniques. In Martini, L., and Ganong, W. F. (eds.), *Frontiers in Neuroendocrinology*. Raven Press, New York, pp. 25–62.

Zor, U., Kaneko, T., Schneider, H. P. G., McCann, S. M., Lowe, I. P., Bloom, G., Borland, B., and Field, J. B. (1969). Stimulation of anterior pituitary adenyl cyclase activity and adenosine 3′:5′-cyclic phosphate by hypothalamic extract and prostaglandin E_1. *Proc. Natl. Acad. Sci.* **63**:918.

18

Joseph Meites

Joseph Meites was born in Kishinev, Russia, at the end of 1913, and arrived in St. Joseph, Missouri, at the age of 6 in time to begin the first year of grade school. He received the B.S., M.A., and Ph.D. degrees at the University of Missouri, the last degree in 1947. In between (1942–1946) he served in the U.S. Army. He has been in the Department of Physiology at Michigan State University since 1947, as a full professor since 1953. In 1955–1956 he spent a sabbatical year at the Weizmann Institute in Israel. At Michigan State University, he received the Junior (1953) and Senior (1967) Sigma Xi Awards and the Distinguished Faculty Award in 1970.

He served as a member of the Endocrinology Study Section, NIH, from 1966 to 1970; on the National Pituitary Agency, NIH, from 1969 to 1976; on the Subcommittee on Use of Hormones in Domestic Animals, National Academy of Sciences–National Research Council, from 1960; as chairman of the IUPS Commission on Neuroendocrinology since 1971; and on the council of the International Brain Organization since 1976. He was the first president of the International Society of Neuroendocrinology from 1972 to 1976.

Dr. Meites presented the second W. E. Petersen Memorial Lecture at the University of Minnesota in 1973, and was the Pfeizer Traveling Fellow of the Clinical Research Institute of Montreal in 1977. He served on the editorial boards of *Endocrinology* from 1953 to 1956 and again from 1966 to 1974, *Neuroendocrinology* from 1967 to 1976, and *Comparative and General Endocrinology* from 1968 to 1971. At present he serves on the editorial boards of the *Proceedings of the Society for Experimental Biology and Medicine*, *Psychoneuroendocrinology*, and *Endokrinologie*, and is an associate editor of *Cancer Research*. Together with his students and research fellows, he has published more than 400 scientific articles, including several dozen chapters in books, and has been editor or coeditor of several books including *Pioneers in Neuroendocrinology*.

Studies on Neuroendocrine Control of Prolactin and Other Anterior Pituitary Hormones

JOSEPH MEITES

BEGINNINGS

My entry into endocrinology was accidental. While still an undergraduate at the University of Missouri, I became a part-time technician in the laboratory of the late Professor C. W. Turner, and after a year became one of his graduate students. Turner received his training in genetics but also developed an early interest in endocrinology while at the University of Wisconsin. In the 1920s he went to the Dairy Science Department at the University of Missouri, where he soon became immersed in work on endocrine regulation of mammary growth and lactation. His research attracted many graduate students and postdoctoral fellows. Turner not only was an enthusiastic and dedicated investigator but also was a kind and generous man, and a good spirit of camaraderie prevailed in the laboratory. The atmosphere was stimulating, and there was a general belief in hard work and long hours. Everyone had a maximum of freedom to pursue his own ideas, and Turner always was prepared to give good advice and encouragement. Over the years he developed some pet theories which he held onto tenaciously even when challenged by his students and others, but he never took offense and conceded the right of everyone to their opinions. He retained his scholarly interests to the end of his life in 1975, and

JOSEPH MEITES • Department of Physiology, Neuroendocrine Research Laboratory, Michigan State University, East Lansing, Michigan 48824.

probably contributed more than anyone else to the foundations of modern lactational physiology.

STUDIES WITH DRUGS ON CONTROL OF PROLACTIN RELEASE AND LACTATION

I took my first course in endocrinology in 1938, while still an undergraduate student, and was taught at that time that the endocrine and nervous systems were separate and for the most part operated independently of each other to control body functions. The nervous system was believed to act rapidly in response to different stimuli, whereas the endocrine system was thought to operate more slowly and for a longer duration of time. It was vaguely appreciated that there was some kind of connection between the pituitary and brain, as evident from the action of the milking stimulus on lactation, the effect of coitus in evoking ovulation in the rabbit, the stimulating effect of light on reproductive processes in birds and mammals, etc. However, there was little understanding of the mechanisms involved.

I had become interested as a graduate student in the control of prolactin secretion and lactation, and had studied the effects of different hormones on these processes. For my Ph.D. thesis, I developed a theory to explain the onset of lactation of parturition based on the interactions during pregnancy among estrogen, progesterone, and prolactin (Meites and Turner, 1942). Aside from estrogen and to a lesser extent testosterone, none of the other hormones, including oxytocin and vasopressin, appeared to be capable of stimulating prolactin secretion. It was therefore with great interest that I read the reports of Markee, Sawyer, and Everett on control of ovulation with drugs and by French workers and Felix Sulman and his colleagues at the Hebrew University of Jerusalem of the actions of tranquillizing drugs on the endocrine system. My first actual experience with drugs was with ergot compounds in collaboration with Dr. M. Shelesnyak while on a sabbatical leave at the Weizmann Institute in Israel in 1955–1956, but more about this later. In 1956 I attended a meeting where I was fortunate to see Bob Gaunt, whom I had known for some years because of his significant contributions on the relation of the adrenal cortex to lactation. Bob was then a research director of CIBA and had studied the endocrine effects of reserpine. He offered to make reserpine available to me to see whether it could initiate lactation in rabbits as it had been reported to do occasionally in women. Since I knew that virgin female rabbits had only rudimentary mammary glands, I first injected them with estrogen for 2–3 weeks or made them pseudopregnant to develop their mammary secretory (alveolar) tissue, and then gave them a single intravenous injection of reserpine. This produced lactation in most of the rabbits, which I attributed to the release of prolactin (Meites, 1957). Charles (Tom) Sawyer in the same year independently observed this phenomenon in rabbits. Estrogen-treated female rats also responded with onset of lactation when injected with reserpine or chlorpromazine, and we subsequently used initiation of lactation in estrogen-pretreated female rats as a semiquantitative test for prolactin release.

Interestingly, when pigeons were injected with reserpine or chlorpromazine, crop milk secretion was not elicited, as might have been expected if prolactin were released. This was another indication to us that hypothalamic control of prolactin secretion in birds probably was different from that of mammals. An earlier observation of a difference in control of prolactin secretion in pigeons was that injections of testosterone, estrogen, and progesterone did not increase pituitary prolactin levels in the pigeon as in rats and other laboratory mammals (Meites and Turner, 1947). We subsequently reported in a series of papers that hypothalamic extracts from pigeons, chickens, Japanese quail, and turkeys promoted prolactin release when incubated *in vitro* with pituitary tissue from these species (Meites and Nicoll, 1966). There is as yet no full understanding of the mechanisms involved in control of prolactin release in birds.

C. H. Nicoll, P. K. Talwalker, A. Ratner, and I soon found that many drugs could initiate lactation in estrogen-primed rats or rabbits, whereas others had no apparent effect. Thus serotonin, acetylcholine, pilocarpine, eserine, atropine, meprobamate, morphine sulfate, amphetamine, insulin, several carcinogens (3-methylcholanthrene, 3,4-benzpyrene, and 9,10-dimethyl-1,2-benzanthracene), and a CRF preparation obtained from Roger Guillemin were effective in producing lactation, whereas LSD was ineffective (Meites, 1963). Later, we found that LSD actually inhibited prolactin release and depressed lactation. Two of the most active drugs for evoking lactation were morphine and serotonin, both subsequently shown to elevate serum prolactin levels as measured by radioimmunoassay. In 1964 Mizuno, Talwalker, and I reported that iproniazid, a MAO inhibitor, suppressed postpartum lactation, and later Nagasawa and I (1970) found that it produced regression of carcinogen-induced mammary cancers in rats. We also found by radioimmunoassay that iproniazid significantly reduced serum prolactin levels in rats (Lu and Meites, 1971).

One of the most fascinating group of drugs we worked with were the ergot derivatives. I already have mentioned some collaborative research with M. C. Shelesnyak (see his chapter in *Pioneers in Neuroendocrinology*,

Vol. 1), who was studying the effects of ergot drugs on implantation of fertilized eggs and deciduoma formation in pseudopregnant rats. Since the ergot drugs inhibited implantation and deciduoma formation, "Shelly" correctly deduced that they inhibited prolactin secretion and thereby depressed progesterone secretion by the ovaries. We proceeded to test this by injecting rats with various ergot drugs, and then removed the pituitaries and assayed them for prolactin by the pigeon crop sac assay. To our disappointment, we found no significant change in pituitary prolactin content. The pigeon crop assay was not sufficiently sensitive to detect prolactin in the blood, and hence we were unable to show that it depressed prolactin release until the radioimmunoassay for prolactin was developed.

Years later, Sandoz A.G. in Basel, Switzerland, offered to make some ergot drugs available to me, and together with Dr. H. Nagasawa, a mammary cancer researcher from Tokyo who was on a postdoctoral fellowship in my laboratory, we tested the effects of ergocornine on growth of carcinogen-induced mammary cancers and on serum prolactin concentration by the newly developed radioimmunoassay for rat prolactin (Niswender *et al.*, 1969). We found that ergocornine induced marked regression of mammary cancers in the rats and profoundly depressed serum prolactin levels (Nagasawa and Meites, 1970). We also reported that ergot drugs induced regression of a prolactin-secreting transplantable pituitary tumor and reduced the extraordinarily high serum prolactin of the rats carrying these tumors to normal levels (Quadri *et al.*, 1972). Ergot drugs, with a few exceptions, have not been reported to inhibit growth of breast cancer in human subjects but have produced some improvement in patients with prolactin-secreting pituitary tumors. Ergot drugs also have been shown to be useful for inhibiting postpartum and inappropriate lactation in women.

Wolfgang Wuttke and John Lu, who were in my laboratory during the work with ergot drugs, showed that ergot drugs had no effect on the estrous cycle of rats when given in moderate doses but could inhibit LH release when given in high doses. By use of ergocornine, Wolfgang Wuttke and I were able to determine the purpose served by the prolactin rise on the afternoon of proestrus in the rat and later in the mouse (Meites *et al.*, 1972). Apparently the proestrous surge of prolactin produces luteolysis of the old corpora lutea from the previous cycle every 4–5 days. If the prolactin surge was prevented by giving ergot drugs daily during successive estrous cycles in the rat and mouse, the ovaries became loaded with corpora lutea which did not disappear. We reported that ergot drugs acted directly on the pituitary as well as on the hypothalamus to inhibit prolactin release.

For several years, Karl Nicoll, P. K. Talwalker, and I were engaged in a friendly debate with John Folley, Keith Benson, and Alfred Cowie of the National Institute for Research in Dairying, Reading, England, as to whether oxytocin could stimulate prolactin release from the AP. In 1956 Benson and Folley first suggested that oxytocin stimulated prolactin release, based on their observation that administration of oxytocin into postpartum lactating rats after litter removal retarded involution of the mammary glands. Since prolactin injections previously had been shown to inhibit mammary involution in similarly treated postpartum rats, they concluded that oxytocin acted by promoting prolactin release. This appeared to be an attractive idea since the work of Bargmann (see his chapter in *Pioneers in Neuroendocrinology*, Vol. 1) suggested that nerve tracts from the supraoptic and paraventricular nuclei might deposit some oxytocin and vasopressin into the portal vessels on their way down to the posterior pituitary. Soon there were many reports claiming that oxytocin induced pseudopregnancy and vaginal mucification, presumably indicative of prolactin release and luteal stimulation. L. Martini and others also had reported that vasopressin could induce release of several other AP hormones, and Geoffrey Harris agreed that oxytocin might act similarly on prolactin release. As a graduate student in Turner's laboratory, I had injected a posterior pituitary preparation (Pituitrin, Parke-Davis) and Pitocin (Parke-Davis) into guinea pigs and rabbits, and neither could initiate lactation or alter pituitary prolactin content (Meites and Turner, 1942). Apparently this has been overlooked by the British workers.

Although we confirmed the work by Folley and his colleagues that oxytocin could inhibit mammary involution in postpartum rats after litter removal, we presented evidence that this was due to a direct action of oxytocin on the mammary alveolar apparatus causing evacuation of milk into the ducts. We showed that oxytocin had no effect on mammary involution in the absence of milk in the glands. In hypophysectomized postpartum rats, we reported that oxytocin alone did not retard mammary involution, but when given together with prolactin and ACTH it acted synergistically to prevent mammary regression. Karl Nicoll and I also reported that culture of rat pituitary with oxytocin or vasopressin did not result in enhanced prolactin release, an observation confirmed later by Pasteels. Subsequently it was shown by radioimmunoassay of prolactin that oxytocin did not increase serum prolactin concentrations in several species, including the rat and cow. At the conclusion of the Symposium on Neuroendocrinology in Miami in December 1961, where both John Folley and I gave talks, Roy Greep in his summary of the meeting stated his belief that the controversy had been resolved in our favor (see Nalbandov, 1963).

The work with central-acting drugs has continued to fascinate me, as it appeared then and appears even more now as a promising approach for controlling release of pituitary hormones in animals and man. The evidence is overwhelming that some of the neurotransmitters in the brain are involved in regulating release of pituitary hormones, and many laboratories are engaged in trying to clarify their actions. Many drugs are available to either increase or decrease the activity of these brain neurotransmitters. What is particularly needed are drugs relatively free of side effects that can alter secretion of single rather than several pituitary hormones at once. Several drugs already have been used successfully for controlling various endocrine disorders.

IN VITRO STUDIES: DEMONSTRATIONS OF HYPOTHALAMIC PRF, PIF, AND OTHER RELEASING FACTORS

Several early investigators had cultured pituitary tissue *in vitro* successfully, but the major interest had been in maintaining viability and observing growth rather than in determining whether the pituitary could continue to release hormones. In the few studies in which hormones were measured, gonadotropins, ACTH, or TSH was found to be released into the medium for only a few days after initiation of the cultures. It appeared to us that the cultured pituitary should be able to secrete prolactin continuously, since Desclin and Everett independently reported in 1955–1956 that when a rat pituitary was transplanted underneath the kidney capsule it could maintain corpora lutea and stimulate growth of mammary tissue for weeks or months. This suggested that secretion of prolactin could continue independently of the hypothalamus. However, in the rat preparations of Desclin and Everett, the possibility could not be discounted that some factor(s) from the hypothalamus or elsewhere in the body might be released into the blood to stimulate the grafted pituitary. Consequently, we thought it would be informative to culture rat pituitary *in vitro* and see whether prolactin would be released and whether secretion could be maintained for a relatively prolonged period without hypothalamic influence.

It was my good fortune to see Dr.Raymond Kahn shortly after he arrived at the University of Michigan to begin work in the Department of Anatomy. Ray had become an expert in organ culture techniques and had published a laboratory manual on this subject for a course he was teaching. He agreed to collaborate with us (Karl Nicoll and me), and we cultured small explants of rat pituitary in a physiological medium by a watch glass technique in an atmosphere of air or by constant gassing with 95% O_2/5% CO_2. To our delight, we found that we could detect prolactin by the pigeon crop assay in the medium for as long as 3 weeks or more after replacing the medium every few days with fresh medium (Meites *et al.*, 1961). The amount of prolactin we recovered from the medium was manyfold greater than was initially present in the fresh AP, providing further evidence that prolactin could be secreted by the AP independently of the hypothalamus or of other body stimuli. Cultures of AP from the mouse, guinea pig, rabbit, monkey, and pigeon also were successfully carried out, and the amounts of prolactin recovered in the mammalian species were always greater than the initial amounts present (Meites *et al.*, 1963). Only the amounts of prolactin recovered from the pigeon pituitary cultures were not significantly greater than originally present in the fresh AP, providing further evidence that the avian AP required hypothalamic stimulation to promote prolactin secretion.

Until 1960 the few reports that the hypothalamus contained hypophysiotropic releasing factors dealt almost exclusively with CRF and release of ACTH. We became interested in testing the effects of extracts of hypothalamus on prolactin release, and soon published our first work on this subject (Meites *et al.*,1960). Using the estrogen-primed female rat for our prolactin bioassay we observed that hypothalamic extracts could readily initiate mammary secretion, whereas large doses of cerebral cortex (the control brain tissue used at that time) induced lactation in only a few rats. We concluded in that report that the hypothalamus contained prolactin-releasing activity. This was the first claim for a prolactin-releasing factor (PRF) in the hypothalamus. Later Mishkinsky *et al.* (1968) used a similar estrogen-primed rat preparation to show that hypothalamic extracts had PRF activity. More recently, the laboratories of Nicoll and Reichlin have presented additional evidence for the presence of a PRF in mammalian hypothalamic extracts, but its structure is unknown. TRH can release prolactin, but work in our and other laboratories makes it appear doubtful that this is the principal PRF of the hypothalamus.

We presented our first *in vitro* evidence for the presence of prolactin release inhibiting activity (PIF) in hypothalamic homogenates and extracts at the Symposium on Neuroendocrinology in Miami in December 1961 and at the annual meeting of the Endocrine Society in 1962. A full report was published a year later (Talwalker *et al.*, 1963). Independently, Pasteels also reported *in vitro* evidence for prolactin release inhibiting activity in rat hypothalamus. We named the presumed prolactin release inhibitor "prolactin-inhibiting factor" (PIF), but its chemical identity remains unknown to the present day. The presence of PIF activity in hypothalamic extracts was confirmed by other workers, notably Sulman and colleagues, Schally, and Grosvenor and McCann.

An interesting question was raised by Jack Everett and Tom Sawyer soon after the reports of the presence of PIF and LRF in hypothalamic extracts, namely whether the two were not identical since LH and prolactin release usually do not occur together. In a collaborative study (Schally *et al.*, 1964), we reported that partially purified LRF of bovine and ovine origin did not contain PIF activity, and more recently Schally's laboratory reported that the pure decapeptide, LRH, also had no effect on prolactin release. The role of the biogenic amines and of other neurotransmitters in the hypothalamus on prolactin release was unknown in the early 1960s. Some evidence in recent years has suggested that dopamine and PIF may be one and the same, since dopamine can inhibit prolactin release, but this has not yet been proven, and additional observations have indicated that a separate peptide PIF probably is present in the hypothalamus.

In the early 1960s little was known about possible hypothalamic control of growth hormone (GH) secretion. It had been demonstrated by a number of laboratories, including our own, that when a pituitary was grafted underneath the kidney capsule of young hypophysectomized rats the result was a partial body growth response, suggesting that some GH continued to be released by the deafferentated pituitary. Placement of lesions in the ventral hypothalamus of rats by Spirtos, Ingram, Bogdanove, and Halmi in 1954 and by Reichlin in 1961 apparently resulted in reduced secretion of GH. However, except for one unconvincing report, there was no evidence for the presence of GH-releasing activity in the hypothalamus.

Using the same pituitary culture technique we had employed for prolactin studies, Roger Deuben and I (1964) cultured rat AP explants with and without neutralized acid extract of rat hypothalamus or cerebral cortex for 6–18 days. GH was measured by the tibia bioassay in hypophysectomized rats. The hypothalamic extract increased GH release during 6 days of culture four- to sixfold over the control AP explants, whereas cerebral cortical extract had no significant effect on GH release. When the hypothalamic extract was boiled for 10 min to destroy any GH that might be present, most of the GH-releasing activity was retained. This suggested that GRF was a small molecule. In another experiment, we cultured rat pituitary explants for 9 days, replacing the medium every 3 days. After the first 3 days of culture no significant amounts of GH were released by control pituitary tissue, but if rat hypothalamic extract was added beginning on the last 3 days of culture GH release returned to 78% of that liberated during the first 3 days of culture.

Soon thereafter other investigators confirmed the presence of GRF activity in the hypothalamus in a variety of animal species, but the structure of this factor still remains to be elucidated (see Schally, this volume). In 1968, Krulich, Dhariwal, and McCann demonstrated the presence of a GH release inhibiting factor (GIF) in the hypothalamus, and it appears that GH is controlled by two hypothalamic factors, a GRF and a GIF. In 1973 Roger Guillemin and his colleagues determined the structure of GIF to be a tetradecapeptide (somatostatin).

In 1964 both McCann's laboratory (Igarashi and McCann, 1964) and our laboratory (Mittler and Meites, 1964) independently published evidence for the presence of follicle-stimulating hormone releasing activity (FRF) in the hypothalamus of rats. Although Igarashi and McCann's report appeared before ours, their results were based on an assay method they had developed for measuring FSH in the mouse which had not been validated, whereas we used the Steelman-Pohley method in the rat which had been established as an accurate assay for FSH. We were able to demonstrate a good dose–response relationship between the amount of hypothalamic extract used and release of FSH from rat pituitary tissue *in vitro*. Our assay procedure for FRF activity subsequently was used by a number of other investigators. Since the isolation by Schally's laboratory in 1971 of the decapeptide LRH-FSH, which released both LH and FSH, the existence of a separate FRF has become questionable. However, under a variety of experimental conditions, LH and FSH release appear to occur separately, and the possibility has not been completely eliminated that a distinct FRF is present in the hypothalamus.

During the early attempts to prove the existence of hypothalamic releasing factors, and be among the first to do so, I recall a conversation with Roger Guillemin about this matter. We both agreed that many of the experiments reported were poorly planned, were badly executed with the use of uncertain methodology (particularly some of the bioassay methods developed to measure hypothalamic factors), gave equivocable results that often could not be validated statistically, and, amazingly enough, ended with the claim that the presence of the particular hypothalamic factor sought for had been demonstrated! This certainly was true of some of the early reports, and the question still remains as to which laboratory should be given credit for being the first to demonstrate the existence of any particular hypothalamic factor. I list here (Table 1) the individuals whom I believe first presented presumptive evidence for the presence of hypophysiotropic factors in the hypothalamus.

When the momentous reports appeared in the fall of 1969 from the laboratories of Roger Guillemin and Andrew Schally on the structure and synthesis of TRH, I realized that we had now entered a new era in neuroendocrinology. This was brought home even more forcefully with the subsequent characterization and synthesis of LHRH and GIF (somatostatin) a few years later. The "chemotransmitter hypothesis" of Geoffrey Harris had indeed been vindicated. Andrew Schally and Roger Guillemin had made the isolation of the elusive hypothalamic hormones their major goal, and rightfully were rewarded with many honors for their success. I know that even right up to the time of the announcements of the structure of TRH, there still was considerable skepticism as to whether hypothalamic hormones really existed and whether they would ever be isolated by any of the individuals working on them. It is even doubtful if funding for pro-

TABLE 1. Investigator(s) First Reporting Evidence for Presence of Hypothalamic Releasing Factors

Factor	Year	Investigators
1. Corticotropin-releasing factor (CRF)	1955	Saffran and Schally; Guillemin and Rosenberg
2. Luteinizing hormone releasing factor (LRF)	1960	McCann *et al.*, Harris *et al.*
3. Prolactin-releasing factor in mammals (PRF)	1960	Meites *et al.*
PRF in Avian Species	1965–1966	Kragt and Meites; Nicoll
4. Prolactin-release inhibiting factor (PIF)	1961–1963	Talwalker, Ratner, and Meites; Pasteels
5. Thyrotropin-releasing factor (TRF)	1961–1962	Schreiber[a] and Kmentova; Guillemin *et al.*
6. Growth hormone releasing factor (GRF)	1963–1964	Deuben and Meites
7. Follicle-stimulating hormone releasing factor (FRF)	1964	Igarashi and McCann; Mittler and Meites
8. MSH release inhibiting factor (MIF)	1965	Kastin *et al.*
9. MSH-releasing factor (MRF)	1965	Taleisnik and Orias; Kastin *et al.*
10. Growth hormone release inhibiting factor (GIF)	1968	Krulich and McCann

[a] Note Dr. Schreiber's comment in his chapter that the tripeptide, TRH, failed to stimulate acid phosphatase activity by the adenohypophysis *in vitro*.

grams to identify the hypothalamic releasing factors would have continued much longer if the faith, dedication, and hard work of the relatively few "true believers" had not resulted in success. Personally, I never made any deep commitment toward the chemical characterization of the hypothalamic factors, although I did try to enlist the help of various pharmaceutical companies and of some chemists on our campus. Nevertheless, our own pioneering studies on hypothalamic factors, as well as those of others, particularly Don McCann and Murray Saffran, helped to lay the foundations for much of the subsequent work on hypothalamic factors.

PHYSIOLOGICAL STUDIES ON HYPOTHALAMIC FACTORS

Beginning about 1964 we devoted a considerable amount of effort toward measuring hypothalamic "content" of PIF, GRF, LRF, FRF, and

TRF under different physiological states. When we began this work there was evidence only that releasing factors were present in the hypothalamus, but nothing was known as to whether they were synthesized and released in response to different stimuli. We began with the assumption that they were secreted by the hypothalamus in a manner similar to that of other endocrine glands, and that their secretion could be altered by appropriate stimuli. On the whole, we believe this assumption has proven to be correct. All of these activities were measured by incubating rat AP halves with a neutralized acid extract of hypothalamus for 1–4 hr and then assaying the hormone released into the incubation medium by a standard bioassay. We assumed there was a separate factor controlling release of each AP hormone. The importance of hypothalamic biogenic amines and other neurotransmitters on release of AP hormones had not yet been demonstrated. What we were actually measuring was the sum of all the substances in the hypothalamus that influenced release of a specific AP hormone.

We encountered the usual problem of trying to interpret the significance of an increase or decrease in "content" of a hypothalamic factor during a particular physiological state, similar to the problem of interpreting content of any hormone in a gland or tissue. Despite these and other shortcomings of our early studies, they did indicate that the hypophysiotropic factors of the hypothalamus were in a dynamic state and could be altered by different stimuli. Apparently they were synthesized and released as other hormones secreted by endocrine glands. It is of interest to mention briefly a few of the changes we reported in hypothalamic "content" of different factors as influenced by a variety of stimuli (Meites, 1970). Albert Ratner and I observed in rats that the suckling stimulus during lactation was associated with reduced PIF content of the hypothalamus and increased prolactin release. We also reported that estrogen or reserpine administration decreased hypothalamic content of PIF, and presumably thereby accounted for the increased prolactin release. On the other hand, administration of prolactin ("short-loop feedback") or of ergot drugs reduced prolactin release and increased PIF content in the hypothalamus.

We also reported that during postpartum lactation in rats the suckling stimulus by the pups decreased hypothalamic LRF and PIF, and thus presumably reduced LH and increased prolactin release (Minaguchi and Meites, 1966). We observed that the prototype of the contraceptive steroids, Enovid, depressed both PIF and LRF in the hypothalamus of female rats, apparently accounting for the increase in prolactin and inhibition of LH release and ovulation by this drug (Minaguchi and Meites, 1967).

We and others reported that although hypothalamic factors could not normally be detected in the systemic blood of intact rats, they did appear in the systemic blood one or more months after hypophysectomy. This appeared to be logical, since any negative feedback from the AP (short-loop feedback) and target organs (long-loop feedback) on the hypothalamus would presumably disappear after hypophysectomy. This problem deserves further investigation. We also presented evidence that a number of stimuli (constant light, drugs, hormones) could induce release of sufficient hypothalamic factors into the systemic circulation of hypophysectomized rats to activate grafted rat pituitaries (Meites, 1970).

Karl Nicoll, Chandra Talwalker, and I first began to determine the effects of different kinds of stress on prolactin release in the rat in the late 1950s. As with our early drug studies, we used initiation of lactation in the estrogen-primed female rat as a semiquantitative test for prolactin release. We treated rats daily for 5 days with injections of 10% formalin, restraint stress, electrical stimulation of the uterine cervix, and warm or cold temperatures and observed variable degrees of lactation in these rats. These results were published in the *American Journal of Physiology* (Nicoll *et al.*, 1960) and constitute the first evidence that stress can evoke prolactin release. We also reported that electrical stimulation of the nipples, skull, lumbar area, and nasal mucosa initiated lactation in estrogen-primed rats (Maqsood and Meites, 1960). In a review of lactation (*Neuroendocrinology*, Vol. 1, edited by Martini and Ganong, 1966), I expressed the view that stress can evoke prolactin as readily as ACTH release. Since then, the ability of stressful stimuli to induce prolactin discharge has been widely confirmed by radioimmunoassay in animal and human subjects. Recent work from our laboratory suggests that in at least one form of stress, immobilization, the increase in prolactin release is associated with enhancement of serotonin turnover in the hypothalamus (Mueller *et al.*, 1976).

Another type of stimulus that has interested us since the mid-1960s is the effect of continuous light on reproductive function. Béla Piacsek and I undertook a study to determine whether constant illumination could stimulate the ovaries of hypophysectomized female rats carrying two pituitary grafts from rats of the same strain and sex. After 3 weeks of constant light, the rats with the pituitary grafts showed ovaries with large, well-developed follicles in contrast to the tiny follicles in the ovaries of control hypophysectomized–pituitary grafted rats on a normal light-dark schedule. The grafted pituitaries of the rats under constant light were more differentiated and weighed more than the grafted pituitaries of the rats under a normal light-dark schedule (Piacsek and Meites, 1967). In a follow-up study while in Roy Greep's laboratory, Béla showed that grafted pituitaries in hypophysectomized rats under constant light contained 125% more FSH than grafted pituitaries from control rats under normal light-dark conditions. These studies suggested that continuous light stimulates the hypothalamus to release sufficient LHRH into the systemic circulation to activate the grafted pituitaries of hypophysectomized rats to release gonadotropins.

In 1952 we suggested that circulating levels of prolactin might help regulate pituitary prolactin secretion, based on observations that a variety of body tissues could readily inactivate prolactin *in vitro* (mammary tissue, ovaries, liver, kidney, adrenal, pigeon crop sac) (Sgouris and Meites, 1952). We believed that such inactivation and subsequent lowering of blood prolactin levels might stimulate the pituitary to increase prolactin secretion. Although high serum levels of prolactin subsequently were demonstrated to inhibit pituitary prolactin release, low serum prolactin levels have not been shown to stimulate prolactin release. The reports by Luciano Martini and his colleagues on the "short-loop feedback" of gonadotropins and ACTH on their own release prompted us to investigate whether high blood levels of prolactin could alter pituitary prolactin secretion. Apparently, Bob MacLeod had the same idea, since we both reported independently that rats carrying a prolactin and GH-secreting pituitary tumor (originally obtained from Jacob Furth) had a low *in situ* pituitary prolactin content. We also found that the PIF content in the hypothalamus was increased, suggesting that this was the mechanism by which prolactin content was decreased (Chen *et al.*, 1967). Subsequently, Fuxe and Hökfelt reported that prolactin injections in rats activated the tuberoinfundibular dopaminergic neurons, suggesting that an increase in dopamine activity was associated with the reduction in pituitary prolactin content. Recently, we (Gudelsky *et al.*, 1976) have confirmed the observations of Fuxe and Hökfelt, and have observed that prolactin injections increased dopamine turnover in the median eminence and anterior hypothalamus but not in the posterior hypothalamus of the rat.

Jim Clemens, David Chen, Jim Voogt, and I reported many studies in which we implanted small amounts of prolactin in the median eminence of rats. We found that this decreased pituitary prolactin content and increased hypothalamic PIF activity and that such implants induced marked mammary regression in cycling or ovariectomized rats, inhibited lactation in postpartum rats, and terminated pseudopregnancy or early pregnancy in rats, all indicative of inhibition of prolactin release (see Meites and Clemens, 1972). When radioimmunoassays became available, we showed that median eminence implants of prolactin not only decreased serum prolactin levels but also significantly increased serum LH and FSH. Thus the suppression of pituitary prolactin secretion by prolactin resulted in enhanced secretion of gonadotropins. This reciprocal relationship between prolactin and gonadotropin secretion is observed under many physiological states and continues to excite our interest and study.

One of the most interesting observations that resulted from our "short-loop feedback" studies on prolactin was the finding that either daily injections of prolactin or an implant of prolactin in the median eminence of prepubertal rats hastened the onset of puberty by 6–7 days. This has been confirmed by a number of other laboratories and suggests that prolactin may have a role in controlling the onset of puberty in rats. When we measured normal pituitary and serum prolactin in prepubertal female rats, the levels were relatively low until the onset of puberty, when they rose significantly. Apparently these low levels of prolactin in prepubertal rats exert an inhibitory action on release of gonadotropins. Wolfgang Wuttke recently has provided evidence that prolactin does have a role in the onset of puberty in the rat.

An important question that continues to interest us is whether the negative feedback of prolactin on its own secretion has any physiological significance. We have suggested that it may serve to prevent rapid depletion of pituitary prolactin content during acute release evoked by such stimuli as stress or suckling. For example, during postpartum lactation in rats, Clark Grosvenor reported that a single short period of suckling by a litter can reduce pituitary prolactin content by 50% or more. Can the initial outpouring of prolactin from the pituitary in response to stress or suckling feed back on the hypothalamus to reduce further prolactin release? We have been testing this hypothesis by injecting prolactin prior to administering a stress or before permitting suckling in lactating rats and find that this significantly reduces the release of prolactin produced by these stimuli (Advis *et al.*, unpublished). However, further work is necessary to prove that the "short-loop feedback" of prolactin is a physiological mechanism that helps control prolactin secretion.

MAMMARY AND PITUITARY TUMORS

The possible relationship of prolactin to development and growth of mammary and pituitary tumors had fascinated me for many years. The realization of the important role of the hypothalamus in controlling prolactin secretion and indirectly in controlling estrogen secretion via the gonadotropins prompted us to explore the relation of the neuroendocrine system to development and growth of mammary tumors. We used the method developed by Dr. Charles Huggins and his colleagues at the University of Chicago to induce a high incidence of mammary adenocarcinomas in Sprague-Dawley rats, by administering carcinogenic chemicals. This method had great appeal for me since, besides the relative speed at which these tumors appeared, they bore some resemblance in their origin, morphology, and hormone dependency to human breast cancers. We also worked with spontaneous mammary tumors which arise in relatively large numbers in aging female Sprague-Dawley rats. Transplantable pituitary tumors were kindly provided by Dr. Jacob Furth, who developed them in inbred rats by prolonged treatment with estrogens or by radiation. These tumors usually secreted large amounts of prolactin and growth hormone and some also secreted ACTH.

In one of our initial studies, we wanted to know whether we could induce mammary tumors by a carcinogen in ovariectomized or ovariectomized-adrenalectomized rats if we gave them only pituitary hormones. Chandra Talwalker and I ovariectomized Sprague-Dawley rats and gave them 7,12-dimethylbenz[*a*]anthracene (DMBA). No tumors developed in the ovariectomized rats given only DMBA, but when estrogen, a combination of prolactin and GH, or prolactin alone was injected, mammary tumors developed as readily in the rats given pituitary hormones as in the rats injected with estrogen (Talwalker *et al.*, 1964). Since adrenalectomized rats could not tolerate DMBA, which is toxic, Kaleem Quadri and I administered DMBA subcutaneously over individual mammary glands of ovariectomized-adrenalectomized rats and injected them with prolactin and GH. We observed mammary tumors only in the DMBA-treated mammary glands. These experiments and related ones by Olaf Pearson suggested that pituitary hormones were more important than ovarian hormones in development of mammary tumors in rats.

We also reported that the hypothalamus had a direct role in development and growth of mammary tumors. Jim Clemens, Cliff Welsch, and I showed that placement of lesions in the median eminence dramatically increased growth of existing DMBA-induced mammary tumors in Sprague-Dawley rats. However, when median eminence lesions were made prior to DMBA administration, this resulted in inhibition of mammary tumor development. Hormonal stimulation of normal mammary growth apparently protected the mammary epithelium against the action of carcinogens. Lesions placed in the preoptic or in the amygdaloid area of rats with existing DMBA-induced mammary tumors resulted in a reduction in mammary tumor growth, perhaps due to a decrease in prolactin or estrogen secretion. Cliff Welsch and I also reported that placement of bilateral lesions in the median eminence of otherwise untreated Sprague-Dawley female rats resulted in early development and increase in total number of spontaneous mammary tumors as compared with control sham-operated rats. These were among the earliest demonstrations that the hypothalamus could influence development and growth of mammary tumors, although earlier reports by Leo Loeb, Mühlbock and Boot, and others, had shown that pituitary grafts in mice could induce mammary tumor development. We suggested that alterations in hypothalamic function might give rise to spontaneous development of mammary and pituitary tumors.

We also reported that central-acting drugs that increase prolactin secretion (reserpine, haloperidol, chlorpromazine, etc.) hastened development of spontaneous mammary tumors and increased growth of existing DMBA-induced mammary tumors, whereas drugs that decreased prolactin release (L-dopa, iproniazid, pargyline, ergot derivatives) induced regression of mammary tumors in rats. Drugs that either increased or decreased prolactin secretion inhibited development of DMBA-induced tumors, presumably by stimulating or inhibiting normal growth of the mammary epithelium and thereby rendering the carcinogen less effective on the mammary tissue. Our early reports that L-dopa and ergot drugs depressed mammary tumor growth in rats were subsequently tested in human breast cancer patients with a limited degree of success. An important role for prolactin in human breast cancer still remains to be demonstrated.

We also worked on the relation of the hypothalamus to development and growth of pituitary tumors in rats. Cliff Welsch and I implanted single pituitary grafts underneath the kidney capsule of intact rats, and also implanted a pellet of diethystilbestrol subcutaneously every few months. Tumors developed both in the grafted and in the *in situ* pituitary, but the tumors *in situ* were larger and more numerous than those under the kidney capsule. This suggested that the hypothalamus contributed to the development and growth of the *in situ* pituitary tumors, perhaps by stimulating PRF secretion in addition to the direct action of estrogen on the pituitary. In another study, Quadri *et al.* (1972) reported that administration of a variety of ergot drugs could induce rapid regression of transplantable Furth pituitary tumors. Serum prolactin was reduced from values of up to 100,000 ng/ml to normal levels within 24 hr of administration of the ergot drugs. Dilip Sinha and I found that these rat pituitary tumors could still respond to rat hypothalamic extract when incubated *in vitro*, with a decrease in release of prolactin and increase in release of GH. Apparently, these pituitary tumors do not lose their hormonal responsiveness, except perhaps when they become very large. The role of the hypothalamus in their development still remains to be clarified. It is difficult to understand why pituitary tissue, separated from hypothalamic influence, continues to secrete GH or ACTH in supranormal amounts. Possibly genetic alterations produced by prolonged estrogen or radiation treatment render the pituitary independent of hypothalamic control.

REPRODUCTIVE SENESCENCE

Our work on the relation of the neuroendocrine system to reproductive senescence in rats began in the mid-1960s. In part it was a logical extension of our work on mammary and pituitary tumors which we commonly observed in aging female rats. We became interested in determining why, as female rats aged, most of them stopped cycling and began to develop mammary and pituitary tumors. In 1961 we had observed that the pituitaries of old female rats released significantly more prolactin *in vitro* than pituitaries from young sexually mature female rats. Did changes occur in the hypothalamus with aging, and, if so, could they be reversed by appropriate neuroendocrine treatments?

The major patterns that we and others observed in aging female rats were irregular cycles, constant estrus, pseudopregnancies of extended length, and, in the oldest rats, anestrus. The largest single group of aging rats showed the constant-estrus syndrome, characterized by cornified vaginal smears and ovaries with well-developed follicles but no corpora lutea. We found we could induce ovulation in these rats by direct electrochemical stimulation of the preoptic area of the hypothalamus or by administration of progesterone or epinephrine in oil (Clemens *et al.*, 1969). We therefore hypothesized that old female rats probably were deficient in hypothalamic catecholamines. More recently we have found this to be correct. Old rats also exhibited increased turnover of serotonin in the hypothalamus (Simpkins *et al.*, 1977).

Jim Clemens, Kaleem Quadri, Henry Huang, and I were able to reinitiate cycling in old constant-estrus rats by administering L-dopa, iproniazid, epinephrine in oil, progesterone, or ACTH and even by subjecting rats to ether stress twice daily. This indicated that, whatever the deficiencies in the hypothalamus, pituitary, or ovaries of old rats, they could be overcome by appropriate central stimulation. Pituitary and serum LH and FSH generally were lower and prolactin was higher in old than in young rats, and the capacity to release gonadotropins in response to stimuli such as ovariectomy or gonadal hormone administration was greatly reduced in old as compared to young rats. The anestrous old rats had almost undetectable amounts of serum gonadotropins and very high serum prolactin levels. This largely explained why aging rats ceased to cycle and developed spontaneous mammary tumors. Later we found that aging male rats showed a similar decline in ability to secrete LH and FSH and increased capacity to secrete prolactin (see Meites *et al.*, 1976).

Unlike the ovaries of postmenopausal women, which become inactive and unresponsive to gonadotropins, the ovaries of aging rats apparently remain capable of function until the end of life. In some respects, cessation of reproductive functions of aging rats represents a return to the prepubertal state. As in the pre- and postpubertal states, the hypothalamus of aging rats appears to be the chief agent involved in regulating reproductive functions.

SOME REFLECTIONS ON A RESEARCH CAREER

Scientific research is not without its problems and frustrations, but there also are many compensations. Among the greatest satisfactions for me has been the opportunity to be associated with many young, bright, and motivated graduate students and postdoctoral fellows, and, outside of my laboratory, with many outstanding scientific colleagues throughout the world. A total of 28 graduate students have received their doctorate degree with me and about 23 postdoctoral fellows have come to my laboratory for additional training. These associations almost always resulted in reciprocal benefits, since I have received much from my students. Without their skills, hard work, and insights into problems, most of whatever we shared and accomplished in neuroendocrinology would not have been possible. I have followed their careers after they left my laboratory with great interest and have taken pride in their accomplishments. I always enjoy the opportunities to get together with them at scientific or other meetings, and to exchange personal and scientific notes. We have remained good friends (Fig. 1).

Another great personal pleasure has been the opportunity to meet and to make friends with scientific colleagues throughout the world. From them I have received much stimulation, encouragement, understanding, unfailing kindness, and hospitality which would be difficult to ever repay in full. Their approval or criticisms of our work have provided a strong motivation for our research. All the men and women I have known in the field of neuroendocrinology have been and remain strong personalities, as befits those who enter untrodden paths. They all have displayed highly individual styles in their approaches to their research and in their personal lives outside of the laboratory. Sometimes there have been sharp disagreements, but I do not believe I have ever permitted an honest difference of opinion to interfere with my admiration, respect, and friendly feelings for my scientific colleagues. I consider the international scientific fraternity (and sorority) to be one of the marvels of the world.

Like others, I feel deeply indebted to Geoffrey Harris in many ways. I looked up to him as a teacher who inspired many of our first efforts in neuroendocrinology, and as a friend. Because of my early interest in the control of prolactin and other pituitary hormones, I was highly stimulated

FIGURE 1. Neuroendocrine Research Laboratory, Michigan State University, 1975. Left to right, front row, seated: W. Chen, G. Campbell, J. Meites, J. Bruni, M. Kurcz, J. Chen, J. Simpkins (standing); second row: H. Huang, B. Yazejian, C. (Bradley) Cukier, G. Kledzik, K. (Kowalski) Robin, C. Twohy, G. Mueller, C. Vandenberg, L. Grandison; third row: J. Advis, J. Geneau, C. Hodson, R. Mioduszewski, L. Warshawsky, S. Marshall.

by his article in *Physiological Reviews* in 1948. His 1955 book *Neural Control of the Pituitary Gland* became a virtual bible of neuroendocrinology in our laboratory, as Karl Nicoll and Chandra Talwalker must well remember. I consider myself particularly fortunate to have spent many hours with him at the last meeting at which I saw him, the International Congress of Physiological Sciences in Munich in 1971, since he died a few months later. At the Congress, he participated with enthusiasm in the formation of the new International Society of Neuroendocrinology, and undoubtedly would have been its first president had he survived. He was a warm and sensitive human being, and the recognized leader of neuroendocrinology. All of us will continue to miss his presence.

Among the highlights of my career was the period in 1965–1970 when I served on the Endocrinology Study Section of the National Institutes of Health (NIH). Although it was mostly a great deal of hard work, I greatly enjoyed the opportunity of working with the many distinguished colleagues on the committee. Morris Graf did an extraordinarily fine job as executive secretary, and ran its affairs with aplomb and efficiency. I was proud of the work of the Study Section, which dealt fairly and conscientiously with the numerous applications it reviewed. Most of my own financial support for our work in neuroendocrinology has come from NIH, for which I am very grateful. I also have received funding from the Michigan Agricultural Experiment Station, from the American Cancer Society, the Michigan Cancer Foundation, the National Research Council, the Upjohn Company, Eli Lilly Co., and other sources. For many years I operated on a relatively modest budget in a small laboratory, but was fortunate enough to be able to increase my funding just at the time we were entering our most productive phase of neuroendocrine research.

I have had the privilege of working on some of the early studies on hypothalamic releasing factors and of applying neuroendocrine approaches to the solution of basic problems in reproduction, lactation, mammary and pituitary tumors, nutrition, aging, and environmental stimuli. There is no question in my mind that the period from the early 1950s to the present, when my major focus has been on neuroendocrinology, has been the most exciting, fruitful, and enjoyable of my career. The exhilaration of having been able to participate in development of a new branch of science will remain with me to the end.

REFERENCES

Chen, C. L., Minaguchi, H., and Meites, J. (1967). Effects of transplanted pituitary tumors on host pituitary prolactin secretion. *Proc. Soc. Exp. Biol. Med.* **126**:317.

Clemens, J. A., Minaguchi, H., Storey, R., Voogt, J. L., and Meites, J. (1969*a*). Induction of precocious puberty in female rats by prolactin. *Neuroendocrinology* **4**:150.

Clemens, J. A., Amenomori, Y., Jenkins, T., and Meites, J. (1969*b*). Effects of hypothalamic stimulation, hormones and drugs on ovarian function in old female rats. *Proc. Soc. Exp. Biol. Med.* **132**:561.

Deuben, R. R., and Meites, J. (1964). Stimulation of pituitary growth hormone release by a hypothalamic extract *in vitro*. *Endocrinology* **74**:408.

Gudelsky, G. A., Simpkins, J., Mueller, G. P., Meites, J., and Moore, K. E. (1976). Selective actions of prolactin on catecholamine turnover in the hypothalamus and on serum LH and FSH. *Neuroendocrinology* **22**:206.

Kledzik, G. S., and Meites, J. (1975). Reinitiation of estrus cycles in light-induced constant estrous female rats by drugs. *Proc. Soc. Exp. Biol. Med.* **146**:989.

Lu, K. H., and Meites, J. (1971). Inhibition by L-dopa and monoamine oxidase inhibitors of pituitary prolactin release; stimulation by methyldopa and *d*-amphetamine. *Proc. Soc. Exp. Biol. Med.* **137**:480.

Maqsood, M., and Meites, J. (1961). Induction of mammary secretion in rats by electrical stimulation. *Proc. Soc. Exp. Biol. Med.* **106**:104.

Meites, J. (1957). Induction of lactation in rabbits with reserpine. *Proc. Soc. Exp. Biol. Med.* **96**:728.

Meites, J. (1963). Pharmacological control of prolactin secretion and lactation. In Guillemin, R. (ed.)., *Proceedings of the First International Pharmacological Meeting*, Vol. 1, Pergamon Press, New York, pp. 151–181.

Meites, J. (1966). Control of mammary growth and lactation. In Martini, L., and Ganong, W. F. (eds.), *Neuroendocrinology*, Vol. 1, Academic Press, New York, pp. 669–707.

Meites, J. (1970). Direct studies on the secretion of the hypothalamic hypophysiotropic hormones (HHH). In Meites, J. (ed.), *Hypophysiotropic Hormones of the Hypothalamus*, Williams and Wilkins, Baltimore, pp. 261–278.

Meites, J., and Clemens, J. A. (1972). Hypothalamic control of prolactin secretion. *Vit. Horm.* **30**:165.

Meites, J., and Nicoll, C. A. (1966). Adenohypophysis: prolactin. *Annu. Rev. Physiol.* **28**:57.

Meites, J., and Turner, C. W. (1942). Studies concerning the mechanism controlling the initiation of lactation at parturition. *Endocrinology* **30**:711, 719, 726.

Meites, J., and Turner, C. W. (1947). Effect of sex hormones on pituitary lactogen and crop glands of common pigeons. *Proc. Soc. Exp. Biol. Med.* **64**:465.

Meites, J., Talwalker, P. K., and Nicoll, C. S. (1960). Initiation of lactation in rats with hypothalamic or cerebral tissue. *Proc. Soc. Exp. Biol. Med.* **103**:298.

Meites, J., Kahn, R. H., and Nicoll, C. S. (1961). Prolactin production by rat pituitary *in vitro*. *Proc. Soc. Exp. Biol. Med.* **108**:440.

Meites, J., Nicoll, C. S., and Talwalker, P. K. (1963). In Nalbandov, A. V. (ed.), *Advances in Neuroendocrinology*, University of Illinois Press, Urbana, pp. 239–277.

Meites, J., Lu, K. H., Wuttke, W., Welsch, C. W., Nagasawa, H., and Quadri, S. K. (1972). Recent studies on functions and control of prolactin secretion in rats. *Recent Progr. Horm. Res.* **28**:471.

Meites, J., Huang, H. H., and Riegle, G. D. (1976). Relation of the hypothalamo-pituitary-gonadal system to decline of reproductive functions in aging female rats. In Labrie, F., Meites, J., and Pelletier, G. (eds.), *Hypothalamus and Endocrine Functions*, Plenum, New York, pp. 3–20.

Minaguchi, H., and Meites, J. (1966). Effects of suckling on hypothalamic LH-releasing factor and prolactin inhibiting factor, and on pituitary LH and prolactin. *Endocrinology* **80**:603.

Minaguchi, H., and Meites, J. (1967). Effects of norethynodrel-mestranol combination (Enovid) on hypothalamic and pituitary hormones in rats. *Endocrinology* **81**:826.

Mishkinsky, J., Khazen, K., and Sulman, F. A. (1968). Prolactin releasing activity of the hypothalamus in post-partum rats. *Endocrinology* **82**:611.

Mittler, J. C., and Meites, J. (1964). *In vitro* stimulation of pituitary follicle-stimulating-hormone release by hypothalamic extract. *Proc. Soc. Exp. Biol. Med.* **117**:309.

Mizuno, H., Talwalker, P. K., and Meites, J. (1964). Inhibition of mammary secretion in rats by iproniazid. *Proc. Soc. Exp. Biol. Med.* **115**:604.

Mueller, G. P., Twohy, C. P., Chen, H. T., Advis, J. P., and Meites, J. (1976). Effects of *l*-tryptophan and restraint stress on hypothalamic and brain serotonin turnover, and pituitary TSH and prolactin release in rats. *Life Sci.* **18**:715.

Nagasawa, H., and Meites, J. (1970). Suppression by ergocornine and iproniazid of carcinogen-induced mammary tumors in rats; effects on serum and pituitary prolactin levels. *Proc. Soc. Exp. Biol. Med.* **135**:469.

Nalbandov, A. V. (1963). *Advances in Neuroendocrinology*, p. 515, University of Illinois Press, Urbana.

Nicoll, C. S., Talwalker, P. K., and Meites, J. (1960). Initiation of lactation in rats by non-specific stresses. *Am. J. Physiol.* **198**:1103.

Niswender, G. D., Chen, C. L., Midgley, A. R., Jr., Meites, J., and Ellis, S. (1969). Radioimmunoassay for rat prolactin. *Proc. Soc. Exp. Biol. Med.* **130**:793.

Piacsek, B. E., and Meites, J. (1967). Stimulation by light of gonadotropin release from transplanted pituitaries of hypophysectomized rats. *Neuroendocrinology* **2**:129.

Quadri, S. K., Lu, K. H., and Meites, J. (1972). Ergot-induced inhibition of pituitary tumor growth in rats. *Science* **176**:417.

Ratner, A., and Meites, J. (1964). Depletion of prolactin-inhibiting activity of rat hypothalamus by estradiol or suckling stimulus. *Endocrinology* **75**:377.

Schally, A. V., Meites, J., Bowers, C. Y., and Ratner, A. (1964). Identity of prolactin inhibiting factor (PIF) and luteinizing hormone releasing factor (LRF). *Proc. Soc. Exp. Biol. Med.* **117**:252.

Sgouris, J. T., and Meites, J. (1952). Inactivation of prolactin by body tissues *in vitro*. *Am. J. Physiol.* **169**:301.

Simpkins, J. W., Mueller, G. P., Huang, H. H., and Meites, J. (1977). Evidence for depressed catecholamine and enhanced serotonin metabolism in aging male rats: Possible relation to gonadotropin secretion. *Endocrinology* **100**:1672.

Talwalker, P. K., Ratner, A., and Meites, J. (1963). *In vitro* inhibition of pituitary prolactin synthesis and release by hypothalamic extract. *Am. J. Physiol.* **205**:213.

Talwalker, P. K., Meites, J., and Mizuno, H. (1964). Mammary tumor induction by estrogen or anterior pituitary hormones in ovariectomized rats given 7,12-dimethyl-1,2-benzanthracene. *Proc. Soc. Exp. Biol. Med.* **116**:531.

19

Seymour Reichlin

Seymour Reichlin was born on May 31, 1924, in New York City. After graduation from Washington University Medical School in 1948, he trained as an internist through internship and residencies in medicine at the New York Hospital and Barnes Hospital. Two years as a research fellow of Dr. Geoffrey W. Harris at the Maudsley Hospital followed, and in 1954 Dr. Reichlin returned to St. Louis to establish a research laboratory in the Department of Psychiatry at Washington University. Between 1955 and 1960 he served as chief of the Psychosomatic Division, Department of Medicine, held joint appointments in medicine, psychiatry, and preventive medicine, and conducted an active research program. In 1961 Dr. Reichlin joined the faculty of the University of Rochester, School of Medicine, as associate professor (later professor) of medicine and chief of the Endocrine Division. In 1969 he became professor and chairman of the Department of Medical and Pediatric Specialties at the newly opened University of Connecticut School of Medicine, and after 2 years relinquished this job to become the first chairman of the Department of Physiology at the school.

In 1972, upon the retirement of Dr. Edwin B. Astwood, the directorship of the Endocrine Division of the New England Medical Center fell free, and Dr. Reichlin assumed this position as professor of medicine of Tufts University and head of the Clinical Research Unit.

Dr. Reichlin is a member of the American Physiological Society, the American Society for Clinical Investigation, Association of American Physicians, the Endocrine Society, the British Society for Endocrinology, and the American Psychosomatic Society, among others.

He has served on the editorial boards of *Endocrinology*, *New England Journal of Medicine*, and *Psychoneuroendocrinology*, has been on the council of the Endocrine Society, and has served on several advisory boards of the NIH.

His honors include the Eli Lilly Award of the Endocrine Society, Lawson Wilkins Award of the Society for Pediatric Research, Upjohn Award of the American Society for Fertility, and Distinguished Lectureships of the Western Society for Clinical Investigation, the Society for Investigative Gynecology, the Italian Endocrine Society, and the Mexican Society for Endocrinology. He served as president of the Endocrine Society from 1975 to 1976 and as president of the Association for Research in Nervous and Mental Disorders from 1976 to 1977.

Formative Years as an Investigator of Hypothalamic-Pituitary Physiology

SEYMOUR REICHLIN

As I see it now, my interest in neuroendocrinology arose from two elements, the first being a general and intense curiosity in natural history and about how living things work and the second a specific concern with the mechanism by which stress causes illness. Both elements were well established by my early teens, the first by exposure to high school courses in biology and the second by my concern over my father's angina, which clearly seemed related to life stress. These factors determined my choice of medicine as a career, sensitized me to the significance of the stress response, and established my outlook in experimental science as that of a physician looking for solutions to disease in disturbances of central nervous system function. In college, I learned of Cannon's view of homeostasis, and of his formulation of the "fight or flight" description of the response to stress. I also became aware of voodoo death and other popularized psychosomatic phenomena. My first effort in neuroendocrine research logically followed this theme, for at Antioch College in 1943 I chose to repeat the original chemical synthesis of epinephrine as my junior year thesis project in organic chemistry. It was this early fascination with the mind–body problem that made me so susceptible to the neuroendocrine virus when I heard about Hans Selye and the general adaptation syndrome (Selye, 1946). Selye's views were set forth in a lecture to the sophomore medical school class in 1945 by Dr. Irwin Levy, an instructor in neurology at Washington

SEYMOUR REICHLIN • Endocrinology Division, New England Medical Center Hospital, Boston, Massachusetts 02111

University Medical School in St. Louis. That lecture gave an insight into the way that the brain might influence bodily functions and cause disease. The strong impact of that lecture on my entire career is a constant reminder to me of the influence that a teacher may have on his students.

In the light of three decades of research, it is hard for those who were not exposed to the early Selye excitement to recognize the impact of his views. McCann, Fortier, and Guillemin are but three of the individuals who have told me that they were "turned on" to neuroendocrinology by Selye. The essential point of this work was that a variety of physical and psychological stresses may produce a relatively nonspecific bodily response largely mediated by the adrenal gland, which if intense and prolonged enough could (in Selye's words) be derailed and cause structural disease.

But how did stress activate the adrenal gland? The role of the adenohypophysis in this response had just been worked out by the Sayerses (Sayers and Sayers, 1947, 1948), who used the adrenal ascorbic acid depletion method, and thus the problem had been defined as a brain-pituitary problem. Again, I attended a crucial lecture by Dr. Peter Heinbecker, then a young professor of surgery, who spoke on the neural origin of Cushing's disease. Heinbecker had worked with White and Rolf in the Physiology Department at Washington University on diabetes insipidus. Their work, now considered a classical study, had shown, among other things, that the anterior lobe of the pituitary was not essential for the development of diabetes insipidus, but that urine volume depended on the presence of an intact adenohypophysis (Heinbecker *et al.*, 1947). Heinbecker later studied the histology of the anterior pituitary gland after the production of various hypothalamic lesions and observed regression in "basophil cells." For reasons no longer clear to me, he postulated that the neurohypophysis tonically suppressed the adrenal cortex; when the suppression was released (as by hypothalamic sectioning) the adrenal cortex became hyperfunctional and inhibited the adenohypophysis by a feedback effect. In apparent support of Heinbecker's views was his demonstration that the hypothalamus of patients dying of Cushing's disease showed histological damage (Heinbecker, 1946). For want of a more clear hypothesis, Heinbecker's view of Cushing's disease for several years was widely quoted, but was put to rest when Castor *et al.* (1951) showed that high doses of ACTH or cortisone in the rat produced analogous hypothalamic lesions. Thus the hypothalamic lesion was secondary and not primary. Much new experimental and clinical data have given continued support to the concept of Cushing's disease as a neurogenic disorder (Krieger, 1976).

By the middle of my second year in medical school the lectures of Levy and Heinbecker led me to the decision to spend the summer of 1946 studying the neural basis of ACTH control. A fellow medical student, Frank

Norbury, suggested that this could be done by placing electrolytic lesions in the hypothalamus, a procedure that James O'Leary, professor of neurology, promised to teach me. To study ACTH release I chose to study adrenal ascorbic acid depletion following unilateral adrenalectomy. Dr. Ethel Ronzoni Bishop, an assistant professor in biochemistry, agreed to allow me to work in her laboratory so that I could learn how to measure ascorbic acid.

The summer's research was a disaster. For one thing we had no stereotaxic instrument for the rat and lesions were therefore placed freehand using insulated sewing needles. Lesions were localized by frozen sectioning, a technique which I could not master. The animal quarters in the attic were not temperature regulated, a grave problem in the heat of a St. Louis summer, nor were we aware of the effects on ACTH release of simply moving the animal from one cage to another. Therefore, "baseline" adrenal ascorbic acid concentrations varied widely. The only predictable stress was my own. However, it was this early work on hypothalamic control of ACTH secretion that ultimately led to my working with G. W. Harris. As a historical note it is of interest that S. M. McCann, working in Brobeck's laboratory, succeeded in 1954 in showing that hypothalamic lesions caused adrenal atrophy (McCann and Brobeck, 1954) and that Claude Fortier approached the same problem using pituitary stalk section (Fortier, 1951).

Undaunted by a totally unsuccessful first year of research, I again tackled the neural control of ACTH secretion the next summer, this time testing the hypothesis of Sayers that stress-induced adrenal cortical activation was due to epinephrine release. The effects of several pharmacological blocking agents on adrenal ascorbic acid depletion were studied. One was dibenzyline, recently introduced as an adrenolytic drug, and the other was ergotamine tartrate, long recognized as a blocker of sympathetic nervous activity. Unfortunately, all the drugs used were themselves highly stressful. Therefore, our results were not conclusive (Ronzoni and Reichlin, 1950). Later, Roger Guillemin repeated this work and was successful in showing that none of the drugs then available for blocking histamine, acetylcholine, or epinephrine effects blocked stress-induced ACTH release. He was able to do this by exposing the animals to repeated injections of the agent to be tested so that they were in an "adapted" state when the specific stress was applied (Guillemin, 1955).

Following graduation from medical school in 1948 and after completing 4 years of clinical training at Barnes Hospital in St. Louis and at the New York Hospital, I was still determined to study hypothalamic control of the adrenal gland. One of the high points of this experience was close contact with Drs. Stuart Wolfe and Harold Wolff, founders of the Cornell School of Psychosomatic Medicine. But my ignorance about what was happening in the field of neuroendocrinology at that time was abysmal. For example, during my interview for a National Science Foundation fellowship in 1952 Dr. Sam Clark, Sr., of Vanderbilt University, a distinguished neuroanatomist and former colleague of Ranson's, asked me "What is your opinion about neurosecretion?" and I had to confess to him that I did not know what neurosecretion was. I did not get the fellowship. The year 1952 was a milestone for me anyway, for it was in that year that I heard a lecture by Geoffrey Harris, who was visiting Dr. Chandler McC. Brooks, professor of physiology at Downstate Medical School in Brooklyn. Harris's name had come to my attention because of a paper he had written in the 1948 *Physiological Reviews* (Harris, 1948), which is a landmark review of neural control of the pituitary gland. This paper, which formed the basis for his 1955 book *Neural Control of the Pituitary Gland* (Harris, 1955), summarized the evidence proving that the pituitary gland was under control of the brain, and posed the paradoxical fact that the gland did not receive a direct nerve supply. When I read Harris's review, I knew at once that he was the man who would teach me what was known about this system. The lecture, delivered on a cold, dreary, damp night in Brooklyn, to a rather meager audience, confirmed my conviction. Harris reported his work with Colfer on the effects of electrical stimulation of the hypothalamus on pituitary-adrenal function in the rabbit, as inferred from changes in peripheral lymphocyte counts. This work, contemporaneous with the work of David Hume in the dog (Hume and Wittenstein, 1950) and Fortier (1951) in the rat, established firmly the role of the hypothalamus in ACTH regulation.

After his lecture I spoke with Harris, and he agreed to take me on as a fellow provided I could find my own support. I still remember with pride that he had read my 1950 paper on adrenergic agents and the control of the adrenal gland and mercifully had forgotten how inconclusive it had been.

Fortunately, through the efforts of W. Barry Wood, professor of medicine at Washington University, I was able to obtain a fellowship from the Commonwealth Fund to spend 2 years with Harris. The Commonwealth Fund was interested in this work because at that time it was investing in studies of the nature of mental illness, and the directors felt, as did I, that an understanding of the neural basis of the stress response would in some way answer the riddles of psychiatric disorders. The notion that the "pituitary is the window to the hypothalamus" is alive and strong as evidenced by the recent studies of Sachar (1975) and others. Harris was at that time in Cambridge, but shortly before I joined him in September of 1952 he had taken the Fitzmary Chair of Physiology at the Institute of Psychiatry at the Maudsley Hospital in South London, and had moved into what were called the "huts," which were temporary cement slab buildings on the grounds of the Maudsley. The very name "huts" conjures up an image of the Gulag Archipelago, but this was not the case.

Working in the huts at that time with Harris were Keith Brown-Grant, a Cambridge M.B. then working toward his M.D., and Bernard Donovan, working toward his Ph.D. I replaced Curt von Euler on their research team, and was immediately thrown into the problem of hypothalamic control of the thyroid gland. Donovan was working with Harris on neural control of ovulation by injecting catecholamines into the pituitary. I was quite disappointed to be assigned to work on a thyroid problem. I knew nothing about the thyroid, nor did I think that the thyroid gland was particularly interesting. Moreover, I had come to work with Harris on adrenal control. Nevertheless, that was the way the laboratory functioned. The junior people worked on whatever Harris wanted studied. Only a few projects were in progress at any one time, and Harris himself spent a great deal of time in the laboratory, working side by side with the fellows. He reserved for himself the most challenging technical problems, such as section of the pituitary stalk or placement of electrodes, and the more difficult the problem the more he enjoyed it. Harris managed to avoid most administrative chores. He gave a few lectures a year, put in an occasional appearance at committee meetings, but did not enter very actively into the affairs of the institution. He once counseled me to never do a good job on a committee, otherwise I would be asked to devote more time to administrative chores. He followed his own counsel well in these matters.

The Harris approach to analysis of hypothalamic control of thyroid function can be looked upon as the classical Sherringtonian method of analysis of neural functions as strongly influenced by Verney's (1947) earlier studies of neurohypophysial function and Hess's work on hypothalamic stimulation (Hess, 1948). Harris's approach began with the identification of a particular endocrine function and the development of specific techniques for measuring that function. Physiological factors capable of perturbing the system in a controlled way were then applied, and the effects of section of the pituitary stalk on this perturbed function were defined. Later, after the function of the stalk had been defined, one could consider making specific lesions in the hypothalamus. Finally, because lesions or sections might influence normal physiological responses nonspecifically, Harris felt that it was essential to duplicate the endocrine change by direct electrical stimulation of the hypothalamus. He feared that restraint or anesthesia might block or modify electrically induced hypothalamic-pituitary function, a point of great importance to him; therefore, he insisted that electrical stimulation be carried out by remote control techniques such as the induction coil method which he introduced into the field. Harris intended to use this technique for the systematic study of each of the known anterior pituitary functions.

Work on the thyroid was but a part of the larger program that Harris had planned out which was designed to prove that the hypothalamus controlled the adenohypophysis by way of the portal vessels of the stalk. He had begun his work on neural control by studying gonadotropin regulation, had worked on ACTH release, was then engaged in studying the control of TSH release, and had elaborate plans for the study of growth hormone secretion. The last included the development of electrodes which would change position as an animal grew in order to permit measurement of the long-term effects of hypothalamic secretion on growth.

The main problem faced by neuroendocrinologists at that time was the difficulty in measuring plasma levels of hormones in small amounts of blood so as to detect short-term changes. Gonadotropin regulation was easy to study by means of the marked changes in the ovary, but growth hormone was nearly impossible to measure until 1963 with the advent of radioimmunoassay (Roth *et al.*, 1963). Even after the introduction of immunoassay, Harris had a deep distrust of the technique for many years, a distrust he also had for most things chemical. Harris believed in what he could measure directly or see in front of his eyes, a view derived from his background as an anatomist. Fred Greenwood, who with Landon was one of the earliest workers to measure GH in blood by radioimmunoassay, told me that in 1964, soon after the technique was developed, he had approached Harris to collaborate on the study of neuroendocrine control of GH secretion in the monkey, but Harris refused because he did not have faith that what the immunoassay was measuring really was growth hormone. In the light of the recent studies of Ellis and Grindeland (1974) and Nicoll (1975) on the dissociation of bioassayable and radioimmunoassayable GH activity, perhaps he will prove to have been correct all along.

Harris's qualifications as an experimental physiologist were prodigious. He was an excellent experimentalist, devised succinct and simple experiments, and executed them elegantly. He kept meticulous, detailed laboratory notebooks, each operation or procedure drawn and described beautifully. When he did a new experiment, he planned it to be definitive, not just a pilot experiment. He knew the literature intimately, and insisted in his writing that each contribution be fully placed in context with the work that had gone before. He insisted on flawless histological preparations, all made laboriously by the celloidin impregnation technique, each section stained and mounted individually on slides. He reminded us that during World War II, when he had been an instructor of anatomy at Cambridge, he had been obliged to use salvaged glass slides and coverslips, and experimental animals were located in surrounding farms, transported to the laboratory on a bicycle.

Harris was extremely logical but stubborn in his views, and could not be readily budged; an idea that he might reject out of hand on first hearing might turn up several weeks later in modified form as a compromise position and without attribution. During my stay in London (1952–1954), he and

Sir Solly Zuckerman waged their famous debate over Zuckerman's stalk-sectioned ferrets that could be stimulated by altered lighting (Thomson and Zuckerman, 1953, 1955). Zuckerman used this observation to bolster his arguments against the view that the portal vessels controlled anterior pituitary function. Harris was positive (after personal study of Zuckerman's histological preparations of the ferret brain) that he could see regrowth of hypophysial-portal vessels which he said had not been filled properly by india ink injection.

Although Harris may have been correct about the stalk in estrus in the ferret, he proved to be incorrect in his evaluation of the contemporaneous work of the Scharrers on neurosecretion. Harris did not believe that the neurohypophysis had anything to do with adenohypophysial function. Among the reasons for this belief were the lack of effect of posterior pituitary removal on anterior pituitary function and the anatomical fact that in a number of species of vertebrates the neurohypophysial system is separated from the hypophysial-portal circulation by a connective tissue septum. In the early 1950s the Scharrers had proposed a rather narrow definition of neurosecretion, limiting it to the stainable material in the supraoptico- and paraventricular-hypophysial pathways, which may account in part for Harris's vociferous criticism of their views. As the concept of neurosecretion has been broadened to include neuronal secretions which do not stain by classical histochemical techniques, it is clear that Harris and Scharrer were working on common grounds. The basic similarity between the function of the neurohypophysis and the function of the median eminence and its tuberohypophysial nerves has been emphasized (Reichlin, 1963).

Our problem with the study of pituitary-thyroid control in 1952 was that there was no way of measuring blood TSH levels at that time, even by bioassay, nor was it possible to measure moment-to-moment secretion of the thyroid gland. Von Euler, Brown-Grant, and Harris in 1951 had devised techniques for measuring radioactive iodine uptake and release in the rabbit by external neck counting after injection of a tracer dose. They had determined that the release of iodide from the neck region followed a semilogarithmic curve and had proposed that changes in pituitary-thyroid function might be inferred from changes in the slope of the release curve (Brown-Grant *et al.*, 1954*a*). My task, with Brown-Grant, was to devise a standard cold exposure to stimulate thyroid function, the ultimate plan being to study the effects of section of the pituitary stalk on this response. Unfortunately, it proved to be difficult to demonstrate cold-induced thyroid hormone release in a consistent and reproducible way. We tried everything we could think of including shaving the unfortunate rabbits, washing them in alcohol, and exposing them to a draft. In fact, rabbits exposed to extreme cold showed *inhibition* of thyroid functions and then we recognized that stress-induced inhibition of thyroid function was really the predominating influence. The hypothesis was tested by forced restraint, which proved to be a consistent means for inhibiting thyroid function (Brown-Grant *et al.*, 1954*b*).

We went on to show that stress-induced thyroid inhibition could be more readily elicited by simple restraint, and because this response was so consistent the stress response rather than the cold-induced thyroid response was used as the basic stimulus for evaluation of the pituitary-stalk connection in thyroid regulation. We found that stalk section inhibited thyroid function and also blocked stress-induced TSH release (Brown-Grant *et al.*, 1957). The stalk section work was confirmatory of a neurohumoral control of TSH secretion. We also studied the role of the adrenal cortex and medulla in these responses which led to the analysis of TSH regulation in adrenalectomized rabbits, and the response to replacement therapy. The stalk-sectioned rabbits were utilized by Claude Fortier, who joined Harris to study ACTH release after stress (Fortier *et al.*, 1957).

The observation that stalk section lowered thyroid function without interfering with thyroxine-induced inhibition led me to postulate that the stalk determined the "set point" of feedback control of the pituitary thyroid axis, a view later elaborated in studies in the rat (Reichlin, 1963, 1964).

After 2 years with Harris, I had great trouble in finding a job in the United States in which I could continue to do neuroendocrine research and clinical medicine. It was fortunate that Dr. Edwin Gildea, professor of psychiatry at Washington University, and Barry Wood were able to arrange a Palmer Fund research fellowship for me in psychiatry. In addition to the challenge of conducting an independent program, my job was to see whether the neuroendocrine work could be made relevant to clinical psychiatry. From 1954 to 1960, I held joint appointments in psychiatry and medicine, studying thyroid function in mental disease, mental function in thyroid disease, and the effects of electroshock therapy and of drug-induced sleep therapy on thyroid function. My main clinical responsibility was to run the medical house staff clinic for the Department of Medicine, a constant reminder of the role of stress in the causation of disease.

When I returned to St. Louis to set up my own laboratory, I applied for grant support from the NIH. I recall that the head of the site visit committee from the Endocrine Study Section was Charles Sawyer, who encouraged the project, but who told me that I had not asked for enough money. I am now grateful for the fact that there was no one else around in the field at Washington University, although it did not appear to be an advantage at the time, and I worked completely by myself for several years, rather than in a relatively large team as is the current mode.

The first project tackled in the new laboratory was the search for a neurohormonal inhibitor of TSH secretion. This experiment grew out of our observation in London that stressful stimuli inhibited pituitary-thyroid function in the rabbit and that the effect was not due to epinephrine. I postulated that the effect might be due to the secretion of an inhibitory substance from the hypothalamus and tested the possibility that it was vasopressin, a substance known to be released during stress. To elicit endogenous vasopressin release, rats were deprived of water and controls were pair-fed, and when this was done there was no apparent effect on thyroid function, nor was there an effect of injected Pitressin (Reichlin, 1957). In the light of recent studies utilizing antisera to somatostatin, it now appears very likely that this stress-induced response is in fact due to somatostatin release, and not vasopressin.

Attempts were then made to develop remote-control techniques for stimulation of the hypothalamus in intact rats to identify the TSH-regulating areas. At the same time Monte Greer was attempting similar projects (Greer *et al.*, 1960). Neither of us was successful—a result I attribute to the use of unanesthetized animals and insensitive end points of TSH release. It now appears that intensities of electric current sufficient to release TSH (as later shown by my collaborator Dr. Joseph Martin) (Martin and Reichlin, 1970) are not tolerated by unanesthetized rats. In this respect, insistence on the use of unanesthetized animals proved to be a detriment to the research.

Having failed twice to identify either the TSH-regulating area or the inhibitory TSH factor, I turned to the study of TSH regulation in rats with hypothalamic lesions, a project pioneered by Greer (1951) and later by Bogdanove and Halmi (1953), Florsheim (1958), and others (see Reichlin, 1966). These workers had studied PTU-induced goiter in rats with lesions, a procedure which gave ambiguous results in thyroid function tests. My contribution was to study functional changes after partial thyroidectomy. These studies established the fact that hypothalamic lesions in the region of the paraventricular nuclei depressed thyroid function but that the glands were capable of responding to partial thyroidectomy by an increase in activity and were suppressed by T_4 injection, a finding that further supported the idea that the hypothalamus determined the set point of control of TSH secretion. Later, Dr. Martin and I were to show by precise determinations of plasma thyroid hormone level and TSH (by radioimmunoassay) that this is in fact the case (Martin *et al.*, 1970). In these experiments we interpreted the residual pituitary-thyroid activity of rats with lesions to indicate residual functional autonomy of the pituitary. More recently, studies carried out in collaboration with Dr. Ivor Jackson (Jackson and Reichlin, 1977) have shown that the classical thyrotropic area lesions in the rat reduce hypothalamic and median eminence TRH (as measured by immunoassay) only by two-thirds. Thus there is residual TRH activity in such animals, and I now believe that the minimal TSH secretion seen in experimental hypothalamic lesions is residual. In man, one may on occasion see profound hypothyroidism following intrinsic hypothalamic disease, so-called hypothalamic hypothyroidism, comparable in severity to that seen after complete hypophysial destruction (Pittman *et al.*, 1971).

My search for TSH-regulating factors in hypothalamic extracts led in 1957 to the effort to identify TRH. The existence of this material was predicted from the stalk section work carried out in Harris's laboratory and the lesion work by Greer (1951). As early as 1956, Shibusawa and a number of collaborators (Shibusawa *et al.*, 1956) had reported the presence of a thyroid-stimulating material in dog hypothalamus and in human urine and, by analogy with CRF, had named this compound "TRF". It is a curious sidelight to history that the presence of TRF in urine, first claimed by Shibusawa on the basis of bioassays and later discredited when his observations could not be confirmed (Reichlin *et al.*, 1963), has been followed by a number of studies including those with my collaborator Jackson which claim that radioimmunoassayable TRH is present in urine (Jackson and Reichlin, 1974; Gagel *et al.*, 1976). The ultimate irony is that our claim of radioimmunoassayable TRH has also been disputed by other workers (Vagenakis *et al.*, 1975), although our most recent studies indicate that the urine contains TRH (Jackson *et al.*, 1976).

Historically, the main problem in proving that TRF existed was the insensitivity of bioassays available for measurement of biological activity. It is the dispute over the validity of the assay used which has led to the dispute over priority for first demonstration of this factor. In my view, the first fully reliable assay method was that of Yamazaki and Guillemin in 1960 which utilized radioiodine-labeled treated young rats. Important early work was done by Schreiber *et al.* (1962). My own studies indicated that the Mackenzie mouse preparation released TSH in response to injection of hypothalamic extract provided that only minimal amounts of thyroxine had been given (Reichlin, 1964*a*,*b*). Large amounts of thyroxine block the response completely. Subsequent work by Vale *et al.* (1967) showed elegantly that the effects of TRH are blocked by thyroxine in a dose-dependent manner. More recently, Snyder and Utiger (1972) have demonstrated that at the physiological set point of thyroid hormone level, minimal, barely detectable changes in plasma T_4 and T_3 are capable of blocking TRH responses. The work of Schally's laboratory in isolating TRH depended heavily on the use of the codeine-blocked mouse assay (Redding *et al.*, 1966). By way of emphasis of progress that has been made in this field, TRH biological activity is readily detectable at the present time using plasma radioimmunoassayable TSH as an index of effect in a wide variety of animals, and the tripeptide is demonstrable directly by radioimmunoassay.

One of the by products of my work with hypothalamic lesions in rats was the observation that some rats with extensive hypothalamic destruction failed to grow normally. I later found out that this observation had been made as early as 1938 by Cahane and Cahane, and by Endröczi *et al.* (1956). Bioassay of pituitary content showed that GH levels were reduced, and replacement with target gland hormones failed to restore normal growth (Reichlin, 1960). These observations taken together made a strong case that the hypothalamus controlled GH secretion, a postulation that has subsequently proven to be true, although at that time, and even now, there seems to be no teleologically valid reason why the brain should regulate growth. Subsequently, in collaboration with Daughaday's group, our laboratory showed by the newly developed radioimmunoassay that GH secretion in response to hypoglycemia in squirrel monkeys was blocked by lesions of the hypothalamus (Abrams *et al.*, 1966), a finding that further confirmed the work on stalk-sectioned humans reported by Roth, Glick, Berson, and Yalow in 1963. Subsequently, my collaborators Gregory Brown and Don Schalch showed that psychological stress responses in squirrel monkeys leading to increased GH release were also blocked by lesions in the anterior lip of the median eminence (Brown *et al.*, 1971). The firm recognition that GH secretion was under the control of the hypothalamus gave further impetus to the effort to identify growth hormone releasing factor (see Reichlin and Schalch, 1969, for review of early work). Although a number of laboratories including our own have given good bioassay evidence that this material exists (see Reichlin, 1974, for review), the chemistry of GHRF is still unknown. The main chemical result of the search for GHRF to this point was the discovery of GH release inhibiting factor by Krulich *et al.* (1968), which was later chemically characterized as somatostatin by Guillemin and collaborators (Brazeau *et al.*, 1973).

It is not the purpose of the essays in this series to do more than trace the development of the field of neuroendocrinology through the early experiences of its practitioners. Nevertheless, it is worth emphasizing that for many years neuroendocrinology, especially of the anterior pituitary, was not respectable. The watershed was reached in 1969 with the elucidation of the structure of TRH almost simultaneously by the laboratories of Guillemin (Burgus *et al.*, 1969) and Schally (Bøler *et al.*, 1969). This discovery almost overnight legitimized the field in the eyes of the disbelievers. The isolation of LHRH, somatostatin, and the endogenous opiates has further advanced knowledge, to the point that neuroendocrine research has been established firmly as a discipline of neurobiology and neuroscience and hence can now be approached from the standpoint of the neuron rather than as its reflection in pituitary function. This evolution of knowledge will strongly direct the direction of growth of the field toward the brain, and away from the pituitary, which long served as the focus of early neuroendocrine research.

REFERENCES

Abrams, R. L., Parker, M. L., Blanco, S., Reichlin, S., and Daughaday, W. H. (1966). Hypothalamic regulation of growth hormone secretion. *Endocrinology* **78**:605.

Bogdanove, E. M., and Halmi, N. S. (1953). Effects of hypothalamic lesions and subsequent propylthiouricil treatment on pituitary structure and function in the rat. *Endocrinology* **53**:274.

Bøler, J., Enzmann, F., Folkers, K., Bowers, C. Y., and Schally, A. V. (1969). The identity of chemical and hormonal properties of the thyrotropin releasing hormone and pyro-glutamyl-histidyl-proline amide. *Biochem. Biophys. Res. Commun.* **37**:705.

Brazeau, P., Vale, W., Burgus, R., Ling, N., Butcher, M., Rivier, J., and Guillemin, R. (1973). Hypothalamic polypeptide that inhibits the secretion of immunoreactive pituitary growth hormone. *Science* **179**:77.

Brown, G. M., Schalch, D. S., and Reichlin, S. (1971). Hypothalamic mediation of growth hormone and adrenal stress response in the squirrel monkey. *Endocrinology* **89**:694.

Brown-Grant, K., von Euler, C., Harris, G. W., and Reichlin, S. (1954*a*). The measurement and experimental modification of thyroid activity in the rabbit. *J. Physiol.* (*London*) **126**:1.

Brown-Grant, K., Harris, G. W., and Reichlin, S. (1954*b*). The effect of emotional and physical stress on thyroid activity in the rabbit. *J. Physiol.* (*London*) **126**:29.

Brown-Grant, K., Harris, G. W., and Reichlin, S. (1957). The effect of pituitary stalk section on thyroid function in the rabbit. *J. Physiol.* (*London*) **136**:364.

Burgus, R., Dunn, T. F., Desiderio, D., and Guillemin, R. (1969). Structure moleculaire du facteur hypothalamique hypophysiotrope TRF d'origine ovine: Mise en évidence par spectrométrie de masse de la séquence PCA-HIS-PRO-NH_2. *Compt. Rend.* **269**:1870.

Cahane, M., and Cahane, T. (1938). Sur le rôle du diencephale dans le development somatique. *Rev. Franc. Endocrinol.* **16**:181.

Castor, C. W., Baker, B. L., Ingle, D. J., and Li, C. H. (1951). Effect of treatment with ACTH or cortisone on anatomy of the brain. *Proc. Soc. Exp. Biol.* **76**:353.

Ellis, S., and Grindeland, R. E. (1974). Dichotomy between bio- and immunoassayable growth hormone. In Raiti, S. (ed.), *Advances in Human Growth Hormone Research*, Publ. No. (N.I.H.) 74–612, U.S. Government Printing Office, Washington, D.C., pp. 409–425.

Endröczi, E., Kóvaks, and Lissák, K. (1956). Die Wirkung der Hypothalamus-reizung auf das endokrine und somatische Verhalten. *Endokrinologie* **33**:271.

Florsheim, W. H. (1958). The effect of anterior hypothalamic lesions on thyroid function and goiter development in the rat. *Endocrinology* **62**:783.

Fortier, C. (1951). Dual control of adrenocorticotropin release. *Endocrinology* **49**:782.

Fortier, C., Harris, G. W., and McDonald, I. R. (1957). The effect of pituitary stalk section on the adrenocortical response to stress in the rabbit. *J. Physiol.* (*London*) **136**:344.

Gagel, R. F., Jackson, I. M. D., Deprez, D. P., Papapetrou, P. D., and Reichlin, S. (1976). The significance of urinary thyrotropin releasing hormone (TRH) excretion in man. In

Robbins, J., and Braverman, L.E. (eds.), *Thyroid Research*, Proc. 7th Int. Thyroid Conf., Boston, June 9–13, 1975, pp. 8–10.

Greer, M. A. (1951). Evidence of hypothalamic control of the pituitary release of thyrotropin. *Proc. Soc. Exp. Biol. Med.* **77**:603.

Greer, M. A., Yamada, T., and Lind, S. (1960). The participation of the nervous system in the control of thyroid function. *Ann. N.Y. Acad. Sci.* **86**:667.

Guillemin, R. (1955). A re-evaluation of acetylcholine, adrenaline, noradrenaline and histamine as possible mediators of the pituitary adrenocorticotrophic activation by stress. *Endocrinology* **56**:248.

Harris, G. W. (1948). Neural control of the pituitary gland. *Physiol. Rev.* **28**:139.

Harris, G. W. (1955). *Neural Control of the Pituitary Gland*, Arnold, London.

Heinbecker, P. (1946). Cushing's syndrome. *Ann. Surg.* **124**:252.

Heinbecker, P., White, H. L., and Rolf, D. (1947). The essential lesion in experimental diabetes insipidus. *Endocrinology* **40**:104.

Hess, W. R. (1948). *Die organization des vegetativen Nervensystems*, Karger, Basel.

Hume, D. M., and Wittenstein, G. J. (1950). The relationship of the hypothalamus to pituitary-adrenocortical function. In Proc. 1st Clin. ACTH Conf., Blakiston, Philadelphia, p. 134.

Jackson, I. M. D., and Reichlin, S. (1974). Thyrotropin releasing hormone (TRH): Distribution in the brain, blood and urine of the rat. *Life Sci.* **14**:2259.

Jackson, I. M. D., and Reichlin, S. (1977). Brain thyrotrophin-releasing hormone is independent of the hypothalamus. *Nature* **267**:853.

Jackson, I., Sanchez-Franco, F., Baum, G., Soo Hoo, F., and Reichlin, S. (1976). Is there, or is there not, thyrotropin releasing hormone (TRH) in urine. 4th New Engl. Endocrinol. Conf., Boston (abst).

Krieger, D. T. (1976). Neuroendocrinology. In Ingbar, S. (ed.), *The Year in Endocrinology 1975–1976*, Plenum Medical Book Co., New York, pp. 22–23.

Krulich, L., Dhariwal, A. P. S., and McCann, S. M. (1968). Stimulatory and inhibitory effects of purified hypothalamus extracts on growth hormone release from rat pituitary *in vitro*. *Endocrinology* **83**:783.

Martin, J. B., and Reichlin, S. (1970). Thyrotropin secretion in rats after hypothalamic electrical stimulation or injection of synthetic TSH-releasing factor. *Science* **168**:1366.

Martin, J. B., Boshans, R., and Reichlin, S. (1970). Feedback regulation of TSH secretion in rats with hypothalamic lesions. *Endocrinology* **87**:1032.

McCann, S. M., and Brobeck, J. R. (1954). Evidence for a role of the supraopticohypophysial system in regulation of adrenocorticotrophin secretion. *Proc. Soc. Exp. Biol. Med.* **87**:318.

Nicoll, C. S. (1975). Radioimmunoassay and radioreceptor assays for prolactin and growth hormone: A Critical appraisal. *Am. Zool.* **15**:881.

Pittman, J. A., Haiger, E. D., Jr., Hershman, J. M., Jr., and Pittman, C. S. (1971). Hypothalamic hypothyroidism. *New Engl. J. Med.* **285**:844.

Redding, T. W., Bowers, C. Y., and Schally, A. V. (1966). An *in vivo* assay for thyrotropin releasing factor. *Endocrinology* **79**:229.

Reichlin, S. (1957). The effect of dehydration, starvation and pitressin injections on thyroid function in the rat. *Endocrinology* **60**:470.

Reichlin, S. (1960). Growth and the hypothalamus. *Endocrinology* **67**:760.

Reichlin, S. (1963). Medical progress, neuroendocrinology. *New Engl. J. Med.* **269**: 1246–1250, 1296–1303.

Reichlin, S. (1964*a*). Function of the hypothalamus in regulation of pituitary-thyroid activity: Brain-thyroid relationships. *Ciba Found. Study Group* **18**:17.

Reichlin, S. (1964*b*). Physiological evidence for a thyrotrophin-releasing factor. Proc. 2nd Int. Congr. Endocrinol., London, pp. 499–507.

Reichlin, S. (1966). Control of thyrotropic hormone secretion. In Martini, L., and Ganong, W. F. (eds.), *Neuroendocrinology*, Vol. 1, Academic Press, New York, pp. 445–536.

Reichlin, S. (1973). The physiology of growth hormone regulation: Pre- and postimmunoassay eras. *Metabolism* **22**:987.

Reichlin, S. (1974). Regulation of somatotrophic hormone secretion. In *Handbook of Physiology*, Section 7: *Endocrinology*, Vol. IV, Pt. 2, pp. 405–447.

Reichlin, S., and Schalch, D. S. (1969). Growth Hormone releasing factor. In Gual, C. (ed.), *Progress in Endocrinology*, Excerpta Medica Foundation, Amsterdam, pp. 584–594.

Reichlin, S., Boshans, R. L., and Brown, J. G. (1963). A critical evaluation of the "TRF" of Shibusawa. *Endocrinology* **72**:334.

Ronzoni, E., and Reichlin, S. (1950). Adrenergic agents and the adrenocorticotrophic activity of the anterior pituitary. *Am. J. Physiol.* **160**:490.

Roth, J., Glick, S. M., Yalow, R. S., and Berson, S. A. (1963). Hypoglycemia: A potent stimulus to secretion of growth hormone. *Science* **140**:987.

Sachar, E. J. (1975). Twenty-four-hour cortisol secretory patterns in depressed and manic patients. In Gispen, W. H., Wimersma Greidanus, T. B., Bohus, B., and de Wied, D. (eds.), *Progress in Brain Research*, Vol. 42: *Hormones, Homeostasis and the Brain*, Proc. Vth Int. Congr. Int. Soc. Psychoneuroendocrinol., Elsevier, Amsterdam, pp. 81–91.

Sayers, G., and Sayers, M. A. (1947). Regulation of pituitary adrenocorticotrophic activity during the response of the rat to acute stress. *Endocrinology* **40**:265.

Sayers, G., and Sayers, M. A. (1948). The pituitary-adrenal system. *Recent Progr. Horm. Res.* **2**:81.

Schreiber, B., Rybák, Eckertová, A., Jirgl, V., Kocl, J., Franc, Z., and Kmentová, K. (1962). Isolation of a hypothalamic peptide with TRF (thyrotrophin releasing factor) activity *in vitro*. *Experientia* **18**:338.

Selye, H. (1946). The general adaptation syndrome and the disease of adaptation. *J. Clin. Endocrinol.* **6**:117.

Shibusawa, K., Saito, S., Nishi, K., Yamamoto, T., Tomizawa, K., and Abe, C. (1956). The hypothalamic control of the thyrotroph-thyroidal function. *Endocrinol. Japon.* **3**:116.

Snyder, P. J., and Utiger, R. D. (1972). Inhibition of thyrotropin response to thyrotropin releasing hormone by small quantities of thyroid hormones. *J. Clin. Invest.* **51**:2077.

Thomson, A. P. D., and Zuckerman, S. (1953). Functional relations of the adenohypophysis and hypothalamus. *Nature* **171**:970.

Thomson, A. P. D., and Zuckerman, S. (1955). A wax construction of the pituitary region of a ferret which responded to light after stalk section. *J. Endocrinol.* **12**:2.

Vagenakis, A. G., Roti, E., Mannix, J., and Braverman, L. E. (1975). *J. Clin. Endocrinol. Metab.* **41**:801.

Vale, W., Burgus, R., and Guillemin, R. (1967). Competition between thyroxine and TRF at the pituitary level in the release of TSH. *Proc. Soc. Exp. Biol. Med.* **125**:210.

Verney, E. B. (1947). The antidiuretic hormone and the factors which determine its release. *Proc. R. Soc. Lond. Serv. B* **135**:25.

Yamazaki, E., Sakiz, E., and Guillemin, R., (1963). An *in vivo* bioassay for TSH-releasing factor (TRF). *Experientia* **19**:480.

20

Murray Saffran

Murray Saffran was born October 30, 1924, in Montreal, Quebec, Canada. He was educated at McGill University (B.Sc., 1945; M.Sc., 1946; Ph.D., 1949). He joined the faculty as research fellow and demonstrator in the Department of Psychiatry in 1948, and remained at McGill until 1969. He spent sabbatical leaves at the University of Copenhagen (1953–1954), the University of Edinburgh (1958–1959), and Harvard University (1967–1968).

At McGill University he had a full-time appointment in the Department of Psychiatry (Allan Memorial Institute) until 1959, rising to the rank of assistant professor. In 1959 he was made associate professor in the Departments of Biochemistry and Psychiatry. During the period 1963–1965 he was director of the McIntyre Medical Sciences Building, assisting in the planning, construction, and opening of the building. In 1965 he left the Department of Psychiatry and was made professor of biochemistry at McGill in 1966.

In 1969 he was appointed chairman of the Department of Biochemistry at the Medical College of Ohio at Toledo. Dr. Saffran has served on the editorial boards of *Endocrinology* and the *Journal of Steroid Biochemistry*. He was a member of the Endocrinology Study Section of NIH and is presently on the Neurology A Study Section. He was the recipient of the Ayerst Award of the Endocrine Society in 1967.

20

Corticotropin-Releasing Factor: The Elusive Hormone

MURRAY SAFFRAN

In September 1953 three young men sat on the grass on the grounds of the Allan Memorial Institute of Psychiatry, McGill University, in Montreal, Canada. The Allan Institute is situated on the slopes of Mount Royal (Fig. 1) and the view of the city below was spectacularly beautiful (Fig. 2). September and October are the best months in Montreal, with unexpected warmth, clear skies, and brilliant colors in the trees. But the young men were not looking at the scenery; they were planning experiments.

I was outlining to Edward Schonbaum and Andrew Schally experiments to demonstrate the existence of a hypothetical hypothalamic hormone that had been postulated as the agent that controls the release of ACTH from the adenohypophysis. The plan was simple (Fig. 3). Adenohypophysial tissue from the rat would be incubated in simple salt solution with and without an extract of the hypothalamus. If the extract contained the hypothetical hormone, then the pituitary tissue in contact with the extract would discharge more ACTH than pituitary tissue incubated without the extract. The ACTH discharged into the medium would be measured by placing the medium in contact with adrenal tissue from the rat, and the medium containing the most ACTH would provoke the largest amount of steroid formation by the adrenal tissue.

This was a very direct attack on the problem. The role of the hypothalamus in the control of adenohypophysial function had a short history, and some of it has been recounted in the first volume of this series. However, previous attempts to demonstrate the existence of a hypothalamic hormone had utilized *in vivo* approaches of various kinds whose complexity denied the investigators a simple interpretation of the results.

MURRAY SAFFRAN • Department of Biochemistry, Medical College of Ohio, Toledo, Ohio 43699.

FIGURE 1. Allan Memorial Institute of Psychiatry, McGill University (1955).

FIGURE 2. View of Montreal from the lawn of Allan Memorial Institute of Psychiatry (1955).

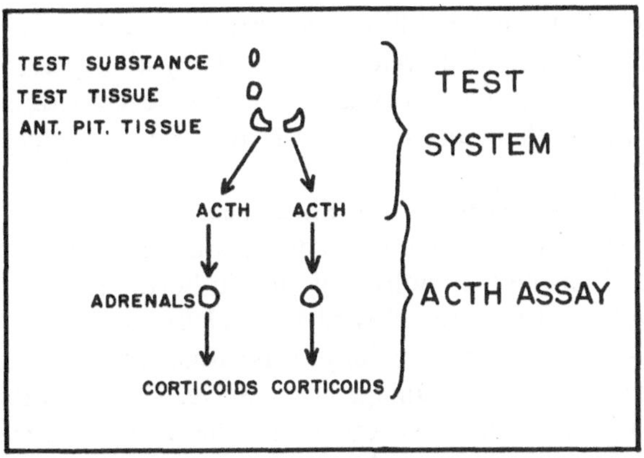

FIGURE 3. Test system for the detection of corticotropin-releasing substances (CRF). From Saffran and Schally (1958), by permission.

The history of science contains many references to happy accidents, or serendipity, but the planned experiments that are successful are seldom as interesting to the reader. In the case of the CRF, whose story is yet to be completely unfolded, the beginning chapters were the result of deliberate planning. I spent many sleepless nights and daylight hours thinking of how CRF might be discovered. The experiment was superficially a very simple one, but the simplicity was the result of my earlier scientific training and experiences.

My first research training was under the late Orville F. Denstedt. Dr. Denstedt was born a Canadian, lived in Canada all his life, and died in Canada. I met him when I was an undergraduate student in biochemistry at McGill University. His kind and friendly manner endeared him to all of the students. When choosing a research director for my graduate degree it was inevitable that I was drawn to Orville Denstedt. I stayed with him for the M.Sc. degree, working on problems related to the preservation of human blood (Saffran and Denstedt, 1951). For the Ph.D. degree he farmed me out to K. A. C. Elliott, for a study in intermediary metabolism, for which Dr. Denstedt did not feel he was equipped.

My training in the use of isolated tissues were received in the hands of K. A. C. Elliott. Dr. Elliott is a South African by birth. He received his Ph.D. degree in the Department of Biochemistry at the University of Cambridge in the great days of Professor F. Gowland Hopkins. His contemporaries numbered among the most famous names in biochemistry. Dr. Elliott went to the United States from Cambridge and after World War II he moved to the Montreal Neurological Institute at the invitation of its director, Wilder Penfield. Although I was nominally a graduate student of O. F. Denstedt, it was Dr. Elliott who directed my work for the Ph.D. on the metabolism of citric acid cycle components by kidney cortex slices (Saffran and Prado, 1949). This work gave me experience with isolated tissues and enabled me to think of studying the function of endocrine tissues *in vitro*.

It was during my work for the Ph.D. degree that I struck up a friendship with Judith Cohen, a fellow graduate student working with the late R. D. H. Heard on the metabolism of steroid hormones. My first contacts with endocrinology were strengthened in 1947 by our marriage. My theoretical knowledge of endocrinology was broadened by lectures from J. S. L. Browne, Eleanor Venning, Hans Selye, C. P. Leblond, R. D. H. Heard, and Lyman Duff and others in the lively endocrine scene in Montreal during my graduate student days.

Even before I was formally granted the Ph.D. degree I was hired in 1948 by Dr. R. A. Cleghorn to assist in his research programs at the Allan Memorial Institute of Psychiatry. Dr. Cleghorn was a pupil of Professor Macleod, who shared the Nobel Prize with Banting and Best for the discovery of insulin. Dr. Cleghorn's area of research had been the adrenal cortex, particularly the cardiovascular and musculoskeletal effects of adrenalectomy. He continued his interest in the adrenal cortex after joining the Allan Memorial Institute of Psychiatry by looking into the relationship between schizophrenia and the adrenal cortex. This line of research led him to the realization that the control of the pituitary-adrenal system resides in the brain, and he very early supported the hypothesis of the hypothalamic control of the pituitary gland. A colleague at the Allan Memorial Institute in those early days was Dr. Bernard F. Graham. Dr. Graham and Dr. Cleghorn worked together in the clinical arm of our research at the Allan Memorial Institute, while I supplied the laboratory supervision. Drs. Cleghorn and Graham educated me in the anatomical and experimental basis of the hypothalamic theory. It was their enthusiastic support for the theory that led me to think in terms of working in the field.

With Drs. Cleghorn and Graham I worked on the effects of anxiety and depression on adrenal cortical function in human subjects (Cleghorn *et al.*, 1950). As a sideline I joined forces with Bernard Grad to show that rat adrenal tissue produces corticosteroids *in vitro* and that the production can be stimulated by the addition of ACTH *in vitro* (Saffran *et al.*, 1952). With Margaret J. Bayliss I found that the stimulation by ACTH could be used as the basis of a bioassay for ACTH (Saffran and Bayliss, 1953). A former fellow graduate student with Dr. Elliott, Marion K. Birmingham, joined the Gerontology Unit of the Allan Memorial Institute. In subsequent years, Dr. Birmingham's work and ours became complementary and I benefitted from her encouragement, assistance, and criticism.

In 1952 I left for a sabbatical leave in the laboratory of Herman Kalckar at the University of Copenhagen. I didn't know at the time, but I was to replace an American postdoctoral fellow named James Watson, who was soon to become famous. While I was in Copenhagen, my laboratory at the Allan Memorial Institute was in the charge of Dr. Frances Elliott, who had worked with me before I left for Denmark. She hired a recently arrived immigrant from England named Andrew V. Schally as her technician. Andrew Schally had obtained his university education at the University of London and had worked for a while at the Institute for Medical Research at Mill Hill, under the direction of the peptide chemist D. F. Elliott.

Frances Elliott (no relation of D. F. Elliott) reported that young Schally was a very vigorous and conscientious worker. He seemed to disregard the clock in his enthusiasm and even worked without interruption while nursing a broken foot. Because of altered circumstances, she suggested that Andrew Schally should transfer upon my return to my laboratory and be my technician.

And so it was, on my return from Copenhagen, that I found a new technician. His energy and capabilities impressed me very much and I readily agreed to his request that he be allowed to become a part-time student in biochemistry at McGill University to augment his bachelor's degree from the University of London with a B.Sc. from McGill University. Within a short time, Andrew Schally managed to fulfill the requirements for the B.Sc. degree, and he applied for and obtained permission to enter graduate school as a candidate for the Ph.D. degree in biochemistry, working under my direction.

Edward Schonbaum was a graduate student in biochemistry at McGill University. His father was a dentist in Jakarta, then called Batavia, when the Japanese overran Java and interned all citizens of the Netherlands. Eddie spent the war years under Japanese custody. After the war he obtained his early education in the Netherlands and came to Canada for graduate work under my direction.

The three young men discussed the plan enthusiastically and divided areas of responsibility among themselves. Eddie Schonbaum would continue to explore the best conditions for the response of adrenal tissue to ACTH *in vitro*. Andrew Schally and Murray Saffran would redesign the assay for ACTH according to statistical principles of bioassay to permit an estimate of the error of the assay in comparison with the response to a stable standard. Then, using the best possible assay for ACTH, the team would apply it to answer the question: does a hypothalamic extract accelerate the release of ACTH from adenohypophysial tissue *in vitro*?

Eddie Schonbaum joined forces with Marion K. Birmingham in an adjacent laboratory of the Allan Memorial Institute and together they systematically surveyed the composition of the medium and a variety of added nutrients for effects on the response of the adrenal to ACTH. Fortunately for the main thrust of the problem, the original conditions for the incubation of adrenal tissue turned out to be virtually optimal (Schonbaum *et al.*, 1956). The survey took up so much of Schonbaum's time that he was unable to participate directly in the search for the elusive hormone. He remained an enthusiastic CRF-watcher throughout his days at the Allan Memorial Institute and remains so to this day in his role as Director of Pharmacology at N.V. Organon in Oss, the Netherlands.

The Saffran-Schally team was complemented by the superb talents of Brigitta U. Zimmermann (later Caplan), who had emigrated to Canada from Germany. Gita was a tireless co-worker. She mastered all chemical and biological aspects of the problem, and remained as a member of the team for many years. Together, the three of us redesigned the ACTH assay on a firm statistical basis and we were now ready to go to the next step (Saffran and Schally, 1955*a*).

My training in biochemistry had given me minimal exposure to neuroanatomy. The first effort to remove hypothalamic tissue was cumbersome. It involved removing the entire lower jaw of the rat's head, chipping away of bone to expose the pituitary, and finally removing the pituitary to expose the hypothalamus. Luckily for our rate of progress, the Allan Memorial Institute was visited by Geoffrey Harris and he very kindly demonstrated a much easier way of getting at the hypothalamus and pituitary. Harris showed us how to open the skull and scoop out the brain with a curved scissors, exposing the hypothalamus and pituitary simultaneously. Access to the tissue was in seconds instead of minutes and the tissue could be obtained unharmed by chips of bone.

Using the ACTH assay, we were able to measure the ACTH released from pituitary tissue incubated for one hour. Coincubation of hypothalamic and pituitary tissue had no detectable affect on the release of ACTH. At about the time of these experiments, C. N. H. Long, at Yale University, had proposed that epinephrine was the stimulant to the release of ACTH from the pituitary. We tried the effect of epinephrine and norepinephrine (then called arterenol), and we did find a slight but consistent increase in the release of ACTH in the presence of the amines, but the increase was not significant. However, when hypothalamic tissue, a sympathetic amine, and pituitary tissue were incubated together, the release of ACTH was statistically greater than from pituitary tissue incubated alone. Surprisingly, hypothalamic tissue could be replaced by brain cortical tissue, provided that norepinephrine was also present. Liver tissue could not be used instead of brain or hypothalamus; neither could acetylcholine or serotonin replace the sympathetic amines.

The most consistent stimulation of the release of ACTH was found with posterior pituitary tissue, plus an amine.

Our first paper, entitled "The Release of Corticotrophin by Anterior Pituitary Tissue *in Vitro*," was published in the *Canadian Journal of Biochemistry and Physiology* in 1955 (Saffran and Schally, 1955*b*).

At about that time, Bruno George Benfey had arrived in the Department of Pharmacology at McGill University and was interested in the purification of peptides of the pituitary gland. He had inherited preparations of vasopression made by R. L. Stehle about 20 years previously in the Department of Pharmacology at McGill University. With his interest in the purification of peptides, Bruno Benfey was a natural partner in the search for the elusive ACTH-releasing hormone. Our interest in the posterior pituitary was increased when we found that norepinephrine apparently stimulated the release of ACTH in the presence of neurohypophysial tissue. The ACTH-releasing activity of the neurohypophysis could be extracted with water and the water extract was active even without norepinephrine. Accetone powders of beef and pig posterior pituitary tissue contained ACTH-releasing activity that could be extracted with dilute acetic acid. The next step was to test the ACTH-releasing activities of preparations of the posterior pituitary hormones then available. Activity was found in a preparation of vasopressin made by Professor Stehle, but not in oxytocin. Stehle's vasopressin was then purified by Benfey by paper chromatography. The purified vasopressin no longer had ACTH-releasing activity! Instead, the CRF activity was recovered from the paper chromatogram in an area clearly separated from vasopressin. The ACTH-releasing activity of this fraction was greater in the presence of norepinephrine, but not statistically so.

In 1954 McCann and Brobeck observed that Pitressin, the Parke-Davis preparation of partially purified vasopressin, reproduced the peripheral effects of ACTH, and they concluded that vasopressin releases ACTH from the adenohypophysis. Their experiments clearly equated vasopressin with CRF. Our results, for the first time, separated the ACTH-releasing activity of the neurohypophysis from vasopressin. As a result of our experiments, a paper was published in *Endocrinology* in October 1955 entitled "Stimulation of the Release of Corticotropin from the Adenohypophysis by a Neurohypophysial Factor" (Saffran *et al.*, 1955). This was the first of our publications on CRF and is the first paper to use this term which we coined for the ACTH-releasing activity. The term "factor" was applied to the activity in keeping with the custom of the 1950s of naming biochemical activities of unidentified substances as factors. Many of the biochemical factors of 1955 were subsequently identified as components of the B complex of vitamins, with roles as coenzymes in intermediary metabolism. The custom has since arisen of calling an unidentified substance a factor, but once the material has been identified it is renamed a hormone, a coenzyme, etc. I should point out that our first CRF was demonstrated in the posterior lobe of the pituitary gland rather than in the hypothalamus.

The purification and identification of CRF looked very simple in 1955. All that the team of Saffran, Schally, and Zimmerman had to do was to subject posterior pituitary extracts to separation by paper chromatography in a variety of solvent systems, assaying material eluted from portions of the paper strip to locate the CRF, until the material was of sufficient purity for chemical analysis. Little did we realize that CRF still would be unidentified more than 20 years later.

Why did we use the posterior pituitary as starting material in the search for CRF? Two reasons. The first—we found the most CRF activity in the posterior pituitary of the rat. The second—in 1955 large amounts of posterior pituitary acetone powder were available (as raw material for the production of oxytocin and vasopressin), but there were no commerical sources of hypothalamic tissue.

We started our search for CRF with a concentrate of posterior pituitary extract, kindly provided by Parke-Davis Company. An amount of approximately 3 g of the extract was subjected to paper chromatography in three systems with painstaking bioassay of eluates of parts of the paper strips for CRF activity. Because of the limited capacity of the paper, the chromatographic steps were repeated hundreds of times. It is a tribute to the energy and patience of Andrew Schally that these experiments were carried out. While Schally did the paper chromatography, Gita Zimmerman and I worked at the assay of the fractions. A year of constant activity resulted in approximately 600 μg of a preparation that showed CRF activity at a dose of half a nanogram.

The preparation of CRF was made possible by the generous donation of posterior pituitary extracts by the Parke-Davis Company and by subsidies from N.V. Organon of the Netherlands. The main financial support for the work was provided through grants to the Allan Memorial Institute by a Federal-Provincial Mental Health Grant whereby the federal government of Canada provided money for distribution by the provinces to medical schools and hospitals for medical research. Additional support came from Nordic Biochemicals Limited of Montreal, manufacturers of ACTH, and from grants by the Foundations' Fund for Research in Psychiatry and from Canada Packers Limited, Toronto. Although the sums were not large by today's standards, the work could not have been carried out without these grants.

A year's work had provided us with 600 μg! What could be done with this small amount?

In return for their generous financial assistance, one-third of our product, 200 μg, was given to the Organon Company for their purposes. They, in turn, provided the material to their medical consultants for clinical trial and CRF was administered to a human subject with the expected results—a large increase in blood cortisol that lasted for over 24 hr. Most of the rest of the material was used for amino acid analysis and for chemical studies. In 1955 micromethods for amino acid analysis were not available. The most sensitive techniques involved paper chromotography with rather crude identification of ninhydrin-positive spots. Using this technique, we found that the amino acids in the preparation resembled those in vasopressin, with the addition of a few extra amino acids. The biological activity, like that of vasopressin, was destroyed by treatment with thioglycollate and performic acid. The material, however, could withstand exposure to 2 N HCl but not to 6 N HCl under conditions of protein or peptide hydrolysis. Our work was published in the *Biochemical Journal* in 1958 under the title "A Corticotrophin-Releasing Factor: Partial Purification and Amino Acid Composition" (Schally *et al.*, 1958).

Some of the purified CRF was tested by A. B. Rothballer in a stalk-sectioned dog (unpublished), and Hokin *et al.* (1958) applied the substance to the pituitary gland and observed an enhanced rate of synthesis of phospholipids.

In 1955 CRF was ripe to be discovered. At the same time as we were working on the problem, completely unknown to us, Roger Guillemin in Texas was also demonstrating the existence of CRF. His story is best told in his own words (see his chapter in this volume). The professor of physiology at McGill University at that time, Dr. F. C. McIntosh, brought a flyer to me announcing a symposium to be held by the Houston Neurological Society at the Texas Medical Center, with Hebbel Hoff as chairman. The topic of the symposium was "Hypothalamic-Hypophysial Interrelationships." Dr. McIntosh knew of our work on CRF and suggested that I send an abstract to Dr. Hoff so that we could present our results at a scientific forum. Dr. Hoff replied to my letter with an invitation to attend the symposium, which was to be held on March 18, 1955. I set out from Montreal by airplane toward New York City to catch a connection for Houston. In 1955 aircraft flew through storms, not over them. The usual 2-hr flight to New York lasted 4 hr because the New York airport was weatherbound and my plane had to fly back to Albany for refueling. By the time we reached New York, the Houston aircraft had left long ago and the next one was not until the next day, too late for the symposium. I returned to Montreal on the same plane and wired my regrets to Dr. Hoff. It was only years later that I learned that the arrival of the abstract describing our work on demonstration of the existence of CRF caused consternation in Houston because they were under the impression that the only demonstration of CRF at that time was by the experiments of Guillemin and his collaborators in Houston. Unfortunately, the March winds prevented a face-to-face meeting of the two groups working on CRF in those early days. The slim volume that arose from the Houston symposium (Fields *et al.*, 1956) contains contributions by the great names of neuroendocrinology, including John Green, Geoffrey Harris, Walter Hild, Arthur Mirsky, and Roger Guillemin. The chapter by Roger Guillemin entitled "Hypothalamic-Hypophysial Relationships in the Production of Pituitary Hormones *in Vitro*" is an excellent summary of the early experiments depicting the hypophysiotropic influence of the hypothalamus on the production of ACTH by organ cultures of the adenohypophysis. In Guillemin's system, in contrast to our *in vitro* incubation, vasopression had no demonstrable activity.

After Andrew Schally had received his Ph.D. degree, he went off to join forces with Guillemin. I felt that the future of CRF was in good hands and that a solution to its structure would soon be forthcoming. My family and I embarked for Edinburgh and a stimulating year with Marthe Vogt. The subsequent history of the Guillemin-Schally partnership in Houston is told in their respective chapters. CRF remained elusive in their hands as well. While we were in Edinburgh, papers began to appear from Texas on the further purification of CRF and other hypothalamic hormones, but no definite identification was made.

During our stay in Scotland, the chairman of the Department of Biochemistry at McGill University, David L. Thomson, made known his intention to retire from the chairmanship, and my former mentor, K. A. C. Elliott, was offered the chair. He wrote to ask me to join him as a half-time member of the Department of Biochemistry. I accepted the offer. When we returned to Montreal, my time was divided between the laboratory at the Allan Memorial Institute and the Department of Biochemistry. The two were located one-quarter of a mile apart horizontally and about 400 ft vertically. Parking was disagreeable in both locations, so I had plenty of exercise commuting on foot between the two. In the laboratory we continued our work on the posterior pituitary by devising methods of separating the large numbers of peptides contained in the tissue. With Seymour Mishkin (a summer student) and Brahm Muhlstock (a graduate student), we percolated posterior pituitary powder with ethanol to obtain extracts rich in peptides and small proteins (Saffran *et al.*, 1962). Enrique C. Preddie isolated a large peptide from one of these extracts and found that it had some oxytocic activity (Preddie and Saffran, 1965*a*). Preddie managed to work out the amino acid sequence of the peptide and demonstrated that a small portion resembled oxytocin in its structure (Preddie and Saffran, 1965*b*).

Enrique C. Preddie was a Trinidadian by birth. He came to the United States for his undergraduate education and found his way to Canada for his graduate work. During the course of his Ph.D. program, he spent several months at the National Institute for Medical Research at Mill Hill in the laboratory of the same D. F. Elliott with whom Schally had once worked, working out the amino acid sequence of the posterior pituitary peptide.

Basudev Shome adopted the method of percolation to acetone powders of hypothalamic tissue, which had become available by this time (Shome and Saffran, 1966). He isolated the prominent peptides in the percolates and obtained an amino acid analysis of one of the major components, which turned out to be a material derived from myelin and is now recognized as the first purification and characterization of the peptide that produces allergic encephalitis in test animals. Basudev Shome was an undergraduate at the University of Dacca in what is now Bangladesh. As the son of a Hindu farmer, he found life and a future career to be very hard in the predominantly Moslem country. He came to Canada for his Ph.D. degree, but is now working at the Harbor General Hospital, UCLA, with Albert Parlow on the structure of the human pituitary hormones.

Ting-chi (Tim) Wuu searched for hormonally related peptides in percolates of posterior pituitary powders. It was his diligence and insight that led him to purify neurophysin and demonstrate for the first time that it was a polypeptide of molecular weight approximating 10,000 instead of a larger protein (Wuu and Saffran, 1969). Wuu was born in mainland China. During the takeover by the Communists, his school was moved to Taiwan, where he completed his schooling and his university training at the National University of Taiwan. He came to McGill University for his Ph.D. degree and worked with me there. I was very impressed with Tim Wuu's abilities, and when the time came for a move to Toledo I invited Tim to move along with me. He has continued his work, independently, on neurophysin at the Medical College of Ohio.

These were side excursions and exercises in the purification and identification of neural peptides, undertaken while we were waiting for Schally and Guillemin to solve the problem of CRF and the other hypothalamic hormones. We didn't have the facilities or the financial support to mount the expensive campaigns needed to compete with other laboratories. However, time passed and the others did not solve the CRF problem. So we reentered the race on a scale compatible with our limited resources. Lorraine T. Chan and Susan M. Schaal joined me in a systematic study of the CRF activity in the rat median eminence (Chan *et al.*, 1969*b*). We determined the best solvents for the extraction of the activity from the tissue, we learned something about the stability of the activity, and we made a few attempts at purification of the active material by chromatography and electrophoresis. We came across a paradox—the purer the peptide became, the less CRF activity it had. There were several explanations for this frustrating finding, among them that the purified peptide became labile in the absence of protecting impurities. However, try as we might, we could make little headway in the purification. We made some progress in sharpening up the bioassay for CRF, first by an attempt to use an *in vivo* method, under the supervision of David de Wied, who showed Lorraine Chan how to place lesions in the rat hypothalamus to inhibit the response to stress in the test animals. Lorraine also provided us with the first clear dose–response curve for the release of ACTH by the rat pituitary gland *in vitro*, using a rat hypothalamic extract as the source of CRF (Chan *et al.*, 1969*a*).

Lorraine Chan was born in Canada of Japanese parents. They lived peacefully in British Columbia until the outbreak of war with the Japanese, when her family, along with other Japanese-Canadian families, was transplanted into more eastern parts of Canada because of fears of a Japanese invasion of the west coast. Lorraine received her bachelor's degree with honors in biochemistry from McGill University and entered graduate school to work under my direction for the Ph.D. degree.

Another development in the bioassay of CRF was the automation of the assay of ACTH. In collaboration with a group at the Ayerst Laboratories, Patricia Rowell, Keith Matthews, and I used the techniques of the clinical laboratory to design a system for the continuous fluorometric measurement of corticosterone that is formed by the rat adrenal gland (Saffran *et al.*, 1967). The automated procedure was far more precise than the old one. Using it, we were able to measure in a day the ACTH released by eight separate pituitary samples in response to CRF *in vitro*.

Before the automated bioassay of ACTH could be used in the CRF problem, I accepted the chair in Biochemistry at the Medical College of Ohio that was being established in Toledo. We moved to Toledo in June 1969 and by September the assay was set up in Toledo for the purpose of returning to the pursuit of CRF. With me in Toledo were Frances Pearlmutter and Eloise Rapino. Frances had come to Toledo as a postdoctoral fellow and Eloise was our very able technician. Eloise had had extensive experience with automated clinical procedures and stepped right in to set up and use the ACTH assay. Together we began to reinvestigate the CRF of the rat median eminence. We soon confirmed the major findings of Lorraine Chan (Saffran *et al.*, 1973).

Frances Pearlmutter is a native New Englander. At an early age she married Fishel Pearlmutter, a rabbinical student, and followed him as he rose to the rabbinate and was called to various pulpits. As a result, her undergraduate education was fragmented among several institutions—Barnard College, the University of Hawaii, and Tulane University. She finally received her Ph.D. degree in physical chemistry at Case Western Reserve University. When her husband moved to Toledo she wrote to the Medical College for a possible opening, and found herself engrossed in the CRF story as a result.

During a discussion with me, Dr. Virginia Upton pointed out the analogy between the peptides that were supposed to release and inhibit the release of MSH and the ACTH system. Peptide fragments derived from oxytocin apparently controlled the release of MSH. Vasopressin was thought to have a role in the release of ACTH and was considered seriously to be a candidate for CRF. MSH and ACTH are chemically related. Oxytocin and vasopressin are chemically related. Therefore, Dr. Upton suggested that we consider a peptide derived from vasopressin as a possible CRF. Through the kindness of Dr. du Vigneaud we received several synthetic peptides derived from the ring portion of vasopressin for testing in our *in vitro* assay. Two of these, pressinoic acid and deaminopressinamide, had small but definite CRF-like activity. Of the two, pressinoic acid was the more potent. This was a very exciting finding and the experiments were repeated many times, until the supply of pressinoic acid was depleted. We hastened to present our findings in a paper (Saffran *et al.*, 1972), but were too quick on the publishing trigger. Subsequent samples of pressinoic acid, obtained from Dr. du Vigneaud and elsewhere, were completely devoid of activity. To this day we cannot explain the differences between the experiments.

We returned to the CRF activity of the rat median eminence and to the puzzle of Lorraine Chan's findings that the purer the peptide the less activity it had.

Gel filtration on Sephadex G-25 is a purification step common to the isolation of all the known hypothalamic hormones. This treatment was applied by Lorraine Chan and was reinvestigated by us in Toledo. A potent extract of the rat hypothalamus, placed on the Sephadex G-25 column, yielded fractions practically devoid of CRF activity. Frances Pearlmutter and Eloise Rapino decided to investigate whether the contact between the extract and the Sephadex gel destroyed the activity. They pooled all the fractions, concentrated the pooled material to a small volume, and reassayed for CRF activity. To their delight, the pooled Sephadex G-25 fractions contained all of the activity of the original extract!

One explanation of this observation was that CRF activity resided in more than one component. The components were separated on the Sephadex G-25 column, to be reunited when the fractions were pooled. The fractions from the Sephadex column were combined in a variety of ways and soon it became apparent that this explanation was true. Two peaks of peptide materials (as detected by ultraviolet absorption) had to be added together to rat pituitary tissue to provoke a release of ACTH. Either peak alone had little, if any, activity. One of the peaks was probably a peptide, because its activity was destroyed by heating in 6 N HCl. To avoid premature publication which plagued us with pressinoic acid, the separation of the activity on Sephadex G-25 was performed literally hundreds of times. Finally the observations were published (Pearlmutter *et al.*, 1975). By this time, however, Frances Pearlmutter had left the field of CRF to return to her first love of temperature-jump and stopped-flow kinetics, while Eloise Rapino retired from science to take care of her family.

Susan Nurrenbern joined the laboratory, learned the details of the CRF activity, and the work continued.

Not only was the CRF activity in the rat hypothalamus a mixture of at least two substances, but also, while most of us looked among the smaller peptides for CRF, the peptide component was large enough to be in the void volume of Sephadex G-25. In preliminary experiments, Marcia van Gemert, who was with us as a guest for a year, determined that the apparent molecular weight of the larger component was in excess of 20,000.

What could we do with this information? Our group and resources were still too small to undertake a large-scale purification of CRF. It occurred to me that it was necessary to join forces with one or more of the laboratories that had collected hundreds of thousands of hypothalamic fragments in the search for the other hypothalamic hormones. I approached my former colleague, Andrew Schally, with the suggestion that we re-form our partnership and that he look in the higher molecular weight residues of the hypothalamic extracts for CRF. We agreed to do the bioassay of these fractions. To our delight, we found CRF activity in the larger molecular weight fractions left over from the purification of the other porcine hypothalamic hormones.

Unlike the CRF of the rat median eminence, that of the porcine hypothalamus seems to be a large molecule that acts without the need for a smaller partner. We are now engaged in collaboration with Dr. Schally and his co-workers in the purification of the peptide CRF and we hope that within a reasonable time this puzzle will finally yield its solution.

It has been more than 20 years since three young men sat on the grass in front of the Allan Memorial Institute of Psychiatry in Montreal and planned the first experiments. Some of the plans have come to fruition, some are still to yield their ultimate results. In the 20 years we have watched progress in neuroendocrinology accelerate to the point that the program of the last meeting of the Endocrine Society was dominated by two subjects, the hypothalamic hormones and hormone receptors.

I look forward to the excitement of the next 20 years, which should bring neuroendocrinology fully into the molecular age.

ACKNOWLEDGMENT

Our recent work on CRF was supported by NIH Grants AM14132, AM18913, and BRS S07-RR-05700-07.

REFERENCE

Chan, L. T., de Wied, D., and Saffran, M. (1969*a*). Comparison of assays for corticotrophin-releasing activity. *Endocrinology* **84**:967.

Chan, L. T., Schaal, S. M., and Saffran, M. (1969*b*). Properties of the corticotrophin-releasing factor of the rat median eminence. *Endocrinology* **85**:644.

Cleghorn, R. A., Graham, B. F., Saffran, M., and Cameron, D. E. (1950). A study of the effect of the pituitary ACTH in depressed patients. *Can. Med. Assoc. J.* **63**:329.

Fields, W. S., Guillemin, R., and Carlton, C. A. (eds.) (1956). *Hypothalamic-hypophysial Interrelationships*, Thomas, Springfield, Ill.

Hokin, M. R., Hokin, L. E., Saffran, M., Schally, A. V., and Zimmermann, B. U. (1958). Phospholipides and the secretion of adrenocorticotropin and of corticosteroids. *J. Biol. Chem.* **233**:811.

Pearlmutter, A. F., Rapino, E., and Saffran, M. (1975). The ACTH-releasing hormone of the hypothalamus requires a co-factor. *Endocrinology* **97**:1336.

Preddie, E. C., and Saffran, M. (1965*a*). Isolation of a large polypeptide from bovine posterior pituitary powder. *J. Biol. Chem.* **240**:4189.

Preddie, E. C., and Saffran, M. (1965*b*). Structure of a large polypeptide of bovine posterior pituitary tissue. *J. Biol. Chem.* **240**:4194.

Saffran, M., and Bayliss, M. J. (1953). *In vitro* bioassay of corticotrophin. *Endocrinology* **52**:140.

Saffran, M., and Denstedt, O. F. (1951). The effect of intravenously injected citrate on the serum ionized calcium in the rabbit. *Can. J. Med. Sci.* **29**:245.

Saffran, M., and Prado, J. L. (1949). Inhibition of aconitase by transaconitate. *J. Biol. Chem.* **180**:1301.

Saffran, M., and Schally, A. V. (1955*a*). *In vitro* bioassay of corticotrophin; modification and statistical treatment. *Endocrinology* **56**:523.

Saffran, M., and Schally, A. V. (1955*b*). The release of corticotrophin by anterior pituitary tissue *in vitro*. *Can. J. Biochem. Physiol.* **33**:408.

Saffran, M., and Schally, A. V. (1958). An *in vitro* system for the study of the neural control of ACTH secretion Curri, A. B., and Martini, L. (eds.), *Pathophysiologia Diencephalica*, Springer, Vienna, p. 780.

Saffran, M., Grad, B., and Bayliss, M. J. (1952). Production of corticoids by rat adrenals *in vitro*. *Endocrinology* **50**:639.

Saffran, M., Schally, A. V., and Benfey, B. G. (1955). Stimulation of the release of corticotrophin from the adenohypophysis by a neurohypophysial factor. *Endocrinology* **57**:439.

Saffran, M., Caplan, B. U., Mishkin, S., and Muhlstock, B. (1962). Use of percolation for the extraction of vasopressin. *Endocrinology* **70**:43.

Saffran, M., Ford, P., Matthews, E. K., Kraml, M., and Garbaczewska, L. (1967). An automated method for following the production of corticosteroids by adrenal tissue *in vitro*. *Can. J. Biochem.* **45**:1901.

Saffran, M., Pearlmutter, A. F., Rapino, E., and Upton, G. V. (1972). Pressinoic acid: A peptide with potent corticotrophin-releasing activity. *Biochem. Biophys. Res. Commun.* **49**:748.

Saffran, M., Pearlmutter, A. F., and Rapino, E. (1973). *In vitro* assays for corticotrophin-releasing factors. In Brodish, A., and Redgate, E. S. (eds.), *Brain Pituitary Adrenal Interrelationships*, Karger, New York, p. 47.

Schally, A. V., Saffran, M., and Zimmerman, B. (1958). A corticotrophin-releasing factor: Partial purification and amino acid composition. *Biochem. J.* **70**:97.

Schonbaum, E., Birmingham, M. K., and Saffran, M. (1956). Metabolism of glucose and steroid formation by rat adrenals *in vitro*. *Can. J. Biochem. Physiol.* **34**:527.

Shome, B., and Saffran, M. (1966). Peptides of the hypothalamus. *J. Neurochem.* **13**:433.

Wuu, T. C., and Saffran, M. (1969). Isolation and characterization of a hormone binding polypeptide from pig posterior pituitary powder. *J. Biol. Chem.* **244**:482.

21

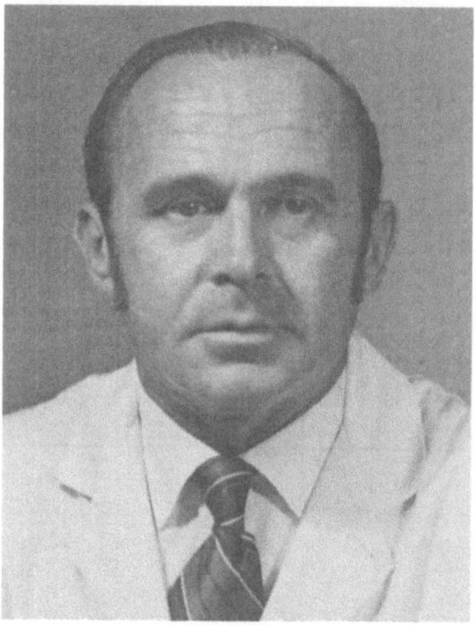

Andrew Victor Schally

Andrew Victor Schally was born on November 30, 1926, in Wilno, Poland, and is presently a U.S. citizen. He received his Ph.D. degree at McGill University in Montreal, Canada. He held research positions at the National Institute for Medical Research, Mill Hill, London, England, from 1949 to 1952, and in the Departments of Physiology and Biochemistry, Baylor University College of Medicine, Houston, Texas, from 1957 to 1962. Since 1962 he has been head of the Endocrine and Polypeptide Laboratory at Veterans Administration Hospital, New Orleans, Louisiana. He also is a professor in the Department of Medicine, Tulane University School of Medicine, in New Orleans.

Dr. Schally has won many awards for his work and that of his colleagues on isolation of hypothalamic releasing factors ("hormones" as he prefers to call them). These include being named a Senior Research Fellow of the U.S. Public Health Service in 1961–1962 and a Senior Medical Investigator in 1973, and being awarded the Van Meter Prize of the American Thyroid Association in 1969, the Ayerst-Squibb Award of the Endocrine Society (U.S.A.) in 1970, the William S. Middleton Award of the Veterans Administration in 1970, the Charles Mickle Award of the Faculty of Medicine, University of Toronto, in 1974, the Gairdner Foundation International Award (Canada) in 1974, the Edward T. Tyler Award in 1975, the Borden Award of the Association of American Medical Colleges in 1975, the Albert Lasker Basic Medical Research Award in 1975, and the Nobel Prize in Physiology and Medicine in 1977. He and his colleagues have written more than 900 publications.

21

In the Pursuit of Hypothalamic Hormones

ANDREW VICTOR SCHALLY

As this book goes to press, several hypothalamic regulatory peptide hormones—thyrotropin-releasing hormone (TRH), luteinizing hormone releasing hormone (LHRH), growth hormone release inhibiting hormone (GHRIH or somatatostatin) and Pro-Leu-Gly-NH_2, proposed as melanocyte-stimulating hormone release inhibiting factor (MIF)—have been isolated, structurally elucidated, and synthesized. These hormones, as well as about one dozen analogues of LHRH and several analogues of somatostatin, are being evaluated clinically for diagnostic and therapeutic uses. This is a reflection of the progress made in this field in the last three decades. Furthermore, during the next few years additional hypothalamic hormones will probably be obtained in a form pure enough to permit determination of their structure and synthesis. Thus the ability to exert a clinical type of control over the endocrine system not previously possible appears to be within our grasp. These practical clinical diagnostic and therapeutic advances and a new field of medical research came into being as a result of endeavors of a few scientists to discover the secrets of the hypothalamus. Thus, in a short span of time, a purely basic research effort has yielded results which bid fair to contribute to the conquest of disease and relief of human suffering, and provide other benefits to humanity, including the development of new methods of birth control.

I have been asked to recount some of the events dealing with my contributions. Although my decision to work on this problem, as will be detailed later, was based on a speculative but logical approach to what appeared to be an important problem in physiology, my endeavors in recent years have been to some extent guided by a practical desire to achieve short-

ANDREW VICTOR SCHALLY • Endocrine and Polypeptide Laboratory, Veterans Administration Hospital, and the Department of Medicine, Tulane University School of Medicine, New Orleans, Louisiana 70112.

term goals in contrast to what might be characterized as my earlier pursuit of idealistic obsessions. This was caused by several factors.

The first factor could be my own character. Having spent a harsh childhood during World War II in the Nazi- and later Soviet-occupied countries of Europe, and having grown up in the atmosphere of necessary general national austerity in postwar England (when, for example, we were asked to reuse one envelope several times), I was perhaps more aware than others of the value of money and of the high cost of medical research. In 1960, when I began to receive my own grants from the Veterans Administration, National Institutes of Health, Population Council, and other agencies, I wanted to use the funds awarded to me wisely and make contributions of a practical nature. I was grateful for the opportunities I was given in the United States, and I strongly wished to be useful to my new country, for which I felt a complete allegiance, having declared my intention to become a U.S. citizen in February 1959 and becoming naturalized in November 1962.

The second factor could be my contacts with many outstanding clinicians interested in using the hypothalamic hormones. Under their influence, I have been concentrating on the isolation especially of those additional hypothalamic hormones which would be useful clinically, as well as trying to modify synthetically the known hypothalamic hormones to impart to them certain physiological properties which would render them more desirable clinically.

The third factor would be my affiliation of 14 years' duration with the Veterans Administration Hospitals and Department of Medicine of Tulane Univerisity School of Medicine, which helped create the right atmosphere of systematic attempts to make contributions of a practical nature to medicine. These practical trends were not in the least disagreeable to me and were in fact clearly self-imposed, as the Veterans Administration, National Institutes of Health, and Tulane University always gave me complete academic freedom in my research.

My own choice of the hypothalamic endocrine field while still an undergraduate student in 1954 was decisively influenced by several events. The most important of these was the formulation by Geoffrey Harris of the basic theories relating to the hypothalamic control of secretion of the anterior pituitary gland and the demonstration in parallel studies by Sawyer, Everett, and Markee of involvement of the central nervous system in the control of gonadotropin secretion. I idolized these great pioneers in my student years (and still do), and it is only because of their brilliant work that my later contributions have been possible. It was clear that, in spite of a strong circumstantial case favoring a hypothalamic control of the pituitary, the proposition would remain speculative until direct evidence for the existence of specific hypothalamic regulatory hormones could be demonstrated.

It was at this stage in 1954 that my own involvement in the hypothalamic field began when I was working in the Department of Psychiatry of McGill University in Montreal. Although I was then only an undergraduate, it was my privilege to work closely with Dr. Murray Saffran in the friendly fraternal atmosphere of his laboratory. I was also immensely helped by the close comradeship and truly friendly and paternal encouragement of Dr. Robert A. Cleghorn, then the chief of the Laboratory for Experimental Therapeutics in the Allan Memorial Institute of Psychiatry, who previously had made important clinical contributions to neuroendocrinology. That period was critical for my formation as a medical researcher and really marked the beginning of my interest in the relationship between brain function and endocrine activity. The work at Allan Memorial was devoted to ACTH and adrenal corticosteroids. At that time, three main theories existed about the control of secretion of ACTH: (1) by systemic levels of epinephrine, (2) control by blood levels of adrenal cortical steroids, and (3) control by a specific hypothalamic neurohumor. The persistence of an increased release of ACTH after stress in adrenalectomized animals and other considerations (such as the ACTH release after strong stress not being blocked by corticoid levels and the response to exogenously administered epinephrine not being rapid enough to explain the sudden elevation of ACTH) ruled out epinephrine of adrenal origin and corticoids as participating in anything but a fine adjustment in the release of ACTH. Dr. Murray Saffran and I independently reached the conclusion that the hypothalamic theory explained most of the experimental facts. This theory was first suggested by Hinsey and Markee as early as 1933 and was strongly supported by the work of Harris and others, who postulated that neurohormones which originate in the median eminence of the hypothalamus and penetrate to the anterior lobe through the portal system could be responsible for the control of anterior lobe secretion. However, this neurohumoral theory was based only on anatomical and physiological considerations, and its final proof had to be the isolation of a hypothalamic neuroendocrine mediator able specifically to stimulate the release of ACTH or indeed of any pituitary hormone.

In the beginning it was not possible to isolate these substances because of a lack of specific methods for the detection of their activity. No importance can be attached to early attempts by various investigators to purify the materials supposedly able to release ACTH and gonadotropins. Murray and I sought to provide evidence for this theory by demonstrating an unequivocal direct effect of hypothalamic substances on the pituitary. Encouraged by the success of the *in vitro* assay system for ACTH based on the use of quartered rat adrenal glands, we devised a test system for measuring pituitary secretion of ACTH using isolated rat pituitary fragments (Saffran and Schally, 1955). This *in vitro* system was exquisitely simple and proved to be of decisive importance not only for demonstrating the existence of hypothalamic substances controlling the release of ACTH but also (after small modifications) for use in subsequent investigations of hypothalamic hormones regulating the secretion of TSH, LH, GH, FSH, prolactin, and MSH. Murray and I were able to prove that hypothalamic (and surprisingly also neurohypophysial) tissue contained a substance which stimulated the release of ACTH. We also obtained evidence that it was a basic polypeptide (Schally *et al.*, 1958). We named the substance "corticotropin-releasing factor" (CRF). Since I think that I made the original suggestion as to this name, I must take the blame for what many regard as an awkward nomenclature. Our finding was the first direct proof of the existence of hypothalamic hormones regulating pituitary function. Murray could have taken the credit for this important discovery himself, since I was then an undergraduate, but he shared it with me. I shall always remember this, since the demonstration of the existence of CRF helped launch my career in the field. At that time, bonds of mutual respect and friendship between Murray and myself were formed, and they have persisted very firmly ever since.

These early findings met with some skepticism, but we were able to withstand the criticism of other investigators at McGill who tried to discourage Murray and myself in our work in many ways, for example, by suggesting that we were dealing with histamine. I am glad that I did not listen to them and later to other investigators in Houston, New Orleans, and Washington, including one who urged me to abandon the project on LHRH on the grounds that there was too much competition in that field—only a few months before our successful resolution of the problem. Reflecting now on it all, I can honestly say that I really rejected without hesitation the advice of these doubting Thomases because their reasoning was generally shallow and their criticisms were usually not constructive and in many instances actually petty and trivial. It was apparent to me very early that my decision to work on CRF, based on the assumption that Harris's theory would prove to be correct, was fully justified and my early success spurred me on to greater efforts.

Despite our initial success with CRF, enthusiasm for our work among other endocrinologists did not last very long. CRF activity proved to be very elusive and difficult to purify, partly because of its instability and partly because of the complexity of the *in vivo* assays and serious interference by substances such as vasopressin, ACTH, and several ACTH-like materials present in the hypothalamic and neurohypophysial extracts. In spite of truly agonizing efforts, I was unable to isolate enough material in pure state for the determination of the structure of CRF.

This work was started with Murray Saffran in Montreal, and continued with Guillemin at Baylor University College of Medicine in Houston, Texas, where I moved in September 1957, having obtained my doctorate at McGill in May of that year. I liked and even admired the progressive spirit prevailing in Texas, but my years in Houston were somewhat discouraging because of serious problems with isolation of CRF and because of my relationship with Guillemin. Our failure to identify CRF cast doubt on the initial findings. We were exposed to the skepticism, sarcasm, and even ridicule and contempt of many scientists and physicians in the endocrine field who had no insight into the technological problems involved in the effort. CRF was called (even in writing) the "monster of Loch Ness," the "abominable snowman of the Himalayas," and various other derogatory names. Other hypothalamic hormones did not fare better, with names such as "family of ghosts" and "watery illusions" being used for them. It is rather ironic that some of the authors of these statements are now using hypothalamic hormones in their basic or clinical work. Some of them later displayed great honesty by publicly admitting their error and have been both friendly and supportive of me after we announced the structure of TRH and LHRH. Despite these years of frustration, I remained unshaken in my confidence that the postulation of the existence of hypothalamic hormones regulating anterior pituitary function would ultimately prove to be correct. Moreover, we were making some headway: methods were being developed for purification, as well as for *in vivo* assays which are essential for any isolation work together with *in vitro* assays, and arrangements were made for procurement of the hundreds of thousands of hypothalami of sheep, pigs, and cattle necessary for any realistic effort to purify useful quantities of hypothalamic hormones. In this respect, after my move to New Orleans, Oscar Mayer and Co. was especially helpful and generous to us in securing over 1 million pig hypothalami. It was also greatly encouraging that some men had confidence in our work and the foresight to appreciate the possible scientific and medical importance and diagnostic and therapeutic potential of hypothalamic hormones.

When Dr. Joe Meyer, then head of VA basic research, offered in June 1962 to set up a VA laboratory devoted to research on the hypothalamus and make me its chief, I quickly accepted, despite the fact that at Baylor I held an NIH Senior Research Career Development Award, since this gave me a clear opportunity to be in complete command of such an effort. For various reasons—among which were the enthusiastic support of Dr. E. H. (Manny) Bresler, then associate chief of staff for research of the New Orleans Veterans Administration Hospital, Dr. George Burch, then chairman of the Department of Medicine of Tulane University School of Medicine, Dr. Cyril Y. Bowers, head of the Tulane Endocrine Clinic, and Dr. William Locke, Ochsner Foundation Clinic, as well as my preference for a warm southern climate (which I learned to love, since it permitted me to do my daily outdoor swimming)—I decided to move to New Orleans. In December 1962 I was appointed chief of Endocrine and Polypeptide Laboratories at the Veterans Administration Hospital and associate professor of Medicine at Tulane and in 1966 professor. After a year and a half in provisional laboratories, we moved in 1964 to a new building which was well equipped by the Veterans Administration. Much of that equipment is still in perfect operating condition and used daily. Immediately after my move to New Orleans, we began to work with great enthusiasm on all the hypothalamic hormones controlling anterior pituitary function, as well as on hypothalamic lipid-mobilizing substances. The earliest members of our 1962 team were Weldon H. Carter and Tommie W. Redding. They have stayed with me all these years, and without their devoted help we could not have solved the thyrotropin-releasing hormone (TRH) problem in 1969, the LHRH problem in 1971, and the porcine somatostatin problem in 1975. Working in a medical environment, I became more aware of patient needs and of materials desirable for better diagnosis and treatment than I had been when in a purely academic setting in Houston. As my position in New Orleans entailed many research responsibilities, I was greatly helped by the experience acquired in my early years of research (1950–1952) at the National Institute of Medical Research, Mill Hill, London, England, where I was fortunate to work with or be exposed to the stimulating influences of such scientists as Dr. Don F. Elliott, Sir Charles Harington, Dr. Rodney R. Porter, Dr. Alan J. P. Martin, Dr. Rosalind Pitt-Rivers, Dr. Jack Gross, Dr. Thomas S. Work, Dr. Heinz L. Fraenkel-Conrat, and Dr. John W. Cornforth. Several of them later won Nobel Prizes for chemistry or physiology and medicine. Although my position was very junior at Mill Hill, my work was appreciated and this was a source of great satisfaction to me, considering that this recognition came from scientists of such caliber. If they only knew how much I learned in those 2½ years, not only technical expertise but also the philosophy of research and approach to investigations. To give just one example, later in my research career I published sometimes nearly 50 publications per year (including abstracts, reviews, chapters in books), but I did not publish anything at Mill Hill partly because I was too busy learning. These years of instruction were indeed decisive in providing inspiration, training, and research discipline, and greatly influenced the course of my research career. In fact, it was at Mill Hill where I endured my "baptism of fire" in medical research and became addicted to it. Being aware of the superior standing of Mill Hill in research, I afterwards considered myself a veteran and member of an elite "research corps."

In my position in New Orleans, I was able to travel and to meet and establish contacts with many leading U.S. scientists and physicians, among them C. H. (Tom) Sawyer, Al Albert, Roy Greep, Paul Munson, Vincent du Vigneaud, J. (Ed) Rall, Al Paulsen, Klaus Hoffman, C. H. Li, S. J. Behrman, "Shelly" Segal, Joe Meites, Si Reichlin, Aaron Lerner, Bernard Horecker, Robert B. Greenblatt, Ralph Hirschmann, Sanford L. Steelman, Al Segaloff, Farahe Maloof, Manny Bogdanove, Ed Rennels, Jim Pittman, Ernie Knobil, Al Parlow, Mort Grossman, and W. F. White, as well as foreign scientists such as Jerker Porath, Reg Hall, G. Mike Besser, Carlos Gual, Arturo Zarate, David Gonzalez-Barcena, Rolf Luft, Victor Mutt, Roberto Mancini, Luciano Martini, Juan Zanartu, Luiz Cesar Povoa, A. Campos da Paz, Vicente Pozuelo-E., Jose Botella-Llusia, Greg M. Brown, Fernand Labrie, Bob G. Edwards, Bernard Donovan, Marian Jutisz, Robert Courrier, M. Fontaine, J. Mathieu, Bruno Lunenfeld, Felix G. Sulman, Zvi Laron, Hans Lindner, Sven J. Nillius, Carl Gemzell, Gerhard Vogel, Rolf Geiger, Shinji Itoh, Romano Deghenghi, Béla Flerkó, Béla Halasz, and more recently Abraham (Tito) Guitelman and Jose Varea-T. I was encouraged, and in some cases even inspired, by some of these remarkable men. I must mention my friendship with and the great admiration I felt for Ed Tyler, Lyman C. Craig, Geoffrey W. Harris, Bernardo A. Houssay, Eric Jorpes, and Arne Tiselius, all of whom are now deceased. They also provided encouragement, help, and inspiration.

Aware of the problem of infertility on the one hand and the necessity on the other to stem the population explosion, I broadened my interest in the brain and endocrine function to include reproduction and birth control. In fact, the central inhibition and stimulation of fertility became one of my main interests after 1962. It had been demonstrated by S. M. (Don) McCann and associates and independently by Geoffrey W. Harris and associates that a LH-releasing activity existed in hypothalamic extracts. I was able to demonstrate with Cy Bowers that purified bovine hypothalamic materials with the properties of peptides stimulate LH release, not only *in vivo* but also *in vitro* (Schally and Bowers, 1964). The latter was the first demonstration that hypothalamic materials release LH by direct action on the anterior pituitary. Soon afterward, my interest in reproductive endocrinology became even deeper, when we confirmed and extended the observations of Igarashi and McCann and Mittler and Meites that hypothalamic materials also release FSH *in vivo* and *in vitro*. I was most fortunate that Dr. Akira Arimura, a brilliant physiologist and endocrinologist, joined us in 1965. Because of his knowledge, enthusiasm, research experience, and exceedingly hard work, Akira made decisive contributions to all phases of our program, which luckily coincided with his own interests. He also broadened our program with many independent contributions and ideas, especially in immunology. We continued work on LHRH and FSHRH (which at that time we considered as separate entities), and, to broaden our knowledge of reproductive processes at the brain level, we started work on the central effects of contraceptive steroids and clomiphene, and on a variety of other projects on reproductive physiology at the central level.

All during this period, I had been also hard at work on TRH with Cy Bowers and Tom Redding. In 1966 we reported the first isolation of porcine TRH and the original correct observation that TRH contains three amino acids (glutamic acid, histidine, and proline) in equimolar ratio. This established for the first time that TRH was at least in part a polypeptide and reestablished equilibrium in a field somewhat confused by an incorrect proposal that TRH and LHRH might not be polypeptides (Guillemin *et al.*, 1966). We even persuaded Merck Sharpe and Dohme Laboratories to make eight synthetic analogues of the possible sequence of these three amino acids, including Glu-His-Pro (Schally *et al.*, 1968, 1969), but unfortunately the correct complete structure was not made. Had they tried to make Glu(NH_2)-His-Pro-NH_2, the TRH problem would have been solved 3 years earlier, since this tripeptide amide would have partially cyclized to pyroGlu-His-Pro-NH_2 after the synthesis. As a matter of fact, because of our deep involvement in reproduction and growth hormone releasing hormone (GHRH), I almost missed solving the TRH problem. However, when Burgus and Guillemin announced at the 1969 Tucson, Arizona, NIH Workshop on Hypothalamic Hormones that they also found the same three amino acids in ovine TRH, I realized that we had the right substance. In an attempt to rapidly solve this project, I formed a temporary partnership with K. Folkers because he had several structural chemists in his laboratory. I provided them with the eight analogues of TRH, including Glu-His-Pro made by Merck Sharpe and Dohme for structural modifications, and with my highly purified materials obtained from biological sources. Although I consider myself less a biochemist[1] and more an all-round endocrinologist, or neuroendocrinologist, with a not inconsiderable interest in clinical endocrine research, I personally carried out the isolation work on TRH, LHRH, somatostatin, PIF, and other hormones. I felt that to get the best and fastest results I would have to do this work myself, since hypothalamic hormones are present in minute amounts and their isolation involves extremely laborious concentration of large amounts of extracts from hypothalamic tissue. I also personally worked on *in vivo* and *in vitro* assays and continue to do so, and on determination of structure, but did not have sufficient time to elucidate the structures all by myself. Such work, in most cases, is routine in the peptide field. However, in the case of TRH, we were puzzled

[1] Biochemistry, as such, especially enzymology, was of little interest to me. In London I became interested in peptide chemistry and in Montreal in endocrinology.

by the absence of end groups, although later we established its correct amino acid sequence in New Orleans (Schally *et al.*, 1969). With expert chemical help from Franz Enzmann, Jan Bøler, and R. M. G. Nair in a parallel effort in my laboratory and in Austin, Texas, we were able to assign a correct structure to porcine TRH (Bøler *et al.*, 1969; Nair *et al.*, 1970). The structure was based on the (1) amino acid sequence of TRH established in my laboratory, (2) comparison of activity of synthetic analogues of Glu-His-Pro in assays carried out separately by Cy Bowers and Tom Redding, (3) mass spectra of natural and synthetic preparations, and (4) physicochemical comparisons of these synthetic analogues and natural TRH, some of which I did myself. Discussing the assays and other results on the phone with Franz Enzmann, we simultaneously exclaimed that TRH was pyro-Glu-His-Pro-NH_2. We immediately notified Folkers of this result, since when this happened he was on a trip to Scandinavia, and he replied in a one-word telegram, "*Wunderbar.*" But I like to think that it was really the huge number of previous TRH assays carried out by my friends and collaborators Tom Redding and Cy Bowers on dozens of fractions from the isolation procedures that enabled us to solve the TRH project by helping provide enough material for a structural identification of TRH.[2] The credit for solving the TRH project had to be shared with Burgus and Guillemin, who elucidated the structure of ovine TRH about that time. With TRH out of the way, I terminated my collaboration with Folkers and Cy Bowers. Cy Bowers has remained very active in this hypothalamic field, and I like to think that strong bonds of respect and friendship remain between us.

The solving of the TRH structure removed much of the skepticism surrounding the work on the hypothalamus and indeed made it appear respectable, desirable, and exciting. We were now able to devote even more of our time to LHRH and FSHRH and to systematic research on reproduction. Dozens of assays on LHRH preparations were done in that and preceding periods, primarily by Weldon Carter using the ascorbic acid depletion method of Parlow. We established that 11 types of estrogen-progesterone combinations commonly used in contraceptive pills and large doses of clomiphene did not suppress the stimulatory effects of LHRH on the release of LH. On the basis of this and other work, we postulated that contraceptive steroids and clomiphene act mainly on the hypothalamus. Later, with Drs. Jessamine Hilliard and Tom Sawyer, and with Dr. Arimura, we were able to prove that these steroids also unquestionably act in part on the pituitary gland. This work helped us postulate later that sex steroids may cause differential release of LH and FSH in response to

[2] In many assays we used up to 200 mice a day, and to avoid ear clipping we devised a system of individual assay cages which we named "Mouse Hilton."

LHRH/FSHRH. Our view, formed after collaborative studies with Dr. Dorsey Holtkamp, that clomiphene also acts mainly on the hypothalamus (Schally *et al.*, 1970) is now generally accepted.

Several other most important developments also occurred in that period. At the 1965 Panamerican Congress of Endocrinology in Mexico City, I met Dr. Carlos Gual, the director of the Endocrine Laboratory of the National Institute of Nutrition in that city. He invited our group to collaborate with him in the clinical testing of hypothalamic hormones in Mexico City. I shall always remember with gratitude this invitation, since it was issued at a time when the hypothalamic hormones were not yet firmly established and also implied confidence on the part of Carlos in the safety and efficacy of our preparations. Cy Bowers and I quickly took advantage of this invitation and showed for the first time that preparations of natural TRH are active in humans. It was also a great stroke of luck that Dr. Abba J. Kastin came to our group from NIH in 1964, to continue his work on MSH and the control of its release and to help us in clinical work. Abba quickly became my best friend and a superbly efficient collaborator. Since he was an accomplished clinician (M.D., Harvard) he became responsible for organizing and carrying out many of our clinical studies. In these studies done in Mexico in collaboration with Dr. Carlos Gual and Dr. Rees Midgley, we established that highly purified porcine LHRH unequivocally released LH and FSH in men and women under a variety of conditions (Kastin *et al.*, 1969, 1970). We realized that LHRH might be useful clinically. This would be in contrast to a hypothalamic GH-releasing decapeptide which we isolated in 1969 after much hard work with Drs. Eugenio Muller, Shinji Sawano, and previously A. Kuroshima and Y. Ishida. This decapeptide appeared to release bioassayable GH, as was later confirmed by collaborative work on synthetic material with Dr. Joe Meites, but had no effect on liberation of immunoreactive GH, as ascertained by RIA in several species. A tripeptide, pyro-Glu-Ser-Gly-NH_2, claimed by the Russians to have GHRH activity also was inactive. Later after beautiful physiological work by Drs. Jurgen Sandow and Jiro Takahara we finally found hypothalamic fractions able to release radioimmunoassayable GH, but at this time the structure of GHRH is still not established.

Also in that period Abba Kastin was hard at work on control of release of MSH. Despite opposition, we established firmly the concept that the control or release of MSH in mammals is exerted primarily by an inhibitory factor (MIF) from the hypothalamus. Having purified bovine MIF some 11,000 times (Schally and Kastin, 1966), Abba and I used this concentrate to carry out various physiological studies and later isolated from it Pro-Leu-Gly-NH_2 and showed that it had MIF properties. Our work agreed with that of Walter, Celis, and Taleisnik, but the nature of mammalian MIF is still obscure.

The clinical success with LHRH encouraged us to continue the agonizing effort involved in the isolation of this hormone. I was able to isolate a small amount (800 μg) of LHRH from 160,000 hypothalami and obtained unequivocal proof from work with proteolytic enzymes and amino acid analyses that it is a polypeptide. I sent two papers to *Biochemical and Biophysical Research Communications* reporting these findings (Schally *et al.*, 1971*a*,*c*). We did not report tryptophan at that time, since amino acid analyses were done at first after acid hydrolysis without thioglycolate. Later using thioglycolate we discovered Trp, but could not revise our paper. I passed this LHRH on to our structural chemists, Dr. Yoshihiko Baba and Dr. Hisayuki Matsuo, for degradation work together with clear instructions for a structural attack. These instructions were based on the effect of proteolytic enzymes and other chemical treatments on LHRH. Since I did not think that the amounts of LHRH on hand would be enough to complete our structural work I decided to continue myself to work on isolation of additional amounts of LHRH. As in the case of TRH and the first batch of LHRH, this involved a great effort, both physical and mental, as well as patience and experience. There were even some personal risks, for example, when working with 100 liters of diethyl ether daily for several days. Dr. R. M. G. Nair and I did this step during Christmas 1970 and the 1971 New Year period, when there were few other people in the lab who could cause or suffer an accident. In the meantime, Dr. Baba and Dr. Matsuo were working very hard and were making good headway toward the elucidation of the structure of LHRH with the very small amount of material then on hand. Dr. Baba was using the Edman-Dansyl procedure and Dr. Matsuo his own selective tritiation method for *C*-terminal analysis. We were able to draw unambiguous conclusions about the amino acid sequence and submitted the first paper (Matsuo *et al.*, 1971), after Dr. Arimura found the synthetic decapeptide made in our laboratory and based on the proposed structure of LHRH to be highly active in releasing LH *in vivo*. Ours was not only the first isolation of LHRH and elucidation of its structure but also its first synthesis. The structure of LHRH was quickly confirmed by Dr. Baba using conventional sequential analyses after cleavage of the *N*-terminal pyro-glutamyl residue by a special enzyme (Baba *et al.*, 1971).

The first paper (Matsuo *et al.*, 1971) reporting the correct sequence of LHRH appeared on June 18, at the time of the Endocrine Society 1971 meeting in San Francisco, at which time I had decided to announce the structure, having sent an abstract in January which was placed on the program. Luckily the issue of *Biochemical and Biophysical Research Communications* containing the structure did not reach the audience, so an air of drama could be maintained. The suspense surrounding that presentation was well captured in an article by Graham Chedd (1971). For example, until the last minute, we did not know if Guillemin's group also was going to announce the structure of this very much sought-after molecule. The groups of Don McCann, Folkers, Harry Gregory, G. W. Harris, and others also were working on this topic, and although we did not think they had achieved the isolation and determination of structure of LHRH, we could not be sure of this. We were absolutely determined to at least share in the discovery. I previously decided not to participate in the International Symposium on Eonadotropins held in May at the New York Hospital Cornell Medical Center which preceded the San Francisco meeting by a few weeks, in order to be able to continue my experimental work on isolating additional amounts of LHRH. However, Dr. Arimura, who represented our lab at that meeting with Dr. Matsuo, was authorized to release the structure of LHRH at this symposium should Guillemin's associates announce it there. They did not, and despite the "needling," "heckling," and sheer provocation from the audience about the structure, Dr. Arimura kept calm and silent. So the drama was reserved for the June meeting of the Endocrine Society in San Francisco. Between March and May, I also gave several lectures at various meetings in the United States and Dr. Kastin gave one in Banff at an international symposium at which we hinted that we might announce the structure at the Endocrine Meeting in San Francisco. We had a distinct feeling that some members of the audiences at these preceding lectures did not quite believe me, but nevertheless the large Imperial Ballroom of the San Francisco Hilton, where I was to give my paper announcing the structure, was very crowded. I listened with tension and suspense during the immediately preceding lecture and discussions, but Guillemin's group did not report the isolation and structure of ovine LHRH. When my time to speak came, I rose to my feet and presented our paper on isolation, structure, and synthesis of porcine LHRH. It was one of the most joyous moments in my life to announce for the first time the solution to the problem which had fascinated me and others for so long, especially inasmuch as the discovery was deemed by others to have so great scientific, clinical, and even social impact, because of the possibilities of stimulation and inhibition of fertility it offers. It should be noted that we won the race to solve the structure of LHRH using only a small fraction of available hormone. By May 1971, I was able to isolate over 11 mg of pure LHRH (Schally *et al.*, 1971*e*) using what I thought was excellent methodology, so that the structure would have been easily completed in any case by standard conventional methods. But Dr. Baba's and Dr. Matsuo's expert structural analyses on the microscale enabled us to achieve our objective about 1 month earlier and our big "reserves" of LHRH were never committed into the race. With Dr. A. Arimura, Dr. Luciano Debeljuk, Dr. Motoi Saito, and others, we continued our work with LHRH with extensive *in vivo* and *in vitro* studies to demonstrate that our decapeptide was indeed the physiological hormone. Since in all these studies natural preparations of porcine LHRH, as well as synthetic ones, stimulated the release of not only LH but also FSH, we put forward a concept that one hypothalamic hormone—our decapeptide—controls the secretion of both gonadotropins from the pituitary gland (Schally *et al.*, 1971*b*). We tried to explain the occasional preferential release of one or the other gonadotropin by the effect of sex steroids, especially estrogens, directly on the responsive pituitary tissue (Schally *et al.*, 1971*d*). This concept still has not been disproved and our later systematic studies showed that even if another FSHRH exists, our LHRH decapeptide does indeed represent the main FSHRH activity, at least in porcine hypothalamic tissue. In clinical studies, too, Dr. Kastin and I continued to carry out a variety of studies in human beings in association Dr. Carlos Gual at the National Institute of Nutrition in Mexico City, and later with Drs. Arturo Zarate, David Gonzalez-Barcena, and Don Schalch at the Hospitals of the Instituto Mexicano de Seguro Social, first using highly purified LHRH of porcine and human origin and then the synthetic hormone. We were also pleased that no adverse side effects of any kind were observed with natural LHRH or with synthetic LHRH. Subsequently, this was also found to be the case for long-acting analogues of LHRH. We felt that these investigations established the principles of the clinical use of LHRH. Induction of ovulation with LHRH in collaboration with Dr. A. Zarate was confirmed and extended by studies in Chile with Dr. Juan Zanartu, and by independent studies of Dr. S. Nillius in Sweden, Drs. Grimes, Taymor, and Acosta in the United States, and Drs. P. J. Keller in Switzerland, R. Nakano in Japan, and F. Huang in Taiwan. Careful and systematic research in Argentina started in 1971 with Dr. Roberto Mancini but now carried out principally in collaboration with Drs. Luis Schwarzstein and Nestor Aparicio indicates that in male infertility, especially oligospermia, LHRH may be useful too (Schwarzstein *et al.*, 1975).

After our announcement of its structure, LHRH was synthesized in many other laboratories based on the structure formulated by us in anticipation of its importance in clinical and veterinary medicine. The activity of various synthetic preparations of LHRH was the same as that of natural LHRH; thus our structure was correct. Numerous basic and clinical studies by others confirmed that LHRH releases LH and FSH in laboratory and domestic animals, and humans. Some work even suggested that LHRH may have an effect on libido, but we feel that such effects—whether exerted directly on the CNS, a general idea for hypothalamic peptides we proposed with Dr. Kastin (Plotnikoff *et al.*, 1971) or through the mediation of sex steroids—are still not clearly proven.

The hypothalamus suddenly rocketed to the top of the endocrine popularity list, even before the subsequent isolation of somatostatin. How long it will remain fashionable is difficult to speculate, but the domination of various U.S. and foreign endocrine meetings by papers on TRH, LHRH, and somatostatin has been truly remarkable. Some investigators who 15 years ago scoffed at the idea of hypothalamic hormones are now proposing the existence of an astronomical number. Needless to say, we did not agree with their earlier views nor do we with their current ones, but we hope to isolate, elucidate, and synthesize a few more hypothalamic hormones.

It is a great tragedy that the untimely death of Geoffrey Harris did not enable him to see the full fruition of his hypothalamic theories. His remarkable scientific insight enabled him to postulate the existence of hypothalamic hormones, and before he died late in 1971 he shared with us the joy of solving the problem of LHRH and saw his life's work proven. His characterization of some of the advances in identification of hypothalamic hormones, "particularly those from [our laboratory] as milestones in the history of endocrinology" (Harris, 1972), will always be regarded by me as the ultimate accolade. The death of Geoffrey Harris was a hard personal loss and left me with a deep sense of grief for some time.

We continued to sustain our momentum in the work on LHRH with immunological studies and synthesis of analogues. Immunological studies by Dr. Arimura in our laboratory and by other groups led to production of several antisera to LHRH and development of a radioimmunoassay (RIA) for LHRH. We were able to show that such antisera caused gonadal atrophy in male animals and blocked the release of LH and FSH and ovulation in cycling rats, and researchers at the Weizmann Institute in Israel made similar findings. Perhaps here lies another potential method of birth control. The RIA for LHRH showed that the peak of serum LHRH coincides with the surge of LH and that in fact LHRH secretion most probably induces this preovulatory rise in blood LH. In addition, these immunological studies confirmed that the LHRH decapeptide is not only the physiological LHRH but also most likely FSHRH.

Fortune favored me again when Dr. David H. Coy, a fine organic chemist, and his wife, Esther, also a remarkable researcher, decided to join our laboratory in 1972. Both were highly trained and skilled in synthetic peptide chemistry and were able to set up in our laboratory a powerful and most productive synthetic section. Because their training was British, as was mine, there was a remarkable understanding between us and we could easily agree on strategies and scientific approaches to various problems. We also quickly formed bonds of friendship. More than 400 analogues of LHRH were synthesized by David and Esther Coy in our laboratory between 1972 and 1977. A large number of analogues were also made by others. This permitted two distinct classes of these compounds to be established:

1. Stimulatory long-acting superactive analogues that cause prolonged liberation of LH and FSH, which should be more effective than LHRH for induction of ovulation in women.

2. Inhibitory analogues, i.e., antagonists of LHRH, which block LH and FSH release and which may eventually form the basis of a new birth control method.

Researchers from many countries continued to join our laboratory to help us in the testing of these analogues. Among them were Dr. Luciano Debeljuk (Argentina), Dr. Jesus Vilchez-Martinez (Venezuela), Dr. Antonio de la Cruz (Peru), Dr. Escipion Pedroza-Garcia (Colombia), and R. Nozomu Nishi (Japan). Particularly the dynamic work of Dr. de la Cruz established for the first time that some of the antagonists of LHRH can completely block ovulation in rats (de la Cruz *et al.*, 1976). Although it is clear that still more active inhibitory analogues of LHRH can be synthesized and although clinical studies indicate that long-acting analogues of LHRH can be useful therapeutically, much work is still needed to make my dream of being able to control reproduction at the central level come true.

The clinical testing of TRH and LHRH (made by various drug houses, Hoechst and Ayerst, and initially Abbott) was carried out by various investigators, not only in the United States but also abroad in many countries including England, Germany, France, Sweden, Italy, Canada, and Japan. We became involved in some of these programs, particularly the evaluation of the analogues, at the official invitation of government and state hospitals in Mexico, Chile, Argentina, Brazil, England, and, later, Spain and Ecuador. In the course of these tests, I decided to learn Spanish and Portuguese in order to be able to pass more directly and more effectively our observations and advice. We had acquired much information based on the physiological studies in animals, as well as information concerning solubility and means to sterilize and preserve hypothalamic hormones and their analogues, and I passed it on to the clinicians. My ability to communicate in these two languages was greatly appreciated by the Latin Americans and the Spanish, and resulted in many beautiful friendships. I now have the pleasure of regularly giving lectures in Spanish, including those of the International Course of Clinical Endocrinology in Madrid, which have been nicknamed "marathons" by the organizer and my good friend Professor Vicente Pozuelo because some of them last 2 hr. My ability to speak Spanish was also exploited in Argentina when the organizers and WHO asked me to lecture in Spanish at the VIIIth Panamerican Congress of Endocrinology and VIIIth World Congress of Fertility and Sterility in Buenos Aires in 1974. The courtesy and enthusiasm which greets me in Latin countries and the great demand for my lectures are a source of personal satisfaction to me, but the greatest reward for learning Spanish and Portuguese came when in 1974, in the course of my work in Brazil, I met a very pretty endocrinologist, Ana Maria de Medeiros-Comaru (M.D.). This charming young clinician did excellent work on the use of long-acting analogues of LHRH in women with secondary amenorrhea (Comaru *et al.*, 1976) and our friendship and romance, reinforced by frequent meetings at various international congresses in North and South America and in Europe, soon deepened into love and led to my second marriage.

Because of my friendship with various clinicians, I became informed of the need for a clinical agent, free of side effects, for the control of galactorrhea. Also of interest were speculations that PIF could possibly be useful in some types of prolactin-dependent breast and prostate tumors, and the prospects of contributing to the treatment of cancer excited me. We had been working on PIF since 1964, but because of a cumbersome bioassay for prolactin could make little headway until a RIA for prolactin was developed. Thus, after solving the LHRH problem, I decided to concentrate on the isolation of PIF. Because we followed the PIF activity by modification of the original *in vitro* pituitary assay, we at first assumed that a substance capable of inhibiting the release of prolactin by a direct action on the pituitary tissue had a good possibility of being the physiological PIF or the prolactin release inhibiting hormone (PRIH). However, 2 years of hard work in which I was enthusiastically helped by a French-Canadian endocrinologist, formerly a surgeon, Dr. André Dupont from Dr. F. Labrie's laboratory, resulted only in the isolation of hypothalamic catecholamines (Schally *et al.*, 1976*b*). Of course, we knew that catecholamines can inhibit prolactin release but assumed mainly because of the work of Kamberi and Porter that they do so by stimulating PIF release from the hypothalamus. There were reports in the literature (McLeod; Meites and colleagues) that catecholamines also can inhibit prolactin release by a direct effect on the pituitary, but somehow we did not attach to them the importance we should have. Thus these years of work led only to a strengthening of the concept that catecholamines may be physiologically involved in the control of release of prolactin, but did not result in the development of any clinical agents. Although the catecholamines appeared to be the most active among fractions inhibiting prolactin release *in vitro*, there were several other materials present with PIF activity. Recently we obtained one of these fractions in pure state and identified it structurally as γ-aminobutyric acid (GABA). It was interesting to find a new effect for this amino acid long known to be present in brain tissue, but the physiological importance of this work is still not clear. Our work on other PIFs is continuing.

In the course of our work on PIF we were measuring not only the prolactin content of the media and plasma but also the GH levels, looking principally for GHRH. We became aware (Schally *et al.*, 1973) that many fractions inhibited GH release, confirming the early work of Krulich and McCann. In our preoccupation with PIF, we did not concentrate on GIF, but after Brazeau *et al.* (1973) announced the isolation and structure of ovine GIF, which they named "somatostatin," we decided that it would be worthwhile to purify porcine somatostatin. This decision was based, among other things, on the fact that after the structure of ovine somatostatin was announced, we were able to synthesize this substance (Coy *et al.*, 1973) and carry out much physiological work (some in collaboration with Dr. F. Labrie) as well as immunological and clinical work which convinced us of its importance. The clinical work was carried out mainly in England. Professor Reginald Hall from Newcastle, who was one of the first to use TRH in patients and to develop the best principles for its clinical use, and Dr. G. Mike Besser, a dynamic clinician with much experience (like Reg Hall) on TRH and LHRH, were our principal collaborators. Other collaborators included Dr. Antonio Gomez-Pan, an enthusiastic Spanish M.D. doing postdoctoral work with Professor R. Hall, Dr. D. Evered, Dr. Chris H. Mortimer, Dr. Steve R. Bloom, and others. These clinical studies, based in part on some of our suggestions, were well organized and performed in England, and resulted in important observations that somatostatin inhibits GH, glucagon, and insulin release, and may thus be clinically useful in the control of acromegaly, gigantism, insulinomas, glucagonomas, and some types of diabetes (Hall *et al.*, 1973). We also made the first original observations (Bloom *et al.*, 1974) that somatostatin inhibits gastrin release in humans, and may therefore be used for control of ulcers. This work on the effect of somatostatin on gastrin release was extended by basic studies carried out in England in collaboration with Dr. Gomez-Pan and in Poland with Professor S. Konturek, which showed that somatostatin also inhibits gastric acid and pepsin secretion and the release of duodenal hormones, secretin, and cholecystokinin. Since our immunological work (Arimura *et al.*, 1975) showed the presence of immunologically active forms of somatostatin in the pancreas, stomach, and duodenum, we then suggested that this substance may be involved in the control of secretion not only of the pituitary but also of the pancreas, stomach, and duodenum. Thus we had strong reasons to continue work on purification of porcine somatostatin, and with the help of Dr. A. Dupont, Dr. Arimura, Dr. D. Schlesinger, and Tom Redding we succeeded in isolating and elucidating its structure (Schally *et al.*, 1976*a*). We were somewhat disappointed that the structure of porcine somatostatin was identical with that of the ovine hormone, although our discovery of long forms of somatostatin which could be prohormones may in the end justify much hard work. Since the clinical effects of somatostatin are short-lived, our approach as well as that of others for developing a more practical clinical agent is to make analogues of somatostatin, some of which may be endowed with prolonged and selective activity.

All this work resulted in many publications. Of more than 900 publications (counting papers, chapters in books, abstracts, reviews) our laboratory produced between 1963 and 1977, about half of these were devoted to LHRH or related reproductive topics. We would have never been able to complete this arduous writing task without the consistent brilliant editorial help of Dr. William Locke of the Ochsner Clinic and occasionally of Dr. Ed Ferguson, Jr., and Dr. Manny Bresler of the VA Hospital. Bill Locke also wrote several reviews with us. We even wrote a clinical book together (Locke and Schally, 1972) which received favorable reviews all over the world. I feel that after writing the first 200 publications I became a better writer, although the reader may be painfully aware of the further need for improvement along these lines.

I have had the satisfaction that my work in the hypothalamus was honored by top U.S. and Canadian awards and recently by the Nobel Prize. I was also made a Senior Medical Investigator by the Veterans Administration, an honor reserved for only a few. However, I do not feel that these prizes will have an adverse effect on my future productivity. My health is excellent,[3] and I am still as desirous as ever to make new discoveries, or at least to continue to make useful contributions to endocrinology. Since I feel that the isolation of new hypothalamic hormones is the hardest and the rate-determining step for any further progress in the field, I will continue to devote much of my own laboratory time to it. Among our present projects are the isolation of all the compounds with PIF activity, GHRH, PRH, and other GIFs. I am also fascinated by the presence in hypothalamic extracts of powerful lipid-mobilizing factors different from known hormones and the possibility that they could be involved in the central control of appetite and obesity. In any case, I wish to continue to work with that "gold mine" of hormones that the hypothalamus has proven to be.

REFERENCES

Arimura, A., Sato, H., Dupont, A., Nishi, N., and Schally, A. V. (1975). Abundance of immunoreactive somatostatin in the stomach and the pancreas. *Science* **189**:1007.

Baba, Y., Matsuo, H., and Schally, A. V. (1971). Structure of porcine LH and FSH releasing hormone. II. Confirmation of the proposed structure by conventional sequential analysis. *Biochem. Biophys. Res. Commun.* **44**:459.

Bloom, S. R., Mortimer, C. H., Thorner, M. O., Besser, G. M., Hall, R., Gomez-Pan, A.,

[3] When in 1955 I was asked for a motto which was to accompany my photo in the book on graduates of McGill University, I wrote "*mens sana in corpore sano*." I believe that I may have accidentally given a true personal motto.

Roy, V. M., Russell, R. C. G., Coy, D. H., Kastin, A. J., and Schally, A. V. (1974). Inhibition of gastrin and gastric-acid secretion by growth-hormone release-inhibiting Hormone. *Lancet* **ii**:1106.

Bøler, J., Enzmann, F., Folkers, K., Bowers, C. Y., and Schally, A. V. (1969). The identity of chemical and hormonal properties of the thyrotropin releasing hormone and pyroglutamyl-histidine-proline amide. *Biochem. Biophys. Res. Commun.* **37**:705.

Brazeau, P., Vale, W., Burgus, R., Ling, N., Butcher, M., Rivier, J., and Guillemin, R. (1973). Hypothalamic polypeptide that inhibits the secretion of immunoreactive pituitary growth hormone. *Science* **178**:77.

Chedd, G. (1971). The switch of fertility. *New Scientist and Sci. J.* (*London*) **51**:758.

Comaru, A. M. deM., Rodrigues, J., Povoa, L. C., Franco, S., Dimetz, T., Coy, D. H., Kastin, A. J., and Schally, A. V. (1976). Clinical studies with long-acting superactive analogs of LHRH in women with secondary amenorrhea. *Int. J. Fertil.* **21**:239.

Coy, D. H., Coy, E. J., Arimura, A., and Schally, A. V. (1973). Solid phase synthesis of growth hormone-releasing inhibiting factor. *Biochem. Biophys. Res. Commun.* **54**:1267.

de la Cruz, A., Coy, D. H., Vilchez-Martinez, J. A., Arimura, A., and Schally, A. V. (1976). Blockade of ovulation in rats by inhibitory analogs of luteinizing hormone-releasing hormone. *Science* **191**:195.

Guillemin, R., Burgus, R., Sakiz, E., and Ward, D. N. (1966). New observations on the purification of the hypothalamic hormone TRF. *C. R. Acad. Sci. Paris* **262**:2278.

Hall, R., Besser, G. M., Schally, A. V., Coy, D. H., Evered, D., Goldie, D. J., Kastin, A. J., McNeilly, A. S., Mortimer, C. H., Phenekos, C., Tunbridge, W. M. G., and Weightman, D. (1973). Action of growth hormone-release inhibitory hormone in healthy men and in acromegaly. *Lancet* **ii**:581.

Harris, G. W. (1972). Humors and hormones: The Sir Henry Dale Lecture for 1971. *J. Endocrinol.* **53**:i.

Kastin, A. J., Schally, A. V., Gual, C., Midgley, A. R., Jr., Bowers, C. Y., and Diaz-Infante, A., Jr. (1969). Stimulation of LH release in men and women by hypothalamic LH-releasing hormone purified from porcine hypothalami. *J. Clin. Endocrinol. Metab.* **29**:1046.

Kastin, A. J., Schally, A. V., Gual, C., Midgley, A. R., Jr., Bowers, C. Y., and Gomez-Perez, E. (1970). Administration of LH-releasing hormone to selected human subjects. *Am. J. Obstet. Gynecol.* **108**:177.

Locke, W., and Schally, A. V. (eds.) (1972). *The Hypothalamus and Pituitary in Health and Disease*, Thomas, Springfield, Ill.

Matsuo, H., Baba, Y., Nair, R. M. G., Arimura, A., and Schally, A. V. (1971). Structure of the porcine LH- and FSH-releasing hormone. I. The proposed amino acid sequence. *Biochem. Biophys. Res. Commun.* **43**:1334.

Nair, R. M. G., Barrett, J. F., Bowers, C. Y., and Schally, A. V. (1970). Structure of porcine thyrotropin-releasing hormone. *Biochemistry* **9**:1103.

Plotnikoff, N. P., Kastin, A. J., Anderson, M. S., and Schally, A. V. (1971). DOPA potentiation by a hypothalamic factor, MSH releasing inhibiting hormone. *Life Sci.* **10**:1279.

Saffran, M., and Schally, A. V. (1955). The release of corticotrophin by anterior pituitary tissue *in vitro*. *Can. J. Biochem. Physiol.* **33**:408.

Schally, A. V., and Bowers, C. Y. (1964). *In vitro* and *in vivo* stimulation of the release of luteinizing hormone. *Endocrinology* **75**:312.

Schally, A. V., and Kastin, A. J. (1966). Purification of a bovine hypothalamic factor which elevates pituitary MSH levels in rats. *Endocrinology* **79**:768.

Schally, A. V., Saffran, M., and Zimmermann, B. (1958). A corticotrophin releasing factor: Partial purification and amino acid composition. *Biochem. J.* **70**:97.

Schally, A. V., Bowers, C. Y., Redding, T. W., and Barrett, J. F. (1966). Isolation of thyrotropin releasing factor (TRF) from porcine hypothalamus. *Biochem. Biophys. Res. Commun.* **25**:165.

Schally, A. V., Arimura, A., Bowers, C. Y., Kastin, A. J., Sawano, S., and Redding, T. W. (1968). Hypothalamic neurohormones regulating anterior pituitary function. *Recent Progr. Horm. Res.* **24**:497.

Schally, A. V., Redding, T. W., Bowers, C. Y., and Barrett, J. F. (1969). Isolation and properties of porcine thyrotropin releasing hormone. *J. Biol. Chem.* **244**:4077.

Schally, A. V., Carter, W. H., Parlow, A. F., Saito, M., Arimura, A., Bowers, C. Y., and Holtkamp, D. E. (1970). Alteration of LH and FSH release in rats treated with clomiphene or its isomers. *Am. J. Obstet. Gynecol.* **107**:1156.

Schally, A. V., Arimura, A., Baba, Y., Nair, R. M. G., Matsuo, H., Redding, T. W., Debeljuk, L., and White, W. F. (1971*a*). Isolation and properties of the FSH and LH-releasing hormone. *Biochem. Biophys. Res. Commun.* **43**:393.

Schally, A. V., Arimura, A., Kastin, A. J., Matsuo, H., Baba, Y., Redding, T. W., Nair, R. M. G., Debeljuk, L., and White, W. F. (1971*b*). The gonadotropin-releasing hormone: One polypeptide regulates the secretion of luteinizing and follicle stimulating hormone. *Science* **173**:1036.

Schally, A. V., Baba, Y., Arimura, A., and Redding, T. W. (1971*c*). Evidence for peptide nature of LH and FSH-releasing hormones. *Biochem. Biophys. Res. Commun.* **42**:50.

Schally, A. V., Kastin, A. J., and Arimura, A. (1971*d*). Hypothalamic FSH and LH-regulating hormone, structure, physiology, and clinical studies. *Fertil. Steril.* **22**:703.

Schally, A. V., Nair, R. M. G., Redding, T. W., and Arimura, A. (1971*e*). Isolation of the LH and FSH-releasing hormone from porcine hypothalami. *J. Biol. Chem.* **246**:7230.

Schally, A. V., Arimura, A., and Kastin, A. J. (1973). Hypothalamic regulatory hormones. *Science* **179**:341.

Schally, A. V., Dupont, A., Arimura, A., Redding, T. W., Nishi, N., Linthicum, G. L., and Schlesinger, D. H. (1976*a*). Isolation and structure of growth hormone-release inhibiting hormone (somatostatin) from porcine hypothalami. *Biochemistry* **15**:509.

Schally, A. V., Dupont, A., Arimura, A., Takahara, J., Redding, T. W., Clemens, J., and Shaar, C. (1976*b*). Purification of a catecholamine-rich fraction with PIF activity from porcine hypothalmi. *Acta Endocrinol.* **82**:1.

Schwarzstein, L., Aparicio, N. J., Turner, D., Calamera, J. C., Mancini, R., and Schally, A. V. (1975). Use of synthetic luteinizing hormone-releasing hormone in treatment of oligospermic men: A preliminary report. *Fertil. Steril.* **26**:331.

22

Vratislav Schreiber

Vratislav Schreiber was born in Prague on June 29, 1924, and qualified in medicine in Charles University, Prague, in 1950. While studying, he worked in the Department of Physiology. Since 1955 he has worked at the Third Medical Clinic of the Faculty of Medicine in Prague, and has been engaged in research work in the Laboratory for Endocrinology and Metabolism of the Clinic since 1957. In 1956 he was awarded a C.Sc. (equivalent of Ph.D.) and in 1963 he became Doctor of Science (in medicine). In the same year he was appointed lecturer (docent) in clinical physiology and in 1968 was made professor.

Professor Schreiber works in the field of experimental neuroendocrinology and was awarded the Prize of the Society of Occupational Medicine (1951), the Prize of the Endocrinological Society (1956, 1960), and the Prize of the Physiological Society (1964). In 1964 he was also awarded the Prize of the Scientific Council of the Ministry of Health and in 1965 the Klement Gottwald State Prize, both for studies in neuroendocrinology.

He is president of the Czech Endocrinological Society and was for many years its secretary. He is also a member of the editorial boards of *Physiologia Bohemoslovaca*, *Endocrinologia Experimentalis* (Bratislava), *Neuroendocrinology* (until 1977), and *Vesmir*. His major research interests have centered on the hypothalamic regulation of the adenohypophysis, particularly on TRH and on the regulation of adenohypophysial growth.

Being an Outsider

VRATISLAV SCHREIBER

PREHISTORY (UP TO 1945)

The Boy Scout system of testing one's knowledge of natural history and technical subjects was my first real contact with the natural sciences and my first attempt at active study, although I had already collected minerals with my father, had gathered beetles, and had made myself a herbarium. I passed my tests in botany, zoology, geology, etc., and even in first aid and general health knowledge. Driven by a desire to shine and by my own ambition, I also became an electrician (how to mend a switch, etc.) and cook (this still comes in handy). However, the main Scout prize, the "three eagle's feathers," escaped my grasp. This test included fasting for 24 hr, remaining in hiding, undiscovered, for 24 hr and observing 24 hr silence. Passing the test, by the way, was rewarded with the handsomest badge, which was worn high on the sleeve, above all the others. Three times I tried and three times I failed, and always on the last lap; if the truth be known, an inability to keep my mouth shut is still one of my main faults. When the Germans occupied Czechoslovakia, the Scout movement was banned and our last summer camp was broken up by the Gestapo, who confiscated everything with the *fleur-de-lis* (the Scouts' emblem) on it. In the case of the belts this produced a serious problem. However, I managed to save my shirt with all my proud badges on it; all through the war it remained hidden in an old stove which was never lit.

During my last years of grammar school, I was fortunate to have Dr. Černohorský for natural history. Dr. Černohorský later became, and still is, professor of botany at the Faculty of Science, Charles University. He knew how to make his students interested in the natural sciences, worked actively in botany himself (on the anatomy of the seeds of Cruciferae), and, what

VRATISLAV SCHREIBER • Laboratory for Endocrinology and Metabolism, Third Medical Clinic, Faculty of Medicine, Charles University, Prague.

was most important, allowed us students plenty of scope for active self-expression. I particularly remember the lectures which we prepared and gave for our own and other classes; one of mine dealt with hormones. Today it is hard to say why my general interest centered on hormones. I did spend several summer holidays with a country doctor uncle and there came into contact with medical literature and with advertisements for all kinds of medicines, among which hormones were beginning to make their appearance. I blush to confess that my very first recollection of hormones is a bundle of such advertisements which had been relegated to the lavatory.

I took my Higher School Certificate in 1943, with honors in the natural sciences. It was actually an easy affair, because some of the examinations had to be done in German and none of the examiners (except the one for the German language) knew the language very well. The universities had been shut down, and we ignored the possibilities of studying in "The Reich," so I put myself down for a 2 years' course in chemistry at technical college. Although I passed the examinations and was accepted, I never started, because the German authorities made me liable to conscription for labor for a year and I was sent to work in a Prague engineering works. I had a job there as a draftsman. We were supposed to produce tanks for the German army, but the supply of materials was already irregular, the attitude of workers and technicians was understandably negative, and production consequently was very low. In all, times by then were very hard; there was not enough food, working hours were very long (12 hr a day for 6 days a week, day and night shifts alternating), and everywhere was the atmosphere of Nazi terror. I had already learned German and English pretty well at school and now I made use of the long hours at "work" to study French and Russian and prepare for my first examinations at the Faculty of Medicine (physics, chemistry, biology). Influenced by my fellow workers, I even embarked on the study of philosophy (the teaching of which was not allowed by the Nazis); it goes without saying, of course, that I studied from banned books. For these reasons I do not regret the 2 lost years. When the war ended, I started my studies at the medical school with plenty in hand, since I was well prepared in all aspects and thus had sufficient time for independent study. Furthermore, I think that the preceding "lean" years had whetted my appetite for learning.

As an ex-grammar school pupil, current draftsman, and future medical student (maybe—nobody could say for certain how things would turn out, even if there was no doubt after 1943 that Nazism would be defeated), I tried to penetrate a little way into medicine. Fortunately, the medical library had been kept open and I was a regular visitor. I also attended the lectures of the Purkyně Society for Study of the Mind and the Nervous System, where Dr. (now professor of neurology) Hrbek submitted a unified system of medical and scientific philosophy. And what was most important, I obtained entry—more or less illegally—to the evening meetings of the Medical Society in Prague. Never since have I made such careful notes as the ones I took from those lectures in 1944–1945. In the spring of 1944, the Society gave a lecture on the accelerated maturation of adolescents. Mention was made in the discussion of the influence of light and of endocrine changes in blindness and Professor Charvát (who had just previously returned from Buchenwald) drew attention to the relationships among light, the pituitary, and the rest of the endocrine system, citing French and American experiments demonstrating the effect of light on development of gonads in birds. This was the primary stimulus which turned my interest toward neuroendocrinology. Since Professor Charvát (formerly Chief Scout of Czechoslovakia) was released from the Buchenwald concentration camp on the intervention of the Crown Prince of Sweden, Gustav Adolf (Chief Scout of Sweden), it can be said that his Highness was also indirectly concerned.

EARLY HISTORY (1945–1955)

In May 1945, after our country had been liberated, I inscribed my name in the books of the Faculty of Medicine, Charles University, Prague. My year was supposed to start with lectures in the autumn (there were 6 years that had been lost by the closing of the universities by Nazis in 1939 and these were given priority). While waiting, I went to Professor Charvát, who had been made head of the new Third Medical Clinic and asked him for a job. Professor Charvát set me to work in his laboratories, and for the whole of the summer and autumn, first with one laboratory assistant and then with two, I carried out standard biochemical tests for the patients of the clinic and tested their basal metabolism.

In the winter of 1945, when the laboratories were already fully staffed, I realized that the clinic's biochemistry laboratories were not the right milieu for my projected study of the effect of light on the endocrine system and asked Professor Charvát to let me go and work at the faculty department of physiology. With a recommendation from Professor Charvát I went to the head of the Physiology Department, Professor Laufberger (a leading Czech neurophysiologist, the discoverer of ferritin, and now known for his work on spatiocardiography) and was accepted. In turn I worked as a demonstrator, assistant scientific worker, and auxiliary assistant (as the last, I crammed students for their practical examinations) and remained there until 1950. Although the remuneration was very small, these were all paid functions and, in addition I worked in Dr. Šilink's biochemistry laboratories (later Dr. Šilink was made director of the Institute of Endocrinology in Prague). When I look at today's students of medicine (my son included), I do not know how I managed to find time for it all—at least 4 hr a day in the physiology department, two 3-hr periods per week of work for Dr. Šilink, a little physical exercise (dancing included), and, of course, my own studies. The reasons were probably that I had learned a lot beforehand, attendance at lectures was not compulsory, and the amount of subject matter was substantially smaller than it is today. The result, however, was that I was not an excellent student of medicine as far as grades are concerned.

At that time Laufberger's activities in the field of neurophysiology and psychophysiology were at their height; he was investigating the mechanisms of nerve impulses and had formed the theory that memory was based on circling impulses. He allotted me to his assistant Holubář, who worked in neurophysiology. In time I was able to persuade Holubář to let us record spikes in the pituitary stalk of frogs whose retina had been exposed to flashes of light. The results were not a success, but Holubář went on studying retinal neurophysiology with similar techniques until he was killed in a car accident in 1967. I was neither gifted in nor liked neurophysiology and psychophysiology, and about a year later I transferred to the Department of General Physiology under Professor Karásek, who already had a group of young workers, led by O. Poupa, specializing in endocrinology and metabolism. They were fortunate in that their work was partly supported by the pharmaceutical industry, for which, in turn, hormone assays were performed.

Under Laufberger, or rather Holubář, I had already begun to study the effect of light. Unable to wait, I started with oat shoots (coleoptile), illuminated them from the side with different colored lights, and watched them grow. I had not studied botanical literature very thoroughly and so after a few months I formulated the theory that "photosynthesis is an antagonist of growth in length." The botanists whom I afterward consulted were not very enthusiastic and it was clear that I had better change over to animals. Because frogs were used on a large scale for practical exercises in physiology, *Rana esculenta* and *R. temporaria* tadpoles were the most easily available. Surprisingly, I found that blinded tadpoles grew a trifle faster than controls if they were left in the light, although there was no difference in the onset of metamorphosis in the two groups (Schreiber 1948*a*). Exposure to artificial light during the winter months raised testicular weight in male frogs (Schreiber, 1948*b*) and caused earlier mating (Schreiber and Tyšerová, 1949). Additionally, the articles of Hans Heller inspired me to carry out experiments which showed that the water metabolism of newborn rats altered characteristically at the time they opened their eyes (Schreiber,

1950). Animals with closed and open eyes had different regression lines for the increase in the percentage of dry body solids in correlation to age. Further studies were largely inspired by those of J. Benoit and R. Collin and a little later they were influenced by the work of G. W. Harris.

In 1946 I spent 2 months at Copenhagen University, where, because of my work in Charvát's clinic, I mainly studied biochemical methods used in hospital laboratories. In 1948 I went for 2 months to France, where I visited French endocrinological and neuroendocrinological laboratories. I was given the greatest help and the largest amount of information in the laboratories of Professor R. Courrier in Paris, by Professor R. Collin in Nancy, and by Professors Max Aron and J. Benoit in Strasbourg. In Paris, Mme. Cologne spent hours teaching me how to hypophysectomize rats. All of them presented me with their books, among them Collin's *L'hypophyse—Deuxième Série*, with his collected works from 1933 to 1936. I do not know whether Collin at that time still held his erroneous view on neurosecretion (see Stutinsky, *Pioneers in Neuroendocrinology*, Vol. 1, p. 279), but my conversation with him reinforced my conviction that adenohypophysial function was regulated by the hypothalamus. I used a quotation from Collin (1937); "*Les régulations hormonales sont insèparables des régulations neurovégétatives et l'endocrinologie classique cédera bientot le pas à une discipline nouvelle, la neuroendocrinologie,*" as the motto for a short review (Schreiber, 1949).

In my view, the role of Rémy Collin in the development of neuroendocrinology is not fully appreciated. His papers particularly merit appreciation for the objective way in which they acquaint the reader with the findings of Scharrer, Roussy, and Mosinger, although he wrongly did not agree with their interpretations. He set forth conflicting views with great accuracy and clarity, and thus gave the reader an opportunity of forming his own conclusions. One example of this exact, informative, and elegant style is the article "*Deux Conceptions de la Circulation Porte Hypophysaire*" (Collin and Fontaine, 1936), although in it the authors expressed the hypothesis that information flowed from the pituitary to the hypothalamus.

My next visits to France came after an interval of over 15 years and again I profited mainly from the kindness of Professors Courrier and Benoit in Paris. Professor Benoit found in his records some notes I had made in 1948 of my future plans; how few of them have come true!

In the paper mentioned above (Schreiber, 1949), I emphasized in the discussion that the most probable basis of hypothalamohypophysial relationships was the one postulated by Harris (1948), i.e., that adenohypophysial function was controlled by a hypothalamic humoral factor (or factors). I shall never forget the excitement which this hypothesis roused in me and my surprise that neither Harris nor anybody else had tested it immediately by classic endocrinological methods, i.e., by extraction and injection. It soon became apparent that this was by no means a simple matter because of the low factor concentration in the hypothalamus, the difficulty of purification, the presence of adenohypophysial hormones in the tuberoinfundibular area, and a number of other reasons. Nevertheless, in the spring of 1949 I began to produce rabbit hypothalamic extracts, inject them into rats and mice, and incubate pituitaries together with the extracts *in vitro*. This work was broken off in 1950, when I left the Department of Physiology; the bulk of the results, plus a few new ones obtained under the most unusual conditions (while doing national service), was summed up in a paper (Schreiber, 1956) which—if it shows anything at all—demonstrates the presence of adenohypophysial tropins in extracts of the rabbit tuber cinereum. The only thing which might possibly testify to the presence of a gonadotropin releasing factor would be a *decrease* in the gonadotropin content of the adenohypophysis after injection of hypothalamic extracts. Anyway, others who began testing hypothalamic extracts at the same time did not get much better results (e.g., Slusher and Roberts; Hume and Wittenstein; Buchanan, Ottaviani, and Azzali).

I was graduated from the Faculty of Medicine, Charles University, in April 1950, without any special distinctions but with a few more or less presentable publications and the satisfaction of having earned my way through most of my studies. My main gain, i.e., a clear plan for the study of hypothalamic extracts, was useless to me, however, since my agreement with the Physiology Department came to an end in June 1950, and, after a brief period in the district hospital in Děčín, I spent the next 5 years doing military service—the personal price I had to pay for the cold war.

THE MIDDLE OF THE STORY (1955–1970): THYROLIBERIN (TRF)

In the summer of 1955 I returned to the Medical Faculty of Charles University, Prague, to work as an intern in the Third Medical Clinic under Professor (now Academician) Charvát. The job was a full-time one, with 20 beds, night duties, and all the rest, and, most of all, the need to learn the fundamentals of the whole of internal medicine. Shortly before my return the clinic had moved to an adapted Jesuit monastery (built in 1659), which had become a military hospital after World War I and had been severely damaged at the end of World War II. Modern adaptations of the spreading building, with its wide corridors, handsome Renaissance ceiling, etc., provided the clinic with large and well-equipped laboratories. This was of the greatest importance for further development of the clinic and further training of its workers. In the laboratories I was once again given the opportunity for experimental work, but the first essential was to complete my studies in internal medicine. In 1956 I qualified as a grade I specialist in this field (later supplementing my qualifications by an endocrinology diploma). In the same year I was awarded a C.Sc. (Candidate of Science) degree (equivalent to a Ph.D.) on the basis of a thesis summing up my previous studies on the effect of light on the pituitary. Fate ordained that I should remain a hybrid between a clinical and an experimental endocrinologist. In 1963 I won my Dr.Sc. (Doctor of Science) degree, in the same year became a "docent," and in 1968 was nominated professor of clinical physiology. Teaching the pathophysiology of the endocrine system and the whole of clinical physiology became my permanent duty.

In the 1950s, after the most serious consequences of World War II had been corrected, scientific research began to receive increasing support in Czechoslovakia. The Czechoslovak Academy of Sciences was established, with a number of large institutes. Research institutes of the Ministry of Health were established (including the Institute of Endocrinology in Prague), and financial and material support for research at the universities was made available. In 1957 the Ministry of Education established the Laboratory for Endocrinology and Metabolism as an adjunct to the Third Medical Clinic and equipped it well with staff and material requirements. Its director, Academician Josef Charvát, created good conditions for work and systematically promoted (and for that matter still promotes) the development of both clinical and experimental research. On his retirement as director of the clinic (which he handed over in 1970 to his pupil Professor Pacovský), he retained his directorship of the Laboratory for Endocrinology and Metabolism and good conditions still prevail there for the workers of the integrated clinic and laboratory.

The research side is endowed by the Ministry of Education, the clinical side (including clinical research) by the Ministry of Health. The endowments of the Ministry of Education are of the institutional type; i.e., the total amount allocated to Charles University is divided according to the requirements of the research plan among the various faculties (the long-term, 5-year plans are revised and adjusted every year). In a similar manner, the faculties divide their share among their various departments and laboratories. The economic commission of the faculty decides where the money shall go, taking into account the productivity of the laboratory, the requirements of the research plan, and the needs of society. The endowments for salaries are separated: after an abrupt increase in the number of workers in the 1950s and 1960s, the number of posts in the individual laboratories is now stabilized and filled as circumstances require. Although it repeatedly has been discussed, the grant system has not been introduced in

Czechoslovakia. As well as having disadvantages, the institutional type of endowment has distinct advantages—a feeling of greater stability and social security for the workers, the possibility of long-term planning, and surprisingly good flexibility (laboratories needing further funds can receive from the faculty funds which have not been exhausted by other laboratories). In my opinion, this way of financing science is the one best suited to a country with a planned economy. During the whole of my period of office as deputy dean for research in our faculty (1966–1969), I never came across a case of a good research program having to be abandoned or reduced in size for lack of funds. This applies absolutely to current materials (experimental animals, chemicals, glassware, etc.). The availability of expensive equipment (e.g. apparatus, especially if imported from the West) is more limited, but even here the standard of equipment has attained a good level.

The clinical environment, the diagnostic needs of patients, and my teaching duties all helped to scatter my interests. From the moment I started work at the clinic, however, I kept returning to the problem of hypothalamic regulation of the adenohypophysis. There was no readily available means of assaying pituitary hormones, and after a few months of frustration with McKenzie's method (the iodine-free diet was a great problem) I posed a simple question—if the adenohypophysis were exposed to the action of hypothalamic extracts *in vitro*, would it be possible to detect any changes by *simple biochemical methods*? The advantage of putting the question this way was the immediate possibility of test. Its chief disadvantage was its manifest nonspecificity, although any changes demonstrated could at least help to explain the mechanism of the action of hypothalamic factors in the adenohypophysis. Thus in 1957 I drew up a program which formed the basis of the work of my laboratory for the next 12 years.

In 1957 I became a scientific worker of the Laboratory for Endocrinology and Metabolism. I gave up my work on the wards and, in clinical medicine, kept only the endocrinological outpatient service. This situation remains today. I was fortunate that at the same time (1957) I acquired an excellent technical assistant, Miss V. Kmentová, who worked with me up until 1969. In the meantime, while working, she studied science at Charles University, took her degree (R.N.Dr., Dr. Rerum Naturalium) and shortly afterward won her C.Sc.

The first method we used was the determination of ascorbic acid. We found (Schreiber and Kmentová, 1958) that incubation of the pituitary with hypothalamic extract was followed by a decrease in the ascorbic acid concentration in the gland. As far as I am aware, this finding has been neither confirmed nor refuted. I likewise do not know why the adenohypophysis is so rich in ascorbic acid (after the adrenals and corpus luteum it has the highest ascorbic acid concentration). Then, at a laboratory consultation, Professor Charvát suggested that we should have a look at alkaline phosphatases. Miss Kmentová already had some experience with the phosphatase assay in clinical biochemistry and knew that it was usual to determine both alkaline and acid phosphatases in clinical material. She therefore prepared reagents for both methods and it was a lucky thing she did so, because when adenohypophysial tissues were incubated with hypothalamic extract, nothing happened to alkaline phosphatase but acid phosphatase activity rose abruptly (Charvát *et al.*, 1958). It was also found that while the factor activating adenohypophysial acid phosphatase was present in the nonprotein fraction of hypothalamic extracts, it was also present in the same fraction of rat temporal lobe extracts, although in a much lower concentration. We determined whether acid phosphatase activity in the adenohypophysis was related to secretion of any of the adenohypophysial hormones by testing the activity of this enzyme in castrated, adrenalectomized, and thyroidectomized rats. Since acid phosphatase activity in the adenohypophysis rose markedly only after thyroidectomy (Schreiber and Kmentová, 1959*a*), we formulated the working hypothesis that adenohypophyseal acid phosphatase is related to TSH secretion and that TRF is its possible activator. This was borne out by the finding that the amount of the factor activating acid phosphatase in the adenohypophysis *in vitro* diminished in the hypothalamus of rats subjected to chemical thyroidectomy (Schreiber and Kmentová, 1959*b*).

Then followed a series of biochemical studies aimed at further fractionation of hypothalamic extracts and the purification, characterization, and, if possible, isolation (*sancta simplicitas!*) of the active factor. The fractionation experiments made increasing demands on the amount of material, and we therefore changed from rat and rabbit to bovine hypothalami; this entailed traveling to the Prague abattoir and collecting the hypothalami in ice-cold acetone. On looking back, this was one of the most difficult phases of the whole study—if I stayed at the abattoir from 6 a.m. to 1 p.m., on good days I would obtain no more than 60–80 hypothalami.

In the first phase of the experiments, acid phosphatase was my only criterion for the presence of the active factor; at the same time we accumulated indirect evidence of the existence of a relationship among this factor, TSH secretion, and TRF activity. Apart from the findings mentioned above, we found reduced sensitivity of the adenohypophysis of rats given thyroid hormones (Schreiber *et al.*, 1961). Stimulation of TSH release *in vitro* was shown in another series of papers (Schreiber *et al.*, 1961*a*, 1962). Of course, we did not isolate pure TRF, but the results were the ones cited by Locke and Schally (1972) when they wrote: "Schreiber probably demonstrated in some of his experiments true TRF effects." The positive aspect of our relatively low degree of purification was that we never expressed the conclusion that TRF was not of a peptide nature. It was not until 8 years later that the actual structure of TRF was elucidated (Folkers *et al.*, 1969; Burgus *et al.*, 1969).

In the autumn of 1967 I had the opportunity of visiting neuroendocrinology laboratories in the United States. Apart from the quality of the results, the main differences between their teams and ours can be summed up under three headings: (1) their much higher professional biochemical level, (2) the efficiency of their system for obtaining frozen hypothalami (tens of thousands a month), and (3) their large laboratory space, large staff size, and the elaborate equipment available. It was brought home to me very forcefully that I was now an outsider in a main line of research (the isolation of releasing factors and elucidation of their structure) and that I should have to look for roads with less heavy traffic for the further program of our laboratory. In any case, I had been most interested in releasing factors at a time when their existence was still held more or less in doubt. Predictability of the results, the ever fiercer competition among the various research workers, and signs of jealousy stripped research during those years of much of its aesthetic and ethical attraction. (The declaration that other teams putting forward the same claims earlier were wrong became a part of scientific reporting on our subject.) In 1966 I agreed to a reciprocal exchange of TRF samples with a well-known laboratory abroad. Their sample arrived and was active in my laboratory, and I sent a report to that effect. Apart from being informed that my sample had arrived and that it was hygroscopic and brown, I learned nothing more. A year later, when visiting the laboratory concerned, I inquired about it and was told that the laboratory was not interested in testing TRF samples which were not sufficiently pure. The reader will no doubt appreciate that this was a source of amusement to me rather than of bitterness.

In this phase (1969–1971), the important question for us was whether the correlation between the increase in acid phosphatase activity in the adenohypophysis and the TRF content in the hypothalamic extracts was a sign that TRF stimulates acid phosphatase activity at the same time as TSH secretion (as I thought). Experiments with synthetic TRF yielded negative results—TRF did not stimulate acid phosphatase activity in the adenohypophysis *in vitro*. The existence of another hypothalamic factor, present in the "TRF" fractions in the initial purification operations and activating adenohypophysial acid phosphatases, is one possible explanation. However, there are others also—e.g., the presence of cofactors—but an explanation must be found for the trophic (growth) and metabolic (enzymic) alteration of the adenohypophysis after thyroidectomy.

Our attempts to demonstrate the existence of TRF produced a series of collateral enzymological, biochemical, and histochemical studies. When we found that there was a correlation between adenohypophysial acid phosphatase activity and TSH secretion, we investigated the enzymes of the adenohypophysis for several years and were fortunate in obtaining the collaboration of Professor Z. Lojda (also of the Medical Faculty) on the histochemical side. When summing up a series of histochemical studies of 17 adenohypophyseal enzymes (Lojda and Schreiber, 1964), it was found that they were activated the most after thyroidectomy, little or not at all after adrenalectomy, and moderately after castration. The adenohypophysis is thus evidently able to produce ACTH without any great metabolic activation, whereas TSH secretion requires maximum activation. The same applies to the growth reaction (Schreiber and Kmentová, 1964), i.e., it is maximal after thyroidectomy, moderate after castration, and minimal or absent after adrenalectomy. On the basis of these findings and a study of the possibilities of inhibiting compensatory hypertrophy of the endocrine glands after hemiablation, I formulated the theory that TSH secretion occupied "a special and privileged position" (Schreiber, 1964, 1971*a*). In the presence of a deficiency of several of the peripheral hormones, the thyrostatic servomechanism is preserved even if it means sacrificing the gonadostatic and adrenostatic servomechanism. The differences observed in the growth reactions of the adenohypophysis under different experimental conditions also got us interested in the regulation of growth of the adenohypophysis in general. These studies have formed a steppingstone to the later stages of our work (e.g., Schreiber and Přibyl, 1972, 1976).

THE "GALLERY OF HYPOTHALAMOLOGISTS"

The "gallery of hypothalamologists" in my laboratory is another byproduct. At the beginning of the 1960s, neuroendocrinologists from other countries began to visit our laboratory, and (inspired by examples I had seen in France) I at once conceived the happy idea of asking them for their photographs. The first visitor was a coeditor of this book, Bernard Donovan of London. The gallery means more to me than a souvenir; apart from reminding me of the faces of distinguished personalities in neuroendocrinology, it also brings back their ideas and their work and in this way acts on me as a mental stimulant—or so I hope.

REFERENCES

Burgus, R., Dunn, T. F., Desiderio, D., and Guillemin, R. (1969). Structure moléculaire du facteur hypothalamique hypophysiotrope TRF d'origine ovine: Mise en évidence par spectrométrie de masse de la séquence PCA-HIS-PRO-NH_2. *C. R. Acad. Sci. Paris* **269**:1870.

Charvát, J., Schreiber, V., and Kmentová, V. (1958). Influence of a crude hypothalamic extract on the acid phosphatase activity of rat pituitaries. *Nature* **182**:62.

Collin, R. (1937). *L'Hypophyse—Deuxième Série*, Thomas, Nancy.

Collin, R., and Fontaine, T. (1936). Deux conceptions de la circulation porte hypophysaire. *Rev. Franc. Endocrinol.* **12**:4.

Folkers, K., Enzman, F., Bøler, J., Bowers, C. Y., and Schally, A. V. (1969). Discovery of the synthetic tripeptide-sequence of the thyrotropin releasing hormone having activity. *Biochem. Biophys. Res. Commun.* **37**:123.

Harris, G. W. (1948). Neural control of the pituitary gland. *Physiol. Rev.* **28**:139.

Locke, W., and Schally, A. V. (1972). *The Hypothalamus and Pituitary in Health and Disease*, Thomas, Springfield, Ill., p. 232.

Lojda, Z., and Schreiber, V. (1964). Activités enzymatiques de l'adénohypophyse du rat dans diverses conditions expérimentales. *J. Physiol. (Paris)* **56**:559.

Schreiber, V. (1948*a*). Vliv světelných podnětů na činnost hypofysy. *Cas.Lek. Cesk.* **87**:409.

Schreiber, V. (1948*b*). Stimulation testiculaire provoquée par l'éclairage artificiel rhythmique. *C. R. Soc. Biol.* **142**:1055.

Schreiber, V. (1949). Fysiologie systému diencefalo-pituitárního. Thomayerova sbírka, Supp. *Cas.Lék. Cesk.* **47**:1.

Schreiber, V. (1950). Two phases of the water metabolism in newly born rats. *Nature* **166**:77.

Schreiber, V. (1956). Zur Frage des hypothalamischen Hormones mit adenohypophysotropher Wirkung. *Endokrinologie* **33**:259.

Schreiber, V. (1964). La position "privilégiée" de la fonction thyréotrope de l'adénohypophyse du rat. *J. Physiol. (Paris)* **56**:559.

Schreiber, V., and Kmentová, K. (1958). Mechanisms of hypothalamus-pituitary relationships: Effect of brain tissue suspensions on the amount of ascorbic acid in the pituitary gland *in vitro*. *Physiol. Bohemoslov.* **5**:437.

Schreiber, V., and Kmentová, V. (1959*a*). Bedeutung der Phosphatasenaktivität in der Hypophyse. *Acta Biol. Acad. Sci. Hung.* **9**:285.

Schreiber, V., and Kmentová, V. (1959*b*). Effects of methylthiouracil or thyroidectomy on the activation of pituitary acid phosphatases *in vitro* by whole hypothalamic extract. *Nature* **184**:728.

Schreiber, V., and Kmentová, V. (1964). L'hypertrophie de l'hypophyse du rat après surrénalectomie, castration, thyroidectomie et leurs combinaisons. *C. R. Acad. Sci. Paris* **258**:4151.

Schreiber, V., and Přibyl, T. (1972). *Adenohypophysial Growth and Thyroxine Binding: Effects of Steroid Hormones*, Acta Univ. Carol. Medica, Mongraphia LI, Praha.

Schreiber, V., and Přibyl, T. (1976). Dopaminergic control of adenohypophysial weight and serum ceruloplasmin level in rats. *Neuroendocrinology*. **21**:58.

Schreiber, V., and Tyšerová, M. (1949). Vliv světla na pohlavní aktivitu žab. *Biol. Listy (Praha)* **30**:255.

Schreiber, V., Charvát, J., Kmentová, V., and Rybák, M. (1959). Fractionation of brain extracts for acid phosphatase-elevating activity. *Nature* **183**:473.

Schreiber, V., Eckertová, A., Franz, Z., Kočí, J., Rybák, M., and Kmentová, V. (1961*a*). Effect of a fraction of bovine hypothalamic extract on the release of TSH by rat adenohypophyses *in vitro*. *Experientia* **17**:264.

Schreiber, V., Rybák, and Kmentová, V. (1961*b*). Effect of thyroid hormones on the sensitivity of the adenohypophysis to the hypothalamic factor, activating adenohypophysial acid phosphatases and TSH secretion. *Physiol. Bohemoslov.* **10**:376.

Schreiber, V., Rybák, M., Eckertová, A., Jirgl, V., Kočí, J., Franc, Z., and Kmentová, V. (1962). Isolation of a hypothalamic peptide with TRF (thyreotrophin releasing factor) activity *in vitro Experientia* **18**:338.

23

David de Wied

David de Wied was born on January 12, 1925, in Deventer, the Netherlands. After high school, he started to study medicine at the University of Groningen and received his M.D. from that University in 1955. In 1950 he joined the Department of Pharmacology of the University of Groningen, where he worked for a Ph.D. in pharmacology under Professor J. H. Gaarenstroom. The thesis was completed in 1952; it dealt with the effect of ascorbic acid on adrenal activity during cold. In 1957 he joined Dr. I. Arthur Mirsky, Department of Clinical Science of the University of Pittsburgh, for 1 year as a research fellow.

He was appointed associate professor of experimental endocrinology of the University of Groningen in 1958 and full professor in 1961. Also in 1961 he worked in Montreal at the Allen Memorial Institute of Psychiatry of McGill University with Dr. Murray Saffran on a NATO fellowship. In 1963 he was appointed professor and head of the Department of Pharmacology of the University of Utrecht, and was appointed honorary professor of pharmacology in the Medical Faculty of the University of Toronto in 1971.

His work has been concerned with hypothalamus-pituitary-adrenal relationships, and at present his interest lies mainly in the field of the effects of pituitary and other peptides on behavior.

Pituitary Peptides and Adaptive Behavior

DAVID DE WIED

In January 1947 I began my medical studies in Groningen, a nice provincial town in the north of the Netherlands. When I had completed the first part of my studies in September 1948, I was asked to participate in the organization of the lustrum festivities of the University which were held in the summer of 1949. It was the first one after the war and a great number of alumni were expected. At that time, the organization was dominated by the activities of the main student corporation "*Vindicat atque Polit*" of which I was a member and acted as the treasurer. The planning and execution of the activities took a great deal of my time. It determined, however, an important part of my future, since the girl whom I invited for the festivities became my wife in 1952. Although this in itself can be regarded as a fringe benefit, I felt that I should use my time a bit more effectively and become more serious with respect to my medical studies. I thus decided that I should spend some time in a laboratory, and for this purpose I inquired whether there would be a possibility of doing some research in bacteriology—bacteriology because my best friend, H. H. Cohen, who presently is one of the directors of the Institute of Public Health in the Netherlands, specialized in bacteriology and seemed to be very excited about the subject. I did not make an appointment with the professor, nor did I mention to my friend that I had in mind to do some research in this area. When I entered the department, I met Cohen's brother-in-law, M. Bruining, who already was a bacteriologist. When he saw me, he asked what I came for. I told him that I was interested in doing some research and that I wanted to explore the possibilities in the Department of Bacteriology. He immediately responded by saying that bacteriology was not the most interesting area of research. He thought that endocrinology would be a field

DAVID DE WIED • Rudolf Magnus Institute for Pharmacology, Medical Faculty, University of Utrecht, Vondellaan 6, Utrecht, the Netherlands.

of the future. I should therefore go to the recently appointed professor of pharmacology, J. H. Gaarenstroom, who worked mainly in this field. Accordingly, I went to see Gaarenstroom, who offered me some space to start a small project. Gaarenstroom had been appointed as head of the department in 1948. He was a charming and courteous man who found an old-fashioned laboratory when he came to Groningen. Within 2 years he had changed it into a lively department filled with a group of young associates. Working under him meant a great deal of freedom. He was highly critical and spotted every illogical line of thought in a presentation. He had received his training first in Amsterdam from the famous, mainly endocrinologically oriented E. Laqueur and subsequently joined S. E. de Jongh when the latter became head of the Department of Pharmacology in the Medical Faculty of the University of Leyden. Gaarenstroom worked with de Jongh on the function of the gonads, and when he was appointed in Groningen decided to change the subject of his research to another endocrine organ, the pancreas. This was probably not a coincidence. Gaarenstroom had severe diabetes mellitus and died at the relatively young age of 58 as a result of complications of this disease.

I worked under supervision of W. H. Lammers, who eventually succeeded Gaarenstroom in Groningen in 1965. The project concerned the effect of adrenalectomy on the sensitivity to anesthetic agents. I remember that we constructed a rotating rod to determine the latency of a rat on the rod following injection of an anesthetic. I found that the time was shorter in adrenalectomized rats thus indicating that the effect of the anesthetic was potentiated in the absence of the adrenal cortex. This, as I found some time later, had already been reported in 1928. Anyway, this first study aroused my interest in the pituitary-adrenal system. Lammers left the department in 1950 for Utrecht, and Gaarenstroom asked me whether I would be interested in replacing him as an "assistant." I was surprised and honored. I felt very much at ease in the department and gratefully accepted the offer after I finished the second part of my studies in September 1950. Gaarenstroom requested my assistance in the practical course in pharmacology for medical students. For the rest I was free to do what I would choose as long as it was economical and did not need expensive equipment.

The Canadian workers Dugal and Thérien in 1949 reported that high amounts of ascorbic acid inhibited adrenal hypertrophy in rats exposed to cold. The function of this vitamin, present in high quantities in the adrenal cortex, was not understood. As suggested in the literature, it might be involved in the formation of corticosteroids and an excess of ascorbic acid might facilitate the production of these steroids. This could affect ACTH release and consequently the size of the adrenal cortex. Using liver glycogen as a measure of adrenocortical activity, I found that high amounts of the vitamin partly prevented the decrease in glycogen content of the liver in rats exposed to cold. The same, however, was found in adrenalectomized rats maintained on salt, Doca, and an extract of the adrenal cortex. Thus the effect of ascorbic acid was not mediated by the adrenal cortex (de Wied, 1953) but probably was related to a decreased metabolic degradation of corticosteroids in the liver, as was later shown by others. The results were good enough for a Ph.D. thesis, which I completed in the summer of 1952. During these very pleasant years I received a lot of stimulation from P. Siderius, who was the chief assistant in the department. He also worked on stress and used the circulating eosinophils and ascorbic acid concentration in the adrenal cortex as measures of pituitary adrenal activity. He taught me the newly developed concepts of the "general adaptation syndrome" as described by Selye which around 1950 markedly influenced our studies on adaptive mechanisms and the role of the pituitary adrenal system. Siderius, who at present is the secretary general of the Ministry of Public Health and Environment, markedly influenced my interest and activities but above all developed my critical attitude. We worked hard and generally did not finish before midnight. Often we went into town for a bite or a beer or even better a game of billiards.

After I had obtained my Ph.D., I decided to continue my medical studies. This required nearly 2 years of internship, which I divided over 3 years so that I could continue my research on a part-time basis. I received my M.D. in the beginning of 1955. I had meanwhile succeeded Siderius as chief assistant after he left for Baghdad, where he was appointed professor of pharmacology. My interest gradually became directed toward the regulation of pituitary ACTH release. With a young chemist, C. Olling, I worked on the influence of chlorpromazine on pituitary-adrenal function. According to French authors this neuroleptic inhibited pituitary adrenal activation in response to stress. We found that it stimulated pituitary adrenal activity in conscious rats but in the presence of pentobarbital blocked stress-induced ACTH release (Olling and de Wied, 1956). From these studies we developed an assay for the hypothalamic corticotropin-releasing factor (CRF) in which many endocrinologists had become interested after Harris (1955) postulated the concept of the humoral control of pituitary function.

In 1956 P. Smelik came to Groningen from Utrecht, where he had begun studies in biology which he wanted to finish in Groningen. He took a course in pharmacology and decided to work for a Ph.D. in the department after he had completed his studies. He also was much interested in the pituitary-adrenal system, and in particular in the central nervous control of pituitary ACTH release. At the same time I invited a promising young medical student, P. Bouman, to join the department. The three of us got on very well and decided to cooperate in the study on regulation of the pituitary-adrenal system. We used pharmacological agents, lesions in the median eminence of the hypothalamus, and corticosteroids to block the system, and studied the influence of Pitressin and hypothalamic extracts on stimulation of pituitary-adrenal activity (de Wied *et al.*, 1958). Pitressin appeared to be a potent releaser of ACTH under the various experimental conditions we used. Others, such as L. Martini and S. M. McCann, already had pointed to vasopressin as a possible CRF. Although these years were very exciting, relatively carefree, and fruitful, I began to feel that I should gain more training and experience elsewhere. Like so many others in Europe at the time, I wanted to go to the United States. I met I. Arthur Mirsky, head of the Department of Clinical Science in the Medical Faculty of the University of Pittsburgh when he visited P. Siderius, who had returned from Baghdad via the United States, where he had spent half a year in Mirsky's department. Siderius thought that I might benefit highly from working with Mirsky. Accordingly, I went for a year to Pittsburgh in the fall of 1957 as a research fellow of Pittsburgh University. This was a decisive period which determined the line of my future research. Mirsky was schooled in biochemistry, physiology, medicine, psychoanalysis, and psychology. He was a man of extraordinary intellect and vision and one of the most creative scientific minds I have ever known. He was a tough-minded scientist, a severe critic of himself and of others, an aggressive researcher for new insights in biological phenomena, a literary and scholarly man. Mirsky was mainly interested in carbohydrate metabolism, but he also spent time studying the relation between ADH and ACTH release. Since this was my main interest, I pursued the subject in Pittsburgh. We studied the influence of prednisolone on ACTH and ADH release in response to stress (de Wied and Mirsky, 1959) and showed that blockade of stress-induced ACTH release by high doses of the steroid did not necessarily block ADH release. This suggested that these phenomena could be dissociated. We did not know, however, that high doses of prednisolone block the pituitary response to vasopressin (de Wied, 1964*a*).

Mirsky insisted that I should get some training from R. E. Miller, a psychologist in the department, so as to become familiar with some basic techniques and concepts in experimental psychology, since knowledge in this area might be important for my future studies. In addition, such techniques were hardly known in the Netherlands at the time, and it might be valuable to introduce some of these when I returned to my country. Miller was a marvelous teacher and I enjoyed working with him tremendously.

Mirsky and associates reported in 1953 that ACTH might have a marked effect on conditioned behavior. From a number of experiments in monkeys and rats they had concluded that ACTH might diminish the effectiveness of an anxiety-producing stimulus. Murphy and Miller (1955) subsequently found that daily administration of ACTH during acquisition of shuttlebox avoidance behavior delayed extinction of the avoidance behavior in rats. Since this effect might be mediated by the adrenal cortex, Miller and I decided to extend these observations by studying the influence of a corticosteroid on extinction of shuttlebox avoidance behavior. To our surprise, we found that the steroid (prednisolone) facilitated extinction of the behavior, thus acting in a way opposite to that of ACTH. This suggested that the effect of ACTH was not mediated by the adrenal cortex and could be an extra target action of ACTH. This was confirmed by Miller and Ogawa in 1962 when they showed that the effect of ACTH on extinction of shuttlebox avoidance behavior also could be demonstrated in adrenalectomized rats. We used a high dose of prednisolone which was rather toxic, for the rats lost a lot of weight and some died during the course of the experiment. We submitted a manuscript for publication, but it was rejected because of our toxic dosage of prednisolone. However, I confirmed the data 10 years later (de Wied, 1967). I received a tremendous impetus from working with Mirsky, Miller, and Shapiro (an internist who also worked in the Department of Clinical Science on psychosomatic aspects of hypertension). I was educated in the many diverse subjects in which Mirsky was interested, such as insulin and insulinase, stress-induced pituitary ACTH and ADH release, behavioral effects of ACTH and corticosteroids, metabolic effects of peptides, communication of affect in monkeys, and psychosomatic aspects of hypertension and stomach ulcers, etc.—discussions which we sustained over many years during regular visits to each other's laboratories. I was very pleased that the University of Utrecht acknowledged Mirsky's impact on my work and that of my group by offering him the degree of Doctor Honoris Causa in Medicine in March 1974, several months before he died of a chronic illness from which he suffered very much during the last years of his life. He was greatly honored and insisted on receiving the degree personally, notwithstanding the fact that he was seriously ill. He came to Utrecht but was not able to attend the ceremonies because he had to be rushed to the hospital. He could barely make the return trip and died on September 15, 1974.

When I returned to Groningen in the fall of 1958, I resumed my original line of research on regulation of the pituitary-adrenal system and the role of vasopressin in stress-induced ACTH release. I also developed a CRF assay in rats in which the nonspecific response to stress was blocked by extensive lesions in the median eminence, but in which the release of ACTH could be evoked by vasopressin or hypothalamic extracts (de Wied, 1961*a*). Smelik during my absence had started to study the role of the posterior pituitary, as the storage site of vasopressin, in pituitary ACTH release. Madame C. Mialhe-Voloss, who worked in M. F. Stutinsky's laboratory in Strasbourg, had reported the presence of relatively high amounts of ACTH in the posterior lobe of the pituitary. Emotional stress, she showed, reduced the concentration of ACTH in this lobe. Smelik reasoned that, if this were true, removal of the posterior lobe of the pituitary should result in inhibition of pituitary-adrenal activation in response to emotional stress. Posterior lobectomy was performed by our excellent technician, J. Melchior, who was able to suck away the posterior lobe after exposing the sella turcica via the parapharyngeal approach. He also could remove the two anterior lobes of the pituitary leaving the posterior lobe intact. I never again encountered a man as surgically able as Melchior. His skill was essential to our work. Smelik (1960) indeed found that posterior lobectomy resulted in a reduced pituitary-adrenal activation following emotional stress such as noise, mild electric shocks to the feet, or exposure of the animals to a novel environment, but not to systemic stress such as hemorrhage under ether anesthesia or injection of histamine or nicotine. In 1960 Smelik went to work for a year with C. H. Sawyer in Los Angeles, and I continued the studies with posterior lobectomized rats, using plasma corticosterone as a measure of ACTH release instead of ascorbic acid depletion in the adrenal gland as employed by Smelik. I confirmed the results and in addition demonstrated that chronic treatment of posterior lobectomized rats with long-acting Pitressin tannate in oil not only removed the mild diabetes insipidus of these rats but also restored the pituitary-adrenal response to emotional stress. We postulated that vasopressin might be involved in the release of ACTH in response to emotional stress (de Wied, 1961*b*).

I should mention here that J. Ariëns Kappers was professor of anatomy in Groningen before he became director of the Brain Institute in Amsterdam and studied the influence of the superior cervical ganglion on the morphology of the pineal gland. I remember that Smelik performed the ganglionectomies in rats for him. We talked a lot about the role of the CSF in the transport of hormones, a subject which appears to be of great importance in our present studies. J. Moll was the closest associate of Ariëns Kappers and also was actively involved in the study of neuroendocrine mechanisms. He had worked in E. Scharrer's laboratory in New York. Moll had studied the regeneration of the pituitary stalk into a "miniature" lobe after extirpation of the pituitary gland and was very interested in the fate of the stalk following posterior lobectomy. We found that a similar regeneration took place and that this miniature lobe responded with release of ADH following dehydration but not following hemorrhage (Moll and de Wied, 1962).

I reasoned that the impaired release of ACTH to emotional stress in posterior lobectomized rats also could reflect a behavioral incompetence of these animals to translate emotional stimuli in the same way as normal animals. For this reason I decided to study the behavior of posterior lobectomized rats, employing the techniques that I had learned in Pittsburgh. Avoidance learning of posterior lobectomized rats in the shuttlebox appeared to be similar to that of sham-operated controls. However, extinction of the response was markedly facilitated in posterior lobectomized rats and was much faster than in control rats. Thus the hypothesis that the CNS response of posterior lobectomized rats is abnormal was strongly supported by these results. This behavioral abnormality could be corrected by treatment with a relatively crude extract of posterior pituitary tissue, Pitressin, and with purified lysine-vasopressin as long-acting preparations. Since Murphy and Miller (1955) had demonstrated an effect of ACTH on extinction of shuttlebox avoidance behavior of intact rats, I also investigated the influence of ACTH and αMSH. Both peptides appeared to exhibit effects similar to that of vasopressin, i.e., to normalize extinction of shuttlebox avoidance behavior in posterior lobectomized rats. These results were obtained when I was still in Groningen. The observations were fascinating but puzzling, since structurally unrelated peptides had seemingly similar effects on avoidance behavior. I decided to postpone publication until I had a better understanding of what was going on and had obtained more data.

During my stay in the United States I had become reader in experimental endocrinology and was promoted to full professor in 1961. I heard the news of my appointment during a 3-month stay on a NATO fellowship at the Allan Memorial Institute of Psychiatry in Montreal. I introduced the CRF assay in median eminence lesioned rats there because I wanted to measure the CRF material that M. Saffran and his associates had isolated. The CRF, however, appeared to be as mysterious a factor then as it had been before and still remains. Ever since, Saffran has shown a slide during lectures on the subject which pictures me sitting in an easy chair reading funnies.

Upon my return, I did a study on the influence of dexamethasone on stress-induced ACTH release and deduced that the blocking effect of this synthetic steroid occurred in the anterior pituitary, a conclusion which met with a lot of disapproval from those who maintained that the feedback was located in the hypothalamus (de Wied, 1964*a*). I also resumed the studies on behavior and decided to explore the role of the adenohypophysis on acquisition and extinction of shuttlebox avoidance behavior—the more so since I discovered one report in the literature on a few animals indicating that hypophysectomy might interfere with avoidance acquisition and that ACTH was able to correct this deficiency (Applezweig and Moeller, 1959). I found that adenohypophysectomized rats were far inferior to sham-operated controls in acquiring shuttlebox avoidance behavior. Since the other results on hypophysectomized rats were similar, the learning deficit had to be due to the lack of anterior lobe principles. Chronic treatment with adrenal maintenance doses of ACTH restored the behavior of hypophysectomized rats, as did substitution therapy with thyroxine, cortisone, and testosterone, albeit to a lesser degree.

In 1963 I had enough courage to publish the results. One manuscript on adenohypophysectomized rats was accepted for publication and appeared in the *American Journal of Physiology* in 1964; a second one on posterior lobectomy appeared in the *International Journal of Neuropharmacology*, in 1965 (de Wied, 1964*b*, 1965). I apparently had mastered enough professional vocabulary to describe these first behavioral studies.

At the end of 1962 I was offered the chair in Pharmacology in the Medical Faculty of the University of Utrecht as successor to U. G. Bijlsma. I decided to accept the challenge and thus moved to Utrecht in the spring of 1963. There were lots of opportunities, sufficient money, and positions for staff members to attract good people. Smelik came with me to Utrecht as my closest associate and wanted to pursue his studies on feedback regulation of the pituitary-adrenal system. For me the move demanded a new line of research. For many years I had been interested in the releasing factors and particular in CRF. Although it was tempting to set up a program to isolate this factor from the hypothalamus, the field was already so competitive that I abandoned the idea. The fascinating results that I had obtained in Groningen on behavioral abnormalities in rats in which the pituitary in part or whole had been removed suggested that more extensive research on the implication of pituitary-adrenal system hormones on adaptive behavior might be fruitful. I also wished to study the problem as a multidisciplinary endeavor, i.e., from a behavioral, endocrinological, biochemical, and neurophysiological point of view. I had asked A. Witter, a peptide chemist, to join me when I was still in Groningen. He worked in the Department of Organic Chemistry of the University of Utrecht and moved to the Department of Pharmacology when I began my work there. Within a few years the department bustled with activity. It was filled with Ph.D. students like W. de Jong, who already was a student of mine in Groningen and who succeeded Smelik when the latter was appointed to the chair of Pharmacology in the Medical Faculty of the Free University of Amsterdam in 1970. Others like Tj. van Wimersma Greidanus and W. H. Gispen were Ph.D. students who became close associates in the main line of research on the pituitary-adrenal system and behavior. P. Bouman succeeded me in Groningen as a professor of experimental endocrinology in 1965.

The first 5 years were rather hectic but a multidisciplinary team was established. Rudolf Magnus had been appointed in 1908 to the first chair in Pharmacology in the Netherlands. In the fall of 1968 we celebrated the sixtieth anniversary of the chair of Pharmacology in Utrecht, and on this occasion the University granted us permission to change the name of the department to the Rudolf Magnus Institute for Pharmacology in honor of a great predecessor.

As soon as possible, I resumed behavioral studies. I found that total hypophysectomy like adenohypophysectomy markedly interfered with avoidance learning in the shuttlebox. Again, maintenance doses of ACTH corrected the deficient behavior of hypophysectomized rats. The same effect was found following administration of αMSH, indicating again that the effect of ACTH was not mediated by the adrenal cortex (de Wied, 1969). Small fragments of ACTH virtually devoid of corticotropic activities, such as ACTH 1–10 and ACTH 4–10, were as active as ACTH, in contrast to the *C*-terminal fragment of ACTH (ACTH 11–24). Thus the active core for the behavioral effect probably was located in the *N*-terminal part of the molecule. From these results I postulated that pituitary peptides related to ACTH normally operate in the formation of conditioned and other adaptive responses: "these peptides with neurogenic activities may be manufactured by the pituitary gland. Whether these peptides are represented by ACTH, αMSH, and βMSH or by peptides closely resembling ACTH 4–10 cannot be determined at present. Such peptides which may be released from the pituitary upon adequate stimulation may affect central nervous structures involved in motivational, learning, and memory processes. The effect of ACTH, α- and βMSH may then be regarded as an incidental finding because these hormones share the sequence 4–10" (de Wied, 1969).

In 1965 B. Bohus, who received his training in Pécs, Hungary, in the famous School of Physiology headed by K. Lissák, came for a year to Utrecht as a research fellow. The interest in Pécs in the behavioral effect of ACTH and corticosteroids (Endröczi, 1972) stimulated the exchange between our laboratories. We studied the effect of long-acting αMSH and Pitressin in intact rats in the shuttlebox and found that both compounds, when given during extinction, inhibited extinction of the response. However, if the treatment was given during acquisition, inhibition of extinction was found in Pitressin-treated rats only. From this we inferred that Pitressin preserves a conditioned avoidance response irrespective of the time of treatment, while αMSH inhibits extinction only during the time of treatment. Thus the mechanisms by which Pitressin and MSH affect avoidance behavior are basically different (de Wied and Bohus, 1966). A great number of experiments were done in which ACTH and vasopressin and their analogues were studied. We amply confirmed these differential effects. The long-term effect of vasopressin was interpreted as an effect on memory consolidation (Bohus *et al.*, 1973) and/or retrieval (Rigter *et al.*, 1974), while the short-term effect of ACTH fragments was interpreted as a motivational effect of these polypeptides (de Wied, 1974) or an influence on concentration (visual/attention) (Kastin *et al.*, 1975).

Another very significant observation which Bohus and I made was that substitution of the D-enantiomer of phenylalanine in position 7 of ACTH 1–10 dramatically altered the behavioral effect and led to facilitation of active avoidance behavior (Bohus and de Wied, 1966). This effect was not obtained if neighboring amino acids were replaced by their D-antipodes. This indicated that we were dealing with highly specific substances.

Bohus came back for the second time in 1970 and after another very fruitful year decided to stay in Utrecht as a staff member of the Rudolf Magnus Institute. Studies on the locus of action, mainly performed by van Wimersma Greidanus and Bohus, revealed that the nuclei parafascicularis of the thalamus and the rostral septal and dorsal hippocampal structures are essential for the behavioral effect of ACTH and vasopressin fragments (van Wimersma Greidanus *et al.*, 1975*a*).

I was convinced that the pituitary contains more specific and potent peptides involved in the formation of new behavior patterns, peptides that would stimulate or inhibit acquisition or maintain or facilitate extinction of avoidance behavior. In view of this, I looked around for means of isolating peptides from the pituitary, to test these in the various behavioral tests that we were using. I knew that A. Lerner and S. Lande of New Haven had isolated a great number of peptides related to ACTH and MSH from hog pituitary material. Lande was interested in a joint program, and so Lande, Witter, and I started in 1969 to test the relatively small peptide fractions which had been isolated in New Haven. Various behaviorally active fractions were found, and the first which was isolated in pure form and characterized appeared to be related to vasopressin rather than to MSH. It was desglycinamide^9-lysine^8-vasopressin (DG-LVP), which was very potent in stimulating acquisition of shuttlebox avoidance behavior in hypophysectomized rats (Lande *et al.*, 1971). Its action was comparable to that of vasopressin in other behavioral tests. Although it was somewhat less active behaviorally, it had lost nearly all classical endocrine effects such as pressor, antidiuretic, oxytocic, and CRF activity. This was an exciting finding in itself, and, like the results with ACTH, indicated that behavioral and endocrine effects are dissociated. In subsequent experiments several other peptides were isolated with potent behavioral activities. One peptide which was obtained in pure form probably was related to the *C*-terminal part of lipotropin (βLPH) (Lande *et al.*, 1973), which, as has been shown recently by various authors, possesses opiatelike activity. Indeed, when these materials became available recently, I found that *C*-terminal fragments of βLPH were even more active than ACTH 4–10 (which by the way is present in the *N*-terminal part of βLPH) in delaying extinction of active avoidance behavior (for review, see de Wied, 1976*a*).

A great number of structure-activity studies were performed over the years in collaboration with H. M. Greven from Organon using a one-way active avoidance test. Greven made the peptides which I tested and in this way we have been able to make more potent compounds and to further dissociate the classical endocrine activities of ACTH, MSH, and vasopressin from the behavioral effects. We had learned that introduction of D-enantiomers except that of phenylalanine would potentiate the behavioral effect of ACTH fragments, probably by protecting them against metabolic degradation (Witter *et al.*, 1975). Other substitutions known to decrease the corticotropic activity of ACTH or the MSH activity of MSH, were introduced. These modifications in [Lys^8]-ACTH 4–9 yielded the hexapeptide H-Met(O)-Glu-His-Phe-D-Lys-Phe-OH. This peptide appeared to be a thousand times more potent in avoidance behavior than the parent molecule. It had a thousand times less MSH activity, a markedly reduced ACTH activity, and no affinity for the opiate receptor (Greven and de Wied, 1973). It was in addition orally active (Verhoef and Witter, 1976). The smallest sequence of ACTH which still contained the full behavioral effect appeared to be ACTH 4–7. The fact that only four amino acids in the ACTH molecule bear the essential requirements for the behavioral effect reminded us of observations concerning the tetrapeptide of gastrin, which exerts essentially the same biological activity as the complete molecule (Tracy and Gregory, 1964). "One may speculate that ACTH and MSH are prohormones from which the active peptide is released by specifically localized peptidases" (Greven and de Wied, 1973). The same view had been proposed by Walter for oxytocin as a precursor molecule for releasing hormones which modulate the release of MSH (Walter *et al.*, 1973). The recent observations on the opiatelike activity of *C*-terminal βLPH peptides underscore the significance of pituitary hormones as precursor molecules for potent neuropeptides (Bradbury *et al.*, 1976).

The exact nature of the effect of neuropeptides on the nerve cell is as yet far from elucidated. Whether these act as neurotransmitters or modulators of neuronal activity is a matter of speculation. Biochemical studies on brain neurotransmission, on cyclic nucleotides, or RNA and protein synthesis, and on other molecular events as yet have not revealed the mode of action of these polypeptides (Schotman *et al.*, 1976). However, it is conceivable that neuropeptides interact with a receptor on the surface of the plasma membrane of the effector cell in the brain in the same way as ACTH has been shown to act on the adrenal cortex (Sayers *et al.*, 1974). This results in an increased production of intracellular cAMP, which mediates the information of peptide cell membrane binding. This then triggers the biochemical train of events underlying the functional response of the effector cell.

From 1967, in lectures and at meetings and conferences, I have discussed the concept that the pituitary is involved in motivational, learning, and memory processes, and that these processes are under the control of neuropeptides generated from hypothalamic and pituitary hormones for which the brain is a target. It has been difficult to convince my endocrine colleagues that the pituitary might have such functions. One good friend once commented that I used to do such nice work before I began to study behavioral aspects of pituitary hormones.

How these peptides reach the brain is in this respect an important question, since approximately 0.01% of a systemically administered ACTH fragment reaches the brain. One has to postulate transport of pituitary hormones directly to the brain. Retrograde transport via the pituitary stalk, or via a microcirculation from the pituitary to the brain, or via direct connections of the pituitary with the CSF (tanycytes?) have been suggested. However, proof of these routes of transport has not been provided. The situation is somewhat easier to visualize for neurohypophysial hormones than for adenohypophysial hormones. The former may be released directly into the third ventricle or be transported via a peptidergic pathway from the hypothalamus to limbic midbrain structures.

Times changed since the first reports on the CNS effects of releasing hormones appeared around 1971, although the emphasis was on pharmacological rather than on physiological actions (Kastin *et al.*, 1975). The observations and findings that these hormones are found all over the brain were a much better eye opener than all the studies that we performed during the preceding 10 years. The more recent discovery of the opiatelike activity of *C*-terminal βLPH peptides has also been of assistance in this respect. Our more recent findings that intraventricular administration of specific vasopressin antibodies disrupts memory consolidation (van Wimersma Greidanus *et al.*, 1975*b*), that Brattleboro rats with hereditary diabetes insipidus have marked learning and memory deficits (de Wied *et al.*, 1975), that amounts of vasopressin as low as 25 pg when given intraventricularly exert potent behavioral effects (de Wied, 1976*b*), that ACTH analogues can be developed with behavioral activity in the order of picogram quantities following systemic injection (Greven and de Wied, 1977), etc., reinforced the concept that the formation of new behavior patterns is under the control of hypothalamic and pituitary neuropeptides. Others have begun to study the effect of ACTH and vasopressin fragments, and part of our work has been confirmed. The group of Kastin contributed considerably with their studies on the implication of ACTH/MSH and peptide fragments on brain function. His group has been able to predict the action of these neuropeptides in animals and has verified the observations in man.

I can say now with confidence that the hypothalamic-pituitary complex produces numerous neuropeptides which carry specific information to brain structures involved in the behavioral repertoire of the organism. Findings so far have demonstrated the existence of neuropeptides which affect pain, sleep, learning, memory, motivation, attention and concentration, sexual behavior, thirst, and aggression as well as neuropeptides which affect addiction and development of tolerance and physical dependence on narcotic analgesics. I realize that a lot has to be done. In particular, it is necessary to extend the search for biochemical and electrophysiological correlates of neuropeptides in the brain, to look for specific receptor sites of the various neuropeptides in the brain, to find highly specific and potent neuropeptides of hypothalamic and pituitary origin which affect motivational, learning, and memory processes, and to study the enzymic release of neuropeptides from hypothalamic and pituitary hormones during the formation and maintenance of new behavior patterns. The first clinical observations in man on the effect of a number of neuropeptides already suggest that we have entered a new era in which it may be possible to mend faltering brain functions. I foresee the development of new compounds from naturally occurring peptides which are much less toxic and more specific than the conventional CNS drugs in use today.

REFERENCES

Applezweig, M. H., and Moeller, G. (1959). The pituitary-adrenocortical system and anxiety in avoidance learning. *Acta Psychol.* **15**:602.

Bohus, B., and de Wied, D. (1966). Inhibitory and facilitatory effect of two related peptides on extinction of avoidance behavior. *Science* **153**:318.

Bohus, B., Gispen, W. H., and de Wied, D. (1973). Effect of lysine vasopressin and ACTH 4–10 on conditioned avoidance behavior of hypophysectomized rats. *Neuroendocrinology* **11**:137.

Bradbury, A. F., Smyth, D. G., and Snell, C. R. (1976). *C* fragment of lipotropin has a high affinity for brain opiate receptors. *Nature* **260**:793.

Dugal, L. P., and Thérien, M. (1959). Influence of ascorbic acid on adrenal weight during exposure to cold. *Endocrinology* **44**:420.

Endröczi, E. (1972). *Limbic System Learning and Pituitary-Adrenal Function*, Akadémiai Kiadó, Budapest.

Greven, H. M., and de Wied, D. (1973). The influence of peptides derived from corticotropin (ACTH) on performance. Structure activity studies. In Zimmermann, E., Gispen, W. H., Marks, B. H., and de Wied, D. (eds.), *Drug Effects on Neuroendocrine Regulation*, Vol. 39 of *Progress in Brain Research*, Elsevier, Amsterdam, pp. 429–442.

Greven, H. M., and de Wied, D. (1977). Structure-activity relationships of MSH and related peptides: Localization and identification of an "active core." In Tilders, F. J. H., Swaab, D. F., and van Wimersma Greidanus, (eds.), *Control, Chemistry and Effects of MSH*, Karger, Basel.

Harris, G. W. (1955). *Neural Control of the Pituitary Gland*, Edward Arnold, London.

Kastin, A. J., Sandman, C. A., Stratton, L. O., Schally, A. V., and Miller, L. H. (1975). Behavioral and electrographic changes in rat and man after MSH. In Gispen, W. H., van Wimersma Greidanus, Tj. B., Bohus, B., and de Wied, D. (eds.), *Hormones, Homeostasis and the Brain*, Vol. 42 of *Progress in Brain Research*, Elsevier, Amsterdam, pp. 143–150.

Lande, S., Witter, A., and de Wied, D. (1971). Pituitary peptides: An octapeptide that stimulates conditioned avoidance acquisition in hypophysectomized rats. *J. Biol. Chem.* **246**:2058.

Lande, S., de Wied, D., and Witter, A. (1973). Unique pituitary peptides with behavioral-affecting activity. In Zimmermann, E., Gispen, W. H., Marks, B. H., and de Wied, D. (eds.), *Drug Effects on Neuroendocrine Regulation*, Vol. 39 of *Progress in Brain Research*, Elsevier, Amsterdam, pp. 421–427.

Miller, R. E., and Ogawa, N. (1962). The effect of adrenocorticotropic hormone (ACTH) on avoidance conditioning in the adrenalectomized rat. *J. Comp. Physiol. Psychol.* **55**:211.

Mirsky, I. A., Miller, R., and Stein, M. (1953). Relation of adrenocortical activity and adaptive behavior. *Psychosom. Med.* **15**:574.

Moll, J., and de Wied, D. (1962). Observations on the hypothalamoposthypophyseal system of the posterior lobectomized rat. *Gen. Comp. Endocrinol.* **2**:215.

Murphy, J. W., and Miller, R. E. (1955). The effect of adrenocorticotropic hormone (ACTH) on avoidance conditioning in the rat. *J. Comp. Physiol.* **48**:47.

Olling, Ch. C. J., and de Wied, D. (1956). Inhibition of the release of corticotrophin from the hypophysis by chlorpromazine. *Acta Endocrinol.* (*Kbh.*) **22**:283.

Rigter, H., van Riezen, H., and de Wied, D. (1974). The effects of ACTH and vasopressin analogues on CO_2-induced retrograde amnesia in rats. *Physiol. Behav.* **13**:381.

Sayers, G., Beall, R. J., and Seelig, S. (1974). Modes of action of ACTH. In Rickenburg, H. V. (ed.), *Biochemistry of Hormones* (M.T.P. Int. Rev. Sci., Biochem. Ser., H. L. Kornberg and B. C. Philips, eds.), Butterworth University Park Press, London, pp. 25–60.

Schotman, P., Reith, M. E. A., van Wimersma Greidanus, Tj. B., Gispen, W. H., and de Wied, D. (1976). Hypothalamic and pituitary peptide hormones and the central nervous system. With special references to the neurochemical effects of ACTH. In Gispen, W. H. (ed.), *Molecular and Functional Neurobiology*, Elsevier, Amsterdam, pp. 309–344.

Selye, H. (1950). "*Stress*": *The Physiology and Pathology of Exposure to Stress*, Acta Inc., Montreal.

Smelik, P. G. (1960). Mechanism of hypophysial response to psychic stress. *Acta Endocrinol.* (*Kbh.*) **33**:437.

Tracy, H. J., and Gregory, R. A. (1964). Physiological properties of a series of synthetic peptides structurally related to gastrin I. *Nature* **204**:935.

Verhoef, J., and Witter, A. (1976). In vivo fate of a behaviorally active ACTH 4–9 analog in rats after systemic administration. *Pharmacol. Biochem. Behav.* **4**:583.

Walter, R., Griffiths, E. C., and Hooper, K. C. (1973). Production of MSH release-inhibiting hormone by a particulate preparation of hypothalami: Mechanisms of oxytocin inactivation. *Brain Res.* **60**:449.

de Wied, D. (1953). The influence of ascorbic acid on the glycogen content of the liver of normal and adrenalectomized rats exposed to cold. *Acta Endocrinol.* (*Kbh.*) **14**:235.

de Wied, D. (1961*a*). An assay of corticotrophin-releasing principles in hypothalamic lesioned rats. *Acta Endocrinol.* (*Kbh.*) **37**:288.

de Wied, D. (1961*b*). The significance of the antidiuretic hormone in the release mechanism of corticotropin. *Endocrinology* **68**:956.

de Wied, D. (1964*a*). The site of the blocking action of dexamethasone on stress-induced pituitary ACTH release. *J. Endocrinol.* **29**:29.

de Wied, D. (1964*b*). Influence of anterior pituitary on avoidance learning and escape behavior. *Am. J. Physiol.* **207**:255.

de Wied, D. (1965). The influence of the posterior and intermediate lobe of the pituitary and pituitary peptides on the maintenance of a conditioned avoidance response in rats. *Int. J. Neuropharmacol.* **4**:157.

de Wied, D. (1967). Opposite effects of ACTH and glucocorticosteroids on extinction of conditioned avoidance behavior. In *Proceedings of the Second International Congress on Hormonal Steroids, Milan, 1966*, Excerpta Medica Int. Congr. Ser. No. 132, p. 945.

de Wied, D. (1969). Effects of peptide hormones on behavior. In Ganong, W. F., and Martini, L. (eds.), *Frontiers in Neuroendocrinology*, Oxford University Press, New York, pp. 97–140.

de Wied, D. (1974). Pituitary-adrenal system hormones and behavior. In Schmitt, F. O., and Worden, F. G. (eds.), *The Neurosciences: Third Study Program*, MIT Press, Cambridge, Mass., pp. 653–666.

de Wied, D. (1976*a*). Peptides and behavior. Minireview. *Life Sci.* **20**:195.

de Wied, D. (1976*b*). Behavioral effects of intraventricularly administered vasopressin and vasopressin fragments. *Life Sci.* **19**:685

de Wied, D., and Bohus, B. (1966). Long term and short term effects on retention of a conditioned avoidance response in rats by treatment with long acting pitressin and αMSH. *Nature* **212**:1484.

de Wied, D., and Mirsky, I. A. (1959). The action of Δ^1 hydrocortisone on the antidiuretic and adrenocorticotropic responses to noxious stimuli. *Endocrinology* **64**:955.

de Wied, D., Bouman, P. R., and Smelik, P. G. (1958). The effect of a lipide extract from the posterior hypothalamus and of pitressin on the release of ACTH from the pituitary gland. *Endocrinology* **62**:605.

de Wied, D., Bohus, B., and van Wimersma Greidanus, Tj. B. (1975). Memory deficit in rats with hereditary diabetes insipidus. *Brain Res.* **85**:152.

van Wimersma Greidanus, Tj. B., Bohus, B., and de Wied, D. (1975*a*). CNS sites of action of ACTH, MSH and vasopressin in relation to avoidance behavior. In Stumpf, W. E., and Grant, L. D. (eds.), *Anatomical Neuroendocrinology*, Karger, Basel, pp. 284–289.

van Wimersma Greidanus, Tj. B., Dogterom, J., and de Wied, D. (1975*b*). Intraventricular administration of anti-vasopressin serum inhibits memory consolidation in rats. *Life Sci.* **16**:637.

Witter, A., Greven, H. M., and de Wied, D. (1975). Correlation between structure, behavioral activity and rate of biotransformation of some ACTH 4–9 analogs. *J. Pharmacol. Exp. Ther.* **193**:853.

24

Solly Zuckerman

Solly Zuckerman was born in Cape Town, South Africa, in 1904. He was educated in Cape Town and London and qualified in medicine at University College Hospital, London, in 1928. After teaching anatomy in South Africa and England, and spending a year as a Rockefeller Research Fellow at Yale University (1933–1934), he was awarded a Beit Memorial Fellowship at the University of Oxford, where he was also appointed university lecturer and demonstrator (1934–1945). Up to World War II he was deeply involved in studies of the primate menstrual cycle, primate behavior, and the physiology of reproduction, but this was interrupted by his successive appointments as scientific advisor to Combined Operations H.Q., chief scientific advisor on planning to the Mediterranean Allied Air Forces, and chief scientific advisor to the Allied Expeditionary Air Forces. He was appointed to the Sands Cox Chair of Anatomy, University of Birmingham, in 1943, and was head of the Department of Anatomy until 1968.

Lord Zuckerman has served on many government bodies, including the Agricultural Research Council, the British Broadcasting Corporation, and the Defence Research Policy Committee. He was Chief Scientific Advisor to the Secretary of State for Defence (1960–1966) and to the British government (1964–1971).

After 22 years as honorary secretary of the Zoological Society of London (1955–1977), he succeeded the Duke of Edinburgh as President. Elected a Fellow of the Royal Society in 1943, he was knighted in 1956, became a member of the Order of Merit in 1968, and was created a life peer in 1971. He also holds the Medal of Freedom with Silver Palm (U.S.A.) and is a Chevalier de la Legion d'Honneur (France).

His writings include *The Social Life of Monkeys and Apes* (Kegan Paul, 1932), *A New System of Anatomy* (OUP, 1961), *Beyond the Ivory Tower* (Weindenfeld and Nicolson, 1970), and *From Apes to Warlords* (Hamish Hamilton, 1978). He is editor of *The Ovary*, Vols. 1 and 2 (Academic Press, 1962; 2nd edition, Vols. 1–3, 1977).

24

A Skeptical Neuroendocrinologist

SOLLY ZUCKERMAN

To the extent that I understand the meaning of the designation, I suppose that I could claim to have become a neuroendocrinologist long before anyone thought that there was any point in coining the term. The etymology of "endocrine" is clear enough. An endocrine organ is a structure whose cells elaborate one or more chemical substances which pass, either directly or indirectly, into the bloodstream, through which they are transported to the cells of some target tissue, where they exercise their effects. The nervous system, from which comes the "neuro" part of "neuroendocrine," is something very different, and consists of a vast network of interacting nerve cells, concentrated mainly in the brain and spinal cord, through whose axons and dendrites currents flow to and from muscle fibers and other tissues, as well as from peripheral sensory cells. My teachers of more than 50 years ago were already talking about the chemical transmission of nerve impulses across synapses. But while we were taught that, apart from the adrenal medulla, endocrine tissues are not controlled by secretomotor nerves, I am sure that we all understood that since both the neural and hormonal systems are involved in the orderly regulation of the organs and tissues, they have to operate in an integrated way. There was nothing very profound in this notion. Why therefore, I ask, prefix the term "endocrinology" with the word "neuro"? At the outset I have to say that I do not quite see what the fuss is all about.

It was my prior interest in the social behavior of wild baboons, creatures which I frequently came across during my childhood and youth in South Africa, which led me to the study of the hormonal basis of reproductive mechanisms. Baboons live in family groups in which mature females and young are "governed" by a dominant male, and it was the realization

SOLLY ZUCKERMAN • Zoological Society of London, and University of East Anglia, England.

that the interrelationships of the mature animals in the group are largely determined by the phases of the sexual-skin cycle of the postpubertal females that started me off in my inquiries into the hormonal control of the menstrual cycle (Zuckerman, 1929, 1932). My early interest in the social behavior of wild animals also made me curious about the control of seasonal reproductive behavior. At the time that I started inquiring seriously into both subjects, all but nothing was known about the endocrine basis of the menstrual cycle. But Rowan (1926) had already reported that seasonal changes in the hours of daylight regulated the migratory behavior of juncos, an observation which Bissonnette (1932) was soon to follow up when he showed that increasing hours of daylight in the spring controlled the onset of estrus in ferrets.

Not long after I had become immersed in these two problems, I also became beguiled by another issue, the possibility of an interaction of the nervous and endocrine systems in the control of the ovarian secretions. At that time, ideas were still current that nervous mechanisms were involved in the control of the menstrual cycle—for example, Pflüger's theory of menstruation had not yet been buried. The year before I started work in Yerkes's laboratory at Yale in 1932, John Fulton, then head of the University's Department of Physiology, and Gertrude van Wagenen (1933) had observed that transection of the spinal cord of monkeys sometimes precipitated menstrual bleeding, after which "regular" cycles were resumed—as Sherrington (1900) had reported. I joined forces with Dr. van Wagenen and found that bleeding does not occur after spinal transection either in immature animals or in mature animals whose ovaries have been removed and that bleeding can be prevented after cord section in mature monkeys if they are given estrogen daily from the time of the operation (van Wagenen and Zuckerman, 1933). The latter observation was independently confirmed by George Corner (1934). The conclusion that I drew was that the bleeding was a consequence of a decline in the level of estrogenic stimulation of the endometrium and that this was caused by some specific change below the level of transection. This inference seemed sufficiently reasonable for me to go a step further and to design an extensive inquiry to test the hypothesis that preganglionic secretomotor pathways to the ovaries are interrupted when the cord is divided. My experiments involved not only cord section and hemisection at different spinal levels but also division or evulsion of the sympathetic chain, the splanchic nerves, and the pelvic parasympathetics. It was exciting work. I transplanted the ovaries to the head region to see if interruption of the sympathetic pathways in the neck or abdominal cavity, or section of the thoracic cord then had an effect. But I was wasting my time—I was chasing a will-o'-the-wisp. The uterine bleeding turned out to be a general phenomenon resulting either from a general vasomotor paralysis or from a paralysis confined to the reproductive organs alone. There were no ovarian secretomotor nerves, or, if there were any, I had certainly not discovered evidence of their existence.

In the only paper in which I reported the results of my arduous experimental inquiries (Zuckerman, 1938), I also referred to the fact that it had long been held that reproductive mechanisms are controlled by hypothalamic nerve centers, remarking that while some of the evidence in favor of this view was acceptable, most of it consisted of uncontrolled experiment and was little better than speculation. John Fulton and I had carried out experiments on three baboons, in two of which we made lesions in the hypothalamus, and in the third merely in the corpus callosum. The sexual-skin cycle in all three was disturbed—but not the menstrual cycle. I began to have doubts about the precise relation of the hypothalamus to reproductive mechanisms. The evidence seemed hazy compared to the precision of that provided by the Chicago school of Fisher *et al.* (1938), which had established a functional and structural connection between the nuclei supraopticus and paraventricularis and the neurohypophysis.

Back in Oxford in 1934 I focused my attention once again on the experimental test of the hypothesis to which I had been led, when working at the Zoological Society of London, about the hormonal control of the phases of the menstrual cycle, and on the reaction to estrogens of vestigial remnants of the Müllerian duct in the male reproductive organs. The latter field of inquiry led me, in due course, to a comprehensive analysis of the histogenesis of tissues sensitive to estrogens (Zuckerman, 1940).

But the problem of breeding seasons was still a preoccupation. Before completing my appointment as Prosector to the Zoological Society in 1932, I had analyzed the breeding records of all mammals that had given birth in the London Zoo from the time of its foundation in 1826. I had also obtained information from zoos in other countries about the times of breeding of creatures which had been moved from the northern to the southern hemisphere or *vice versa.* It was obvious that the breeding habits of mammals are modulated by environmental factors, but the question was "how?" Bissonnette had not only found that ferrets could be made reproductively active in winter by exposing them to artificial light, but had also shown that the response is much impaired after division of the optic nerves (Bissonnette, 1935). Two years before this, Hill and Parkes (1933) had found that artificial illumination in the winter does not stimulate estrus in ferrets whose pituitaries have been removed. *Ergo*, or so it seemed to me, it followed that the impulses started by stimulation of the retina are transmitted along a definable pathway to the adenohypophysis, even though the latter was then not supposed to be richly innervated—an anatomical question to which no final answer has yet been provided. Tom McKeown had come from Canada as a Rhodes Scholar and I was supervising him for his D. Phil. degree. Le Gros Clark, Dr. Lee's Professor of Anatomy in the University, was skilled in operating on the brain. The three of us therefore joined forces to see whether we could track the fibers which we supposed were concerned in the transmission of the impulses between the retina and the pituitary, our technique being to divide the optic pathways of the ferret at different levels. Once again a meticulous set of experiments yielded no firm answer (Le Gros Clark *et al.*, 1939). The estrous response to light occurred in the combined absence of the visual cortex and the superior colliculi. One experiment even suggested that it could occur after the interruption of retinal impulses passing to the dorsal nucleus of the lateral geniculate body, visual cortex, superior colliculi, and the pretectal area. To the best of my knowledge, these experiments have never been repeated, but there is some evidence now that optic fibers do pass from the region of the chiasma to the rostral part of the hypothalamus. Presumably they could be the specific pathway for which we had been searching. However, it was about this time that some rather crude experiments on ducks persuaded Benoit that light had a direct effect on the pituitary, the light waves passing through the surface tissues of the ocular region. Although this story goes on being repeated in the literature, I have always found it fanciful. I would find it easier to imagine a general effect on the body as a whole.

There was another possibility of neural: endocrine interaction in which I became interested at this time. The ferret is an animal which needs the stimulus of coitus for ovulation to occur, follicular rupture being the result of a surge of gonadotropin. McKeown and I therefore designed an experiment to discover the pathways whereby the impulses set up by vaginal stimulation in the estrous ferret are transmitted to the adenohypophysis. The war had started before we were properly launched on this study.

In 1946 I moved from Oxford to Birmingham, and in that year, at a meeting of the Society for Endocrinology, gave a résumé of the considerations which in my view would have to be borne in mind when searching for an answer to the question of how the pituitary reacts to light (Zuckerman, 1947). It was on this occasion that G. W. Harris first put forward the "suggestion" that since a direct neural control of the adenohypophysis—as opposed to the neurohypophysis—could not be established, "one link in the chain of events between the environmental stimulus and the gonadal reaction may be constituted by the hypophyseal portal vessels." My comment on this speculation was that "the crucial experiment of studying the effects of light in animals whose own pituitaries had been replaced by pituitary tissue transplanted to some extracranial site had not been carried out." Many attempts to transplant the pituitary have since been made, but to the best of my knowledge, no one has yet reported success in reconstituting a normal adenohypophysis from transplanted tissue. Nonetheless, Harris's suggestion has been taken up in the literature as Holy Writ. In spite of its speculative character, his thesis goes on being accepted, as if by popular vote, but so far as I am concerned the criticisms to which it ought to stand up have never been answered. Given that the hypothalamus does elaborate specific chemotransmitters or releasing factors—call them what one will—why was it ever necessary to suppose that they needed a specific vascular pathway to get to their target cells a millimeter or so away, when none such is demanded in the body for any other internal secretion that has to travel a longer path? Again, what, in logic, does it mean in the validation of Harris's hypothesis to assert that the flow in the pituitary-portal system is from the median eminence to the adenohypophysis? We first have to establish beyond doubt that there are chemotransmitters specific to the separate functions of the pars distalis, and that has not been done. What is more, the observation is questionable in itself, since the pressure in the capillary network in which the system begins, and in the sinusoids where it ends, is presumably minimal, and the flow would, on hydrodynamic principles, depend largely on gravity. Is it reversed when we stand on our hands?

I have challenged the hypothesis of which the portal vessels are a central feature more than once (Zuckerman, 1952, 1954, 1955). In my most recent review of the problem (Zuckerman, 1970), I pointed out that the theory subsumes the following three separate propositions:

1. The hypothalamus controls, or at least modulates, the functions of the anterior pituitary.
2. A necessary and indeed the only functional connection between the hypothalamus and the pars distalis is the pituitary-portal system, which would have to be a unique anatomical connection between the two.
3. Specifically different chemical substances are produced in the hypothalamus (and, because of the role assigned to the portal vessels, nowhere else), and changes in the function of the pars distalis result from alterations in the secretion of these substances into the portal system.

On the first of these propositions I commented that insofar as it implies a specific and moment-to-moment control of the pars distalis, the thesis is not amenable to experimental verification. Quite apart from the lamentably inadequate nature of the controls in the experiments that have been carried out, the changes in pituitary function which follow stimulation or destruction of different parts of the hypothalamus—which is the basis of the proposition—are not only nonspecific, in the sense that they are associated with profound changes in other physiological functions, but are also highly variable in their nature, mainly because the minute size of the hypothalamic nuclei relative to the instruments which are used in the experiments makes it next to impossible to confine any experimental interference to specified neurons. The belief that the hypothalamus controls the anterior pituitary, whatever the nature of its influence, whether chemical or nervous (given that in the final analysis these are distinct from each other), is hardly likely to be elucidated by the classical approach of stimulation and destruction of groups of nerve cells. Neither technique can be applied with sufficient precision to what are in effect microscopic structures which cannot be approached either with topographic certainty or without seriously damaging or affecting neighboring structures. Most experiments on the subject have been carried out on rats, in which (by analogy from man) the volume of the hypothalamus can barely be more than a cubic millimeter.

I remain equally dubious about the second proposition which derives from the hypothesis as enunciated by Harris, namely that the portal system of vessels constitutes a "necessary" and unique functional-structural connection between the base of the brain and the adenohypophysis. What I fail entirely to see is why the belief that hypothalamic chemotransmitters control the anterior pituitary should depend on the assumption that they have to be transmitted to the gland by a specific vascular pathway. They might, for example, get to the pituitary by simple diffusion through the fluid between the cells of the minute region concerned. And here I repeat an argument that I have already used: the results of experiments in which pieces of pituitary were grafted to areas remote from the hypothalamus could have been interpreted as support for the hypothesis that substances simply diffused from the hypothalamus to the adenohypophysis—had this been the fashionable explanation at the time they were carried out—just as much as they were for the thesis that they were conveyed by the pituitary-portal vessels.

In addition, as I have pointed out, there is a free anastomosis between the vascular beds of all parts of the pituitary gland (Holmes and Zuckerman, 1959). This being so, the critical test of the proposition that the normal function of the pars distalis is uniquely dependent on the integrity of the portal vessels could only be through study of the function of the gland after these vessels, and these vessels alone, had been eliminated. This, in my anatomical judgment, could never be done. The hormonal activity of fragments of anterior pituitary tissue grown in tissue culture could also provide useful evidence. For example, it has been found that the cells of the pars distalis are directly sensitive to gonadal hormones. My own experiments on ferrets showed that while the pituitary gland usually dwindles in size after section of the stalk, due to infarction, atrophy is sometimes only partial. When this happens, the gland may function normally in the light-response even when completely separated from the base of the brain, and when all possibility of a vascular reconnection has been excluded. The fact that others, as well as myself, have obtained negative results in similar experiments is no argument against this positive finding. There is also evidence that the secretion of pars distalis grafts remote from the hypothalamus can alter in response to changes in the level of circulating hormone (this would not be expected to occur if all pars distalis functions depended on variations in the passage of hypothalamic chemotransmitters, via the portal vessels). The second proposition could hardly be shakier than it is.

In examining the third proposition which the hypothesis subsumes, I pointed out that whereas we are dealing with specific anterior pituitary functions—for example, the development and rupture of a Graafian follicle or its transformation into a corpus luteum—we are talking on the other hand about "releasing factors" which mostly seem to be nonspecific in their action, and whose derivation from specific hypothalamic neurons cannot be precisely determined. From the point of view of testing critically a general hypothesis about the control of the anterior pituitary by the portal vessels, experiments with hypothalamic extracts or releasing factors are not only irrelevant; they are also unsophisticated. As a scientist I have to confess to a sense of gloom when I read in more recent literature of experiments in which extracts were made of "ten pooled hypothalami" and when precise physiological functions are then attributed to the extract. Anyhow, if there are chemotransmitters, their existence would no more be dependent on the presence of portal vessels than is the existence of the latter necessary to give credibility to the idea of a chemotransmitter control of the anterior pituitary.

I concluded my 1970 review by referring to the fact that the pineal gland had also by then become involved in the story of the control of the pituitary without much understanding of what this meant. Were I now to bring my critique up to date, I would have to point to other matters which have come to the fore in the past few years, and which add to my confusion. For example, there is now much uncertainty about the specificity of the cells which produce FSH or LH, and about that of these hormones themselves in terms of function. Recent studies have also added to the number of extra-hypothalamic neural structures which are claimed to exercise an influence on the pituitary. To a skeptic such as myself, the whole story of the hypothalamic control of the pituitary, instead of becoming clearer, has become more speculative and more cloudy over the years. Indeed, the conviction is growing in me that the pituitary-portal vessels would never have entered into the picture of the presumed hypothalamic control of adenohypophysial function if the canvas had been painted by physiologists or pharmacologists rather than by anatomists. Whatever part the vessels may play in the economy of the gland, the concept is altogether too simplistic. After all, we do not conjure up special vascular pathways for the transmission of the molecules of other hormones from the sites where they exercise their effects. And what about insect pheromones? We accept that 10^{-10} μg—just a few molecules—of a pheromone released by one insect can exercise its effects on the receptor cells of another insect over a distance of as much as 10 km without supposing that they get there along minute invisible canaliculi (Wigglesworth, 1964; Treherne, 1974). Why should such a mystery have been woven about chemotransmitters traveling 1 or 2 mm in the rat?

As I wrote some time ago, I do not have the slightest idea what the physiologists of the year 2000 will believe about the control of the functions of the pars distalis, and about the interrelation of that gland with other parts of the pituitary—for example, with the anatomically imprecise pars tuberalis, which in mammals has hitherto not been endowed with any function, or with the pars intermedia, which has also not been accredited with any precise role, or indeed with the neurohypophysis, whose physiology is much more clearly understood, as is also its dependence on the nuclei supraopticus and paraventricularis of the hypothalamus. Nor have I any idea about what is likely to be thought of the functions of the pineal (once believed to be the seat of the soul), which in some species is an anatomical vestige of a third eye, and which almost generally is a cerebral appendage whose microscopic appearance hardly suggests a glandular function, although now extracts have been made from this structure which do what the presumed hypothalamic chemotransmitters were supposed to do. Nor, again, have I any idea about the place where the ependyma of the third ventricle will fit in. Will it turn out that the anterior pituitary is controlled chemically by substances transported in the cerebrospinal fluid? Many years ago, Harvey Cushing, that most distinguished of brain surgeons, believed that pituitary secretions made their way up the pituitary stalk to filter through the ependyma into the cerebrospinal fluid of the third ventricle—but his preparations were later shown to be artifacts. I have recently had occasion to review many new papers which deal with the problem. The overwhelming impression they left with me was of experiments that had been embarked upon with little thought to the logic of design, and with bland disregard for the shortcomings of the techniques used.

Yes—I remain a skeptical neuroendocrinologist, at any rate as far as the role of the pituitary portal vessels is concerned. I wish it were otherwise.

REFERENCES

Bissonnette, T. H. (1932). *Proc. Roy. Soc. London Ser. B* **110**:322.
Bissonnette, T. H. (1935). *J. Exp. Biol.* **12**:315.

Corner, G. W. (1934). See Zuckerman, S., in Brouha, L. (ed.), *Les Hormones Sexuelles* (1938), Hermann, Paris, p. 300.
Fisher, C., Ingram, W. R., and Ranson, S. W. (1938). *Diabetes Insipidus and the Neurohormonal Control of Water Balance*, Edwards, Chicago.
Hill, M., and Parkes, A. S. (1933). *Proc. Roy. Soc. London Ser. B* **113**:537.
Holmes, R. L., and Zuckerman, S. (1959). *J. Anat.* **93**:1.
Le Gros Clark, W. E., McKeown, T., and Zuckerman, S. (1939). *Proc. Roy. Soc. London Ser. B* **126**:449.
Rowan, W. (1926). *Proc. Boston Soc. Nat. Hist.* **38**:147.
Sherrington, C. S. (1900). In Schafer's *Text-Book of Physiology*, Vol. II, p. 783.
Treherne, J. E., *et al.* (eds.). (1974). *Advances in Insect Physiology*, Vol. 10, Academic Press, New York.
van Wagenen, G. (1933). *Am. J. Physiol.* **105**:473.
van Wagenen, G., and Zuckerman, S. (1933). *Am. J. Physiol.* **106**:416.
Wigglesworth, V. B. (1964). *The Life of Insects*, Weidenfeld and Nicolson, London.
Zuckerman, S. (1929). The social life of the primates. *Realist* **1**:72.
Zuckerman, S. (1932). *The Social Life of Monkeys and Apes*, Kegan Paul, London, 357 pp.
Zuckerman, S. (1938). In Brouha, L. (ed.), *Les Hormones Sexuelles*, Hermann, Paris, p. 299.
Zuckerman, S. (1947). *Proc. Soc. Endocrinol.* **5**:xv.
Zuckerman, S. (1952). *Ciba Found. Colloq. Endocrinol.* **4**:213.
Zuckerman, S. (1954). The secretions of the brain. *Lancet*, 739, 789.
Zuckerman, S. (1955). *Ciba Found. Colloq. Endocrinol.* **8**:551.
Zuckerman, S. (1970). *Beyond the Ivory Tower*, Weidenfeld and Nicolson, London, Chaps. 4 and 5.

Index